The Evolution of American Ecology, 1890–2000

The Evolution of American Ecology, 1890–2000

Sharon E. Kingsland

The Johns Hopkins University Press
Baltimore

Printed in the United States of America on acid-free paper
9 8 7 6 5 4 3 2 1

The Johns Hopkins University Press
2715 North Charles Street
Baltimore, Maryland 21218-4363
www.press.jhu.edu

Library of Congress Cataloging-in-Publication Data

Kingsland, Sharon E.
The evolution of American ecology, 1890–2000 / Sharon E. Kingsland.
p. cm
Includes bibliographical references and index.
ISBN 0-8018-8171-4 (hardcover : alk. paper)
1. Ecology—United States—History. 2. Botany—United States—History. I. Title.
QH540.83.U6K56 2005
577'.0973—dc22 2004026034

A catalog record for this book is available from the British Library.

To David

Contents

Acknowledgments

Several readers provided comments on portions of the manuscript and saved me from errors: I thank Daniel Bain, Grace Brush, Alice Ingerson, Steward T. A. Pickett, and Peter Taylor, as well as a referee for the Johns Hopkins University Press. My husband, Paul Romney, checked information and gave editorial advice at many stages of the writing. Several graduate students over the years have read and reacted to early versions of some of the chapters. I thank Sheila A. Dean for expert research assistance in the MacDougal Papers in Tucson. Susan Fraser, archivist at the New York Botanical Garden, helped me obtain photographs.

Listening to scientists talk about their work has been an important learning process for me. The final chapter could not have been written without the opportunity to listen to the scientists and educators involved in the Baltimore Ecosystem Study over the course of several years. Jane Maienschein, Gar Allen, John Beatty, and Jim Collins organized three summer courses and workshops on the history and philosophy of ecology, held at the Marine Biological Laboratory at Woods Hole and funded by the Dibner Institute for the History of Science and Technology. These weeks of discussion with historians and philosophers of biology, graduate students, and scientists raised different points of view on the problems I was studying. My citations of their work do not adequately express my indebtedness to this community of scholars.

Charles ReVelle, Justin Williams, and Simon Levin allowed me to sit in on discussions about interdisciplinary approaches to the design of nature reserves. Although this is not the subject of the present book, the problems of interdisciplinary cooperation that they confronted prompted insights relevant to the cases discussed here.

My perspective on the history of ecology and on present-day ecology benefited enormously from my yearlong participation in the Ecological Visions Committee of the Ecological Society of America in 2003–4. The committee was charged with formulating an action plan for the society and the discipline of ecology in general, in light of our pressing global environmental problems. I offer special thanks to Margaret Palmer, chair of the committee, for inviting me to serve on it. I feel a great debt to these committed scientists and educators. Their probing discussions of ecology and its future suggested some of the ideas that have entered into this book.

A grant from the National Science Foundation (SBR-9320310) supported part of the research for this book.

The Evolution of American Ecology, 1890–2000

INTRODUCTION

The Struggle for Place

Ecologists use the metaphor of the "ecological niche" to express the idea that plant and animal species play certain roles in the ecological community. The roles reflect such things as their place in the food chain, the kinds of resources they need to live, and how they act as conduits of matter and energy flowing through the ecosystem. The metaphor also draws attention to the competitive relationship between species. Individuals with similar needs are said to compete and hence would find it difficult if not impossible to occupy the same niche. But the niche does not really exist as a definite place out there in the world, a place that is occupied or fought over. Nor is it something that is fixed in time; the shape or size of the metaphorical niche would depend on the shifting context in which the plant or animal exists. The term *niche* is really a shorthand device for thinking about how organisms relate to their environment and how they compete with other organisms for resources.

Disciplines can be said to occupy niches in a similar metaphorical sense, and as with the ecological niche, the role and place of a discipline evolves over time in competition with other disciplines and in relation to an environmental and social context. I am interested in how ecology came to define its separate niche, that is, how it became a discipline in the United States, and what challenges have faced ecologists over the twentieth century as the disciplinary environment has changed and the niche has been reshaped. In looking at the formation of ecology as a biological discipline, I am also interested in Americans' concurrent discussions about their own adaptation to the land. The ways in which American scientists were interrogating nature mirrored the kinds of questions Americans were asking of themselves as a people, initially as a colonizing people moving across the North American landscape. As they colonized, they adapted the environment to their needs and in so doing changed the relations between themselves and other species. What did it mean for them to fit into a landscape and adjust properly to a new environment? What would guarantee the future strength of the people and the nation? How should human behavior adjust in the light of growing ecological knowledge? As I follow the developing science through the twentieth century, I repeatedly draw attention to the connection between the evolution of the scientific discipline and parallel discussions by Americans about their relationship to the land, the nature of their society, and its directions.

I start with the assumption that the existence of ecology as a discipline should not be taken for granted. Ecology is a broad subject; its scope and definition have vexed its practitioners since its origins about a century ago. Ernst Haeckel, the German Darwinian who coined the term *ecology* in the 1860s, apparently had no intention of trying to create a new discipline. He meant to indicate a set of problems suggested by Darwin's "struggle for existence" that needed further study. As a subject matter, ecology certainly makes sense. Ecological questions and perspectives enter into nearly all branches of biology. It is reasonable to conclude that there are occasions when one should adopt an ecological approach to problems in systematics, genetics, cell biology, embryology, physiology, and other kinds of biological study. Evolutionary biology is also built on the foundation of ecological knowledge, as Haeckel implied, so that detailed understanding of the mechanisms of evolution requires sophisticated ecological analysis. There is no question that the kind of knowledge that we call ecological is a central component of the life sciences and can be applied fruitfully at all levels of analysis.

However, recognizing the value of adopting an ecological perspective on a problem does not imply that there should be a separate discipline called ecology. Does it make unequivocal sense to define a discipline by framing questions about the relationships between organisms and their environment? Does ecology's claim to status as a discipline seem as obvious as the claims of systematics, genetics, cell biology, embryology, or physiology? I would suggest not, and therefore the development of ecology into a discipline needs to be explained.

Over the past century, ecologists—even while expressing faith in their subject's right to exist—have worried out loud about its lack of coherence, its status as a science, and its ability to measure up to other kinds of hypothesis-testing science. Examples of this uneasiness permeate the literature in ecology. The problem is to find an effective way to transcend the anxieties and move forward, if one accepts (as I do) that the ecological perspective is important and valuable as a contribution to understanding.

We must start by discovering the reasons behind the pursuit of ecological knowledge. We must set ecological research in its social, economic, and institutional context and investigate the motives and goals of the scientists who did this research. Our story begins before ecology came to be recognized as a discipline.

Ecology developed along varying paths at different times and in different places. The full study of these paths and diverse contexts is well beyond the scope of this book. I am interested chiefly in the American context and how ecology expressed American goals and aspirations, that is, the ways in which it can be seen as an American science. But this is not a full study of ecology in the United States, either. I have chosen places, people, and episodes that reveal some of the central themes of ecology and address some of the central problems that have to do with ecology's place in the scientific and cultural landscape. Because this study carries us up to the present, the historical analysis can also be used to address a deeper problem that we must consider as well: why ecology *should* be supported now and what it takes to make this claim convincing to society. For me this is an exercise in applied history as well as a scholarly undertaking. In the conclusion I consider current challenges to ecologists as they seek links to other disciplines and to ordinary citizens.

The story begins before the subject of ecology was well articulated, and locates the origins of ecology in a reform movement that transformed American botany in the late nineteenth century. The growth of ecology is linked to a broader American campaign in the 1890s to develop science and make it competitive with, and to some degree independent of, European science.[1] Ecology was one product of this struggle for a place on the world scientific stage. The economic growth of the nation, along with its western development and newfound interest in the Caribbean and Central and South America, also had a crucial bearing on the growth of ecology. This relationship is evident when we look at the allied field of conservation. Samuel Hays, in his important book *Conservation and the Gospel of Efficiency* (1959), insisted that conservation, in its early stages, was the handmaiden of economic development.[2] Hays examined conservation work within the federal government between 1890 and 1920 and argued that conservation was a movement of applied science, not a democratic protest, and that its goal was the rational management of resources. The early conservation movement, as he saw it, was not a "people's movement" set against business interests.

If conservation was an applied science, ecology was the research side of the same coin. Ecology was a science devoted to biological research into the nature of life and adaptation without regard for immediate practical applications. But the same economic imperatives and the same need to rationalize resource use that supported the conservation movement also supported research in ecology. In the first third of the book I explore this connection between the development of a home-grown American science, the economic expansion of the country, and the emergence of this new field of biological research. We should not conflate ecology with environmentalism and the critiques of capitalism that emerged in the 1960s.

Starting with the founding of the New York Botanical Garden and following one of its satellite operations, the Carnegie Institution's Desert Botanical Laboratory, I recount a history that gradually opens onto wider areas of ecological thought. The first four chapters are focused quite closely on a scientific community with connections to the Botanical Garden and the Desert Laboratory. My contention is that this community represents a microcosm of the larger American scientific world in the life sciences. The story of this group of entrepreneurs, which beautifully captures American values and aspirations, is absolutely central to the understanding of American ecology.

Beginning the story in New York, rather than in the midwestern universities where most histories of ecology are situated, may seem unorthodox, but I argue that in crucial respects New York was the birthplace of ecology. The backdrop against which we should view the birth of ecology was a heated campaign for American independence in biological science. That struggle culminated during the 1890s in the building of the New York Botanical Garden, the premier research establishment for botany in the country. Ecology began where wealth was concentrated enough to endow scientific institutions, and where the interests of industrialists matched the goals of scientific research. The story told in chapters 1 through 4 traces the link between the expansion of American science, the break from Europe, the growth of new institutions, the investment of new patrons of science, and the development of ecology as an area of research and a new discipline. I highlight the ways in which the New York Botanical Garden served as midwife for the emergence of ecology in the United States.

I also emphasize the close relationship between ecological research and other lines of investigation that were pursued at the same time, especially in evolutionary biology. In its early stages ecology was not a well-defined subject but a rather eclectic conglomerate of research that included climatic studies, descriptive work on distribution of species, and experimental studies of adaptation that were mainly physiological. Because there were no sharp boundaries that defined and delimited ecology, one can view ecology as growing out of different scientific enterprises, for example biogeography, natural history, or physiology. Thus ecology can be seen as the modern extension of Humboldtian science, or Darwinian science, or laboratory-based physiological research on adaptation.[3] Although I agree with these historical interpretations, I will add to them a context important to the American story that may not have an equally crucial role in the development of ecology elsewhere in the world.

I have found it instructive to see ecological work as part of a larger project to expand research in experimental evolution, an exciting area of study in the early twentieth century. This connection is missing in most histories, or where it is mentioned it has not been properly related to the rise of ecology.[4] It is important to see the relationship between ecology and experimental evolution because the goals of both kinds of research were similar—to understand the nature of life well enough to gain some control over it and over the evolutionary process. Knowing why there was interest in the experimental analysis of evolution helps one understand why there was interest in exploring ecological problems and in funding this kind of research.

Discussion of the origins of ecology leads to broader consideration of some of the central theories and debates that defined ecology from about 1910 through the 1930s. In chapter 5 I consider two competing ways of envisioning what ecological science should be. Ellsworth Huntington thought of it as a broad science that embraced the study of humans and thus was linked to eugenics, public health, and human geography. Frederic Clements thought of ecology as a biological discipline, but one that held important lessons for the organization of society. Chapter 6 examines critiques of ecological ideas and shows their relationship to broader American debates about history, science, and progress in the postwar period.

Chapters 6 through 9 deal with ecological research after the Second World War and develop the themes of the earlier chapters. By the 1950s ecology was a discipline, but one of relatively low status. Debates about the meaning and future of ecology grew from and responded to the theoretical and practical problems of the prewar decades, but the atomic age raised new, urgent problems. Much debate focused on the "growth culture" of the postwar period and the rapid changes that were occurring in the land. I explore how the cold war fostered the emergence of ecosystem ecology and how a split occurred in ecological thought between those who wanted to retain an older, natural-history perspective and those who wanted a new, modern approach that borrowed from physical science. In the 1960s ecology also became a "subversive" science, whose role was partly to critique modern assumptions about technological fixes and economic growth, rather than to promote rational management, as had been the case in the early twentieth century.

After examining how ecologists have grappled with the definition of the ecosystem and the concept of equilibrium in scientific thought, we return to the New York Botanical Garden. From that launching pad we land in Baltimore, the site of a present-day urban ecosystem study. This research project, which is in its early stages and still evolving, provides an opportunity to reflect on one of our main themes: the way ecologists are constantly being challenged to adapt to new circumstances and redefine their science as well as their relationship to the public.

I conclude with some observations about the need for ecologists always to think about the core mission of their science: to teach us all how to meet change with change. Ecologists must be willing to think explicitly and critically about the role of their science and its place in society, and how to communicate to the public the need for adaptive social change. At all times they must respond to changing conditions and be willing to envision their field in new ways. That might mean forming new relationships with other disciplines and revisiting older ideas that may still have important lessons to impart. I argue in the conclusion that ecology has become so bound by its traditions that it is now difficult to adopt a human-centered approach to ecological problems. Yet conveying the lessons of ecology and using ecological knowledge to stimulate social change requires the ability to see things from a human-centered perspective.

Ecology cannot have fixed boundaries and ecologists must not shy away from ambitious goals. They grapple with some of the hardest problems in science and cannot afford to feel oppressed by the claim that theirs is a "soft" science. They must exercise imagination and intellectual flexibility. Ecology exists as a discipline because people one hundred years ago and more had imagination, drive, and flexibility, because they dared to be great and even dared to grate. Their vision of science differed from ours, and aspects of their worldview, especially its racist assumptions, are off-putting to modern sensibilities. But it was a vision they pursued with remarkable perseverance. We are indebted to them for beginning a research tradition that has grown increasingly important for addressing the problems that beset our overcrowded world.

To establish context for the next three chapters, which focus on various reforms that furthered ecological research in the United States, let us first consider how Americans were talking about their land and its future uses in the late nineteenth century and what role they imagined science would play in that future. What was the dominant mood of the time? How were Americans thinking about their social progress in relation to changes occurring in their land? Are there expressions of a common worldview or shared ideology behind the scientific developments that were so crucial for the formation of ecology? Three Americans, writing in a characteristically American vein, expressed the mood that prevailed when ecology was in its infancy: George Perkins Marsh (1801–1882), Nathaniel Southgate Shaler (1843–1906), and Lester Frank Ward (1841–1913). All three wrote about American life with a forward-looking vision, offering prophecies of the world their descendants might live in. All three believed in American power, ingenuity, and progress. As they thought about the future, they raised questions about human foresight, divine providence, the plasticity of human nature, and the relationship between humans and their environment. Their ideas reflected a shared national ideology of American exceptionalism, the belief that the natural resources and republican institutions of the United States would enable the nation to evade the problems that beset European nations and to meet the challenges of the modern age with the help of scientific rationality.[5]

In 1864 George Perkins Marsh published *Man and Nature*, which surveyed how humans had transformed their landscape and argued for a more systematic, scientific analysis of what that transformation entailed.[6] The book appeared in an expanded second edition in 1874 with a new title, *The Earth as Modified by Human Action*. It became a classic of environmental writing, has been periodically reprinted, and is cited often in ecological and environmental literature.[7] Marsh was a man of parts: a businessman, a politician, a diplomat, and a scholar of the English language.[8] He wrote *Man and Nature* during the Civil War, while serving as ambassador to the new Kingdom of Italy, and derived many of his insights from comparisons between the United States and Europe.

Marsh's book drew attention to the ways people everywhere transformed their lands, not always destructively but frequently so. Cutting forests increased erosion and made the land drier. Wherever man colonized, Marsh observed, tenacious weeds and other pests arose to plague him. Many large mammals had been hunted to extinction since the invention of gunpowder (a ferocious pace of killing that picked up speed in Marsh's lifetime, threatening the bison that were still abundant as Marsh wrote). In the United States the people were unsettled, almost nomadic, he complained, and the physical face of the land shared that quality of incessant change. It was time for Americans to settle down a bit and devise ways to restore some of the lands denuded of forests, so that the nation would become "a well-ordered and stable commonwealth, and, not less conspicuously, a people of progress."[9]

Marsh's perspective was a religious one, in line with the viewpoint of natural theology that saw the balance of nature as ordained by God. Nature exhibited order, harmony, and purpose, and humans were transgressing nature's laws, thereby threatening the downfall of all: "The ravages committed by man subvert the relations and destroy the balance which nature had established between her organized and her inorganic creations; and she avenges herself upon the intruder, by letting loose upon her defaced provinces destructive energies hitherto kept in check by organic forces destined to be his best auxiliaries, but which he has unwisely dispersed and driven from the field of action."[10] The earth was becoming unfit for its "noblest inhabitant" and, he thought, might even be reduced to a state that would not support the species. As David Lowenthal argues, Marsh's key insight was that such imbalances did not correct themselves automatically.[11] Humans had to restore what humans had disturbed.

It was not just the changes observable in his time that Marsh had in mind, but also the grand engineering projects that were likely to unfold in the future—major enterprises like the Panama Canal, the draining of the Zuider Zee in the Netherlands, river diversions, excavations for mining—all of which raised the possibility of major disturbances as well as more subtle effects that were hard to trace. He did not disapprove of such great feats of engineering. On the contrary, he devoted a chapter to various land reclamation projects, as in the Netherlands, as examples of necessary and intelligent ways to accommodate a rising population. Nonetheless, these large projects likely would have effects on climate, on ocean currents, and on animal and plant life that were only dimly understood. He foresaw that future projects could have enormous impact on the earth, so much so that human actions "must rank among geological influences."[12] The fact that the effects of each action could not easily be calculated did not mean that they should be ignored.

Marsh asked whether man was "of nature or above her" and concluded that humans were above nature, exerting an influence on nature that was different in character from that of other animals. His purpose was to show that humans were free moral agents working independently of nature, not mere automatons.[13] They did things guided by "self-conscious and intelligent will aiming as often at secondary and remote as at immediate objects." This special status had negative and positive aspects. Animals were limited by the cravings of appetite: only man "unsparingly persecutes, even to extirpation, thousands of organic forms which he cannot consume."[14] But humans also had the ability to design their future intelligently. His book was meant to awaken people to the idea that they should cease to act blindly and should take steps to understand the consequences of their actions and preserve what was needed for the sake of human civilization. Humans had to become co-workers with nature in the reconstruction of the damaged fabric.

Reconstruction of nature's balance involved careful planning, such as reforestation to create a better balance between forest and farmland, or hydraulic projects designed to regulate water flow. Marsh hoped that people would understand the need for scientific study, so that the impacts of human activity could be anticipated and remedied by further engineering and reconstruction. The New World was better able to handle these challenges than the old. Europe seemed to Marsh to be sinking into deeper desolation. He was not convinced that Europe could be rebuilt in the immediate future, at least not in time to keep the "thronging millions" from emigrating.[15] The American continent, southern Africa and Australia, and perhaps a few small oceanic islands were the theaters where Marsh saw humans engaged in transforming the face of nature. As Lowenthal points out, these themes were characteristically American in the way they expressed faith in human power and a commitment to the future.[16] Science was the key to power: more measurements taken at many observation points would allow scientists to gauge the impact of human actions on the land, on climate, and on other organisms.

Nathaniel Southgate Shaler, founder of American geology and for forty years a professor at Harvard University, picked up Marsh's themes and brought them forward in time, adding a racial twist. The goal of his 1891 book *Nature and Man in America* was to trace the probable future of the social and economic development of North America.[17]

Shaler had been a student of Louis Agassiz, the famed comparative anatomist and founder of the Museum of Comparative Zoology in Cambridge, Massachusetts. When Agassiz arrived in America in 1846, he noticed that the forests were far more diverse than those in Europe; he thought they were like the forests that had covered Europe in the Miocene time. There were also many animal species that had vanished from the Old World. In 1860 Agassiz gave Shaler this problem to investigate.

Shaler did not, like Marsh, think that Europe had become impoverished in species as a result of poor treatment of the environment. Shaler decided that the continents were simply at different stages of geologic history. Drawing on the idea that the history of life progressed through stages marked by geologic epochs, he thought Europe must have moved further ahead in its organic history than any other land. North America lagged behind by one geologic period. Asia lagged to the same degree, and Africa and South America lagged still further. Australia was furthest behind, having an assemblage of animals reminiscent of Europe in the Jurassic period.[18]

Following Agassiz's death in 1873, Shaler developed these arguments in an explicitly Darwinian context. Agassiz had never accepted Darwin's theory, but his students all became evolutionists and after Agassiz's death grew bolder in publicly expressing their allegiance to these new ideas.[19] Shaler began to think about the way geography might influence how the Darwinian struggle for existence would be played out. In *Nature and Man in America* Shaler continued this line of thought, focusing on what would happen when geographic changes forced species to migrate and come into contact with new species. He was also interested in what physical conditions were likely to produce the strongest species. The development of life on different continents depended a lot on physical accidents, such as mountain-building. In mountainous regions, where climate changed as the elevation changed, many kinds of species would be packed into a fairly narrow area. The struggle for existence would be intense and therefore promote evolution. These observations were applicable to human evolution as well: mountainous areas were, he noted, cradles of strong peoples.

The way Shaler began to extend the evolutionary thought of his day was in considering what might happen when groups of species were simultaneously affected by geographic or climatic change. Under some conditions, entire groups of species, the "great armies of life," might migrate and invade a new region, rather like the "great migrations which brought the people of northern Asia and northern Europe upon the civilized districts about the Mediterranean."[20] These massive migrations of "organic armies" would result in huge contests between the flora and fauna of different regions. The geologic development of continents had impelled the flora and fauna to "interactive and continual migrations," ultimately producing not just strong species but stronger assemblages by means of the Darwinian combat that ensued. The larger battles of life were not single combats, but "contests between consolidated armies."[21] In these contests, what was destructive to the individual might be good for the assemblage—proof, Shaler thought, that there was a great plan underlying all creation. Always the "whips of necessity" drove organic beings "upward and onward toward the higher planes of being."[22]

This line of argument led to the conclusion that the North American continent lacked the geographic features and resources to serve as the cradle of civilization, which explained why the native Indian populations had not advanced.[23] The Indians, whom Shaler admired as a people, were forever limited by the land they occupied; they could not develop as well as the people of northern Europe and Scandinavia, the true cradles of civilization. Shaler believed that racial differences were markers of the conditions of the geographic "stage" on which people had played their parts. These racial characteristics, once fixed, became stable for long periods, excluding the possibility that American Indians would evolve quickly and catch up to Europeans.

But if North America could not be the cradle of civilization among its native populations, it was "tolerably plain" to him that the continent was "well suited for the development of northern Europeans," who already had achieved a high level of evolution.[24] Transplanted to new ground, they had the ability to use the land as the native peoples could not, although they had to bear many hardships before they could settle the land. He dismissed arguments that the white race might deteriorate or lose the variability that was so important to continued evolutionary advance. He argued that diverse occupations would effectively maintain variation in the character of the people. He even believed that vocations would become hereditary.[25] Given the land's richness and the advanced state of its European population, evolution could only unfold in an ever upward direction. In fact, as Shaler explained in a mammoth survey of the history and resources of the country published in 1894, the very success of "northern Aryans" on the continent was itself proof of "how well suited the land was to their needs, and how well they were themselves suited to the inheritance."[26]

Shaler envisioned that in regions of North America not suitable for development, notably the northern part of Canada, the best use of the land was recreational. Northern Canada could become a vast continental park, preserving the primitive landscape along with the many animals that could only find protection from extinction in undeveloped regions. Here, he thought, the native Indian would find his sanctuary too, a place where he could avoid the vices of civilization, live in a traditional way, and "long remain in his pristine state, the noblest animal of this vast preserve."[27]

Human moral and intellectual advance was linked to the geographic features that had developed through the course of geologic history. This unity of the physical, organic, and moral was evidence to him of divine providence. A divine plan underlay all: the continents were "great engines operating in a determined way to secure the advance of life."[28] As David Livingstone makes clear, Shaler combined a feeling of reverence for nature with a hard-nosed pragmatism about the need to develop the country under the guidance of the Aryan race, which he thought was particularly well adapted to these lands.[29] A spiritual link with nature was needed to encourage moral uplift, but there was no question that nature's resources were there to be used. Nevertheless, he believed that this development should take into account our aesthetic needs and build into the landscape harmonious structures that were pleasing to the eye, not just of practical use.[30]

Where Shaler saw the struggle for existence as a territorial contest between racial groups, a great battle between opposing armies, Lester Frank Ward interpreted the struggle in less racist language as a quest for individual freedom. The result, however, was similar: those transplanted to new lands had greater potential than the native species. Ward is best known as one of the founders of American sociology; he looked to science to provide assurance that orderly social progress could be achieved.[31] As a young man, while his sociological ideas were in gestation, he was also an avid botanist. His botanical studies gave him insights into the mysteries of adaptation, and as he later realized, these meshed perfectly with his thoughts about human society. The botanical problem, explored in an 1876 article, concerned the nature of adaptation and the implications that the new Darwinian logic had for the concept of adaptation.[32]

Naturalists commonly observed that plants were confined to certain environments and inferred that the places where they lived must be especially well suited to their survival. They even proposed that each species was "exactly adapted to the particular habitat where it occurs."[33] Ward became convinced that this putative "law" of adaptation was false. He transplanted assorted wild plants into his garden and observed how they fared in their new surroundings. He found that many species, removed from their native habitats, grew much better when transplanted to a new environment. These little experiments only reinforced what was strikingly evident from common experience. Each time humans disturbed the ground by cultivation, plants emerged from their "obscure natural retreats" and, spreading quickly over farm fields, grew to be "formidable enemies" of farmer and gardener. A plant taken from one country and introduced into a new land "thrives more vigorously than it did at home, and even threatens to drive out indigenous species." Such introduced species could easily break free of the protective influence of humans and "strike out into the forest or spread over the plains, carrying dismay to the native vegetation."[34] If anything, the plant kingdom seemed to Ward to display not attachment to native lands, but a love of change, almost an eagerness to change habitat. It was precisely this plasticity or adaptability that made domestication and improvement of plants possible.

All of these observations suggested to Ward that adaptations were not at all perfect. "Every plant is at all times ready to change its habitat for a better one, and this is actually going on whenever occasion permits."[35] The reason naturalists had not grasped this important truth was that they had not understood the influence of plants on each other. Adaptation had been narrowly construed as meaning adaptation to an inorganic environment. But add the organic surroundings, the various influences of competing vegetation, and one would get a clearer picture of how the distribution of plants was controlled. Ward's insight was a Darwinian one, for Darwin had also emphasized in *The Origin of Species* the "web of complex relations" between organisms that in subtle ways determined the success and failure of species in any given locality.[36] Darwin spoke of the "struggle for existence," but Ward imagined that there must exist a "mutual repulsion" between individuals, as each organism tried to repel the approach of every other one and sought "sole possession and enjoyment of the inorganic conditions surrounding it." This mutual repulsion ultimately produced stasis, a "universal state of forced equilibrium."[37] Plants would end up being confined to certain places depending on how these mutual relations played out.

The fact that plants were so confined did not mean that there existed a "rhythmical (almost pre-established) harmony between species and environment." On the contrary, each plant had a reservoir of vital force, a "potential energy far beyond and wholly out of consonance with the contracted conditions imposed upon it by its environment": "Each individual is where it is, and what it is, by reason of the combined forces which hedge it in and determine its very form. Each species is the perpetual and inexorable antagonist of every other. The 'struggle' is not alone 'for existence,' it is also for *place*. In the plant races, as in the human, there is a recognized hierarchy, the laws of which are as yet to a great extent involved in mystery. But the first principle, as in the rest of Nature, is force." The state of equilibrium found in nature was the result of the many combined forces that locked each plant into its place, where it lived "under constant surveillance" and where "every attempt to overstep its fixed limits is instantly checked." But disturb that equilibrium, as humans so often did, and the forces were set free: "A general swarming begins; some individuals are destroyed, others are liberated; each pushes its advantage to the utmost, and all move forward in the direction of least resistance, till at length they again mutually neutralize each other, and again come, under new conditions and modified forms, into the former state of quiescence."[38] The lesson Ward extracted from his botanical studies was that the way nature might appear was not a criterion of what it was capable of becoming. Liberation from confinement and realization of the organism's greater potential was always possible.

In a 1906 treatise on applied sociology, Ward came back to these lessons from his garden. His botanical results, he found, bore relevance to human society. The context for his extrapolation from plants to people was his disagreement with Francis Galton's analysis of hereditary genius and his studies of what conditions produced scientific eminence. Galton had come down firmly on the side of heredity in asking whether nature or nurture was more important in producing men of talent. Ward thought Galton had overplayed his hand and ignored evidence for the importance of nurture. Ward understood nurture to mean the environment; and as he explained, all biologists recognized the role of the environment, for "life itself is an adjustment of internal to external relations."[39]

Ward drew on his botanical experiences as he reflected on what nature and nurture meant for human progress. There was no better example of the importance of nurture than our improvement of crop species. Nature produced plants such as *Zea mays*, a relatively modest plant compared to the robust and widely cultivated maize that covered the vast cornfields of the West. The difference between the wild and the cultivated plant was a result of nurture by humans, who accomplished something that nature left alone would never accomplish. Ward now grasped the connection between his botanical observations and his social theory. With people as with plants, the forces of nature were chained, the channels of energy choked. He embraced the "new gospel of liberation": "Nurture does not consist in the mere coddling of the weak. It consists in freeing the strong. It is emancipation. It becomes a practical question and not a futile speculation. The important thing is not genius itself but the products of genius, and it becomes evident that these will depend upon the degree of freedom with which genius is allowed to act."[40] For the human mind, just as for the plant in the garden, the environment checked and repressed spontaneous efforts. Remove the cause of opposition, and the possibilities of life would be liberated.

Lest his arguments be mistaken for a form of environmental determinism, Ward added that the environment, being a passive condition rather than an active agent, did nothing to determine or produce civilization. Rather, humans created civilization when they entered into a relationship with the environment. They acted upon the environment, the environment reacted to human activity, and the interaction of humans and environment led to the creation of civilization. Ward called this interaction "synergy" (a term used in medicine denoting the correlated action of bodily organs). Civilization was a result of the progressive mastery of the environment through a synergistic relationship: man transforms and uses the environment and "compels the very powers that at first opposed his progress to serve his interests and supplement his own powers."[41]

The writings of Marsh, Shaler, and Ward provide remarkably clear connections between ecological or proto-ecological perspectives and ideas about power, control, history, and progress. We see there the easy conflation of ideas about nature and about human society and its future. This ability to glide smoothly from nature to society and back was not always logical or consistent, but it was felt to be unproblematic at the time. These writings allow us to glimpse the ideology that underlay the science of ecology in its infancy. It is in the context of such discussions about territorial expansion and progress that ecology took shape as a new science, a hybrid of natural history and experimental study. As Americans probed the natural world and formulated scientific theories about how nature worked, they were also grappling with how their own future would unfold, what their relationship with nature was, and what new things might evolve from the synergistic interactions of people and environments.

Ward reappears in chapter 2 as a polemicist in a campaign for botanical reform in the 1890s that was important for the professionalization of botany. An echo of Shaler's racist ideas can be heard in the rhetoric that accompanied the founding of the New York Botanical Garden. In addition, he inspired human ecologist Ellsworth Huntington, whose ideas about climate and civilization are discussed in chapter 5. Marsh's warnings have been periodically revisited throughout the twentieth century; for sustained discussion of the themes he raised, see chapter 6.

All three men were optimists. Their vision of the future could not be achieved without science, but science had to be remade to meet the challenges of the modern age. Although the inspiration for that reform came from Europe, Americans responded in ways that were adaptive for their environment and expressed faith in America's destiny. The first five chapters of this book tell that story of response and adaptation as the new discipline of ecology was created. The last four chapters examine the development of ecology in the more pessimistic postwar decades as the discipline, still adapting to a changed environment, took on its modern form.

CHAPTER ONE

Entrepreneurs of Science

When young naturalists visited Spencer Fullerton Baird, secretary of the Smithsonian Institution from 1878 to 1887, he would sometimes tease them by pointing out that natural history was commonly regarded as the "domain of women and children and weak-minded persons."[1] But as he well knew, that common perception was fast changing. In the United States, the move toward a higher level of scientific natural history had been started by men like Baird and by Louis Agassiz in zoology and Asa Gray in botany.[2] In the midwestern states, Charles Bessey and John Merle Coulter, following Gray's lead, built botany almost from scratch in the 1870s and 1880s and introduced new laboratory methods inspired by European science. Bessey was a leading educational reformer while on the faculty of the Iowa Agricultural College. After moving to the University of Nebraska in 1894, he built a strong research program focused on the study of Nebraskan flora. Coulter was at Indiana University in the 1880s and eventually joined the faculty of the newly formed University of Chicago in the 1890s. He established the *Botanical Gazette* as a four-page bulletin in 1876; over two decades it grew into the leading journal of botany in the United States.

By the 1890s the professionalization of natural history was well under way and new subjects of biological research were being defined both in the laboratory and in the field.[3] Historian Andrew Denny Rodgers estimated that by 1885 there were almost a dozen "creditable" American botanical laboratories, not a large number but a significant advance in only a decade.[4] Naturalists and collectors who had earlier struggled to support themselves by selling specimens to enterprising dealers now found new opportunities for careers in science.[5]

This period of growth was marked by the founding of new institutions as well as the creation of a network of agricultural experiment stations and colleges across the country. Among the most important of the modern institutions was the New York Botanical Garden, which was founded in 1891 and opened its museum in 1901. As Janet Browne has observed, botany was the "big science" of the mid-nineteenth century.[6] She was referring not only to the importance of museums and herbaria as research centers, but also to their position as hubs of worldwide colonial enterprises. Botany enjoyed huge popularity as recreation and as the basis of lucrative businesses in plant breeding and horticulture. The New York Botanical Garden was the American answer to the "big science" enterprises of the day and as such greatly influenced the subsequent development of biology in the country.

Ecology, often defined as "scientific natural history" in its early years, emerged as a field of study in the United States during this period of expansion. The founding of institutions like the New York Botanical Garden made possible the emergence of ecology by helping to train scientists at the graduate level and also by furthering the professionalization of science. Although the Botanical Garden was primarily a center of taxonomic work, by influencing how botany developed in the country it helped create a place for the new science of ecology. I think of the Garden not so much as the "parent" of this new science, but more as a midwife, helping in the birth process but subsequently not involved in the child's development.

Naturalists of the generation born around 1850 were often self-taught, their love of nature stimulated by parents or older siblings. To have careers in science they had to create the conditions that made such careers possible. They needed societies where serious discussions could be held and whose meetings might one day attract European scientists. At that time Europeans might have heard of Asa Gray and Louis Agassiz but otherwise ignored what was going on in America. American naturalists needed journals that would publish high-quality research and that Europeans would one day feel obliged to read. They needed libraries, taxonomic museums, and experimental laboratories where they could do research and train students. And they needed the wherewithal to continue exploring the world, not just compile lists of species. They wanted to study nature in its full context, to analyze ecological relationships in the field, where the struggle for existence was played out.

They therefore needed to convince the wealthy public that general science was important enough to support, that the philanthropic duties of those who had become rich from America's westward expansion and industrialization extended as much to scientific enterprises as to any other kind of good works. They needed to convince the rest of the public that science had an important role to play in American culture; they believed that through the guidance of science Americans would become more aware of their landscape, its resources, and the need to both develop and protect those resources.

The understanding of what collecting entailed was changing in the light of these goals. In 1891 when John Coulter addressed his fellow botanists at the American Association for the Advancement of Science, he developed the theme that systematic botany needed to be modernized and broadened. He was happy to see an end to the days of the mere "collector," who was interested only in finding new and rare plants and who recorded finds in an isolated way that made the plant "a text without any context." Coulter complained that the discovery of new species had often degenerated into a mania, "and sometimes into kleptomania," pursued by incompetents. But species were not diamonds, things valuable by themselves; rather, botanical information became valuable when plant life was related to its environment.[7] Naturalists wanted to understand why some species were rare and others abundant. They wanted to build on the principles of plant geography that European scientists, following the lead of Alexander von Humboldt, were elucidating.[8] They wanted to probe the dynamic struggle for existence that Darwin had described.[9] In short, they were discovering ecological thinking.

The fulfillment of this vision of a modern, energized natural history required a place to do research and to support a wide variety of research and educational activities. One such place was the New York Botanical Garden, which, after its founding in 1891, began development in 1896 and started its work in earnest around 1900 as its grand museum was nearing completion. Nathaniel Lord Britton was the first scientific director of the Garden and oversaw its scientific work for thirty years.[10] Bringing the Garden into being required enormous energy and single-mindedness, traits that Britton possessed in abundance, albeit disguised within an improbably frail-looking body. It also required teamwork, the willingness of people to agree on a common goal and work together to achieve the goal. And although Britton is the center of the narrative in chapters 1 through 3, the story that follows should not be taken as a story of the "great man" of science and his works. It is the story of a botanical revolution. Britton was a major player, but the revolution depended crucially on a society of individuals who collaborated in many different ways to develop natural history as a respectable scientific pursuit.

The creation of a world-class garden and research center was meant to prove that American institutions could compete with established European taxonomic museums. The garden would combine an educational function, an aesthetic function, and, most important, a scientific function. Britton and his allies built the New York Botanical Garden into a center of taxonomic research whose reputation rested on the credit earned by the discovery, description, and publication of the new species brought to light by the Garden's expeditions around the world. In its early years the Garden was also a center of experimental science where new subjects such as plant physiology and ecology were studied. The Garden extended research into new frontiers in two senses: by exploring and collecting in regions that were unknown botanically, and by developing an experimental program of research in its new, well-equipped laboratories. These activities, involving both field research and experimental science, came to define ecology in the early twentieth century.

Nathaniel Britton was born on Staten Island, New York, in 1859, a few months before Darwin's *Origin of Species* was published. His family was wealthy, with long-established landholdings on Staten Island. His parents had imagined a clerical career for Nathaniel, but he had other ideas. As a young man he formed a friendship with naturalist John J. Crooke, who had studied at Yale University under Benjamin Silliman and was mainly interested in mining and metallurgy. A gentleman collector, Crooke also amassed a good library and natural history collection by purchase from other collectors. He was a friend of John Strong Newberry, the first geologist to explore the Grand Canyon and an active promoter of public interest in scientific research.[11] Crooke also knew the eminent New York botanist John Torrey, one of the central figures of New York's scientific society and a founding member of the Lyceum of Natural History.[12] Crooke nurtured Britton's scientific interests and persuaded his parents to send him to the Columbia School of Mines.

At Columbia Britton came under the influence of Newberry, a professor of geology and paleontology and president of the New York Academy of Sciences. Newberry was a great champion of progress achieved through technological and scientific advance. The progress was occurring literally at "railroad speed" as the country was opened by the newly laid railway tracks. Newberry taught all of the subjects in natural history: botany and zoology in addition to his specialties, geology and paleontology. Britton would have found Newberry sympathetic to modern ideas of Darwinian evolution, although Newberry refused to set these ideas in opposition to religious belief, preferring to see God's hand in the intricate adaptive designs of the evolving world.[13] But in botany, Britton's scientific passion, Newberry's instruction was minimal and had no laboratory component. Students were largely left to educate themselves.[14]

After graduating from Columbia in 1879, Britton worked for five years as an assistant to Newberry and as botanist and assistant geologist for the Geological Survey of New Jersey. That work earned him a Ph.D. from Columbia College and then an appointment as instructor in botany and geology at Columbia, where he took over Newberry's botanical teaching. The college did not have a botany department, but it did have an herbarium that comprised the herbarium of John Torrey, who had lived on the Columbia campus at the end of his life, and the collections of three other botanists.[15] When Torrey died in 1873, he left no botanical successor at Columbia, and his collection languished.[16] Britton tackled the decaying herbarium, organizing and expanding it to form the nucleus of a botany department. By 1891 he was a full professor. He had also made an important ally in the president of Columbia College, Seth Low, who was actively building up the college in the 1890s.[17] Low's financial support proved crucial to Britton's later enterprises.

When Britton first tracked down the herbarium as a student, he also discovered the Torrey Botanical Club, which met in the same building at Madison and Forty-ninth Street. This group had started in 1858 as an informal association of botanists, known simply as "the Club" or the Botanical Club of New York, John Torrey being its most prominent member. In 1870 one member of the club started a bulletin, called it the *Bulletin of the Torrey Botanical Club*, and this name stuck. The club incorporated in 1872 and members voted to accept its charter in 1873.[18] But Torrey died the same year, and with an aging membership the club stagnated. When Britton joined in 1877, along with his friend and fellow student Arthur Hollick, another Staten Islander, they stood out by their youth. As Hollick recalled, "There were no young botanists in those days."[19] Hollick and Britton injected a dose of youthful energy into the club, and they remained lifelong friends.

Hollick's adventurous spirit and enthusiasm for science probably owed more than a little to the influence of his unorthodox father, Frederick Hollick, a physician born in England. The elder Hollick excelled in mathematics and was also an enthusiastic naturalist.[20] He trained in medicine at Edinburgh and by his early twenties was styling himself a "socialist missionary," having absorbed Robert Owen's socialist philosophy in the course of his education. By this time he had a reputation as a "boy-infidel" because of his vigorous defense of socialism, whose essence he took to be the idea that human character was formed largely by external circumstances. Poor environments created poor behavior, and in judging these behaviors due attention needed to be paid to these environmental influences, he believed.[21] Hollick emigrated to the United States in 1842 and continued lecturing on medical subjects and promoting progressive causes, the chief of which was to expand public awareness of the anatomy and physiology of sex and related matters.[22]

Coming from such a progressive household, Arthur Hollick likely had no trouble endorsing one major change in the club's structure that came shortly after he and Britton joined: the admission of women. In 1877, when he joined the club, having women at the meetings would have been "unthinkable," he recalled. The reasons probably stemmed not just from the perceived awkwardness of mixing the sexes in the very male atmosphere of the club, or from inherent biases toward women in science, both very real obstacles. There were also practical obstacles to attending the late evening meetings, especially for those who had to travel any distance. But there were women who were keen on botany, and the club eventually opened to them. The first woman to join, in 1879, was Elizabeth Gertrude Knight; six years later Britton married her. Knight, a New Yorker, had spent much of her childhood in Cuba, where her family had a furniture factory and a sugar plantation, and had developed a love of botany early on.[23] As an adolescent she was sent to live with her maternal grandmother in New York to finish school, which included preparation to become a teacher. She was employed as an assistant in natural science at New York Normal College (later Hunter College).

Knight continued her scientific research as a specialist in the study of mosses, a difficult and at the time relatively unstudied group of plants. Newberry, who thought her the best woman botanist around, encouraged her work. One of her professors at the Normal College thought Knight's marriage in 1885 would be a check to her career in science, preventing her from continuing as a teacher. She did resign her position at the college as soon as she got married, as she would have been expected to do, but she continued to do research and to publish for several years while working closely with her husband. In 1893, when the Botanical Society of America was formed, she was its lone woman member, in recognition of her ongoing scientific contributions.[24] She was remembered as a smart and strong-minded person, formidable and impatient at times but charming and vivacious as well.[25] She had the right mix of traits for the kind of crusade that the creation of a botanical garden demanded.

Henry Hurd Rusby, another young botanist who joined at that time, remembered how the Brittons dominated the Torrey Botanical Club, in particular the field meetings, which were an important aspect of the club's activities. In the 1880s Manhattan offered many places for studying rare plants. The naturalists planned regular field meetings both there and in the region. The idea was to study plants in their native haunts, to follow them "from cradle to grave," watch their struggles for existence and their habits and migrations, and study their usefulness to humans. The work of the club, Rusby noted, was ecological, although the word *ecology* was not then in general use. Rusby first met Britton on one of these field expeditions and recalled that Britton's incisive intelligence "at once attracted me like a magnet."[26] Both Elizabeth and Nathaniel were workaholics. The club was a place to cement social relationships and forge professional alliances, creating ties that lasted a lifetime. As Rusby boasted, the history of the club was the history of botany in New York, for "very little was done that was not directly or indirectly connected with us or, one might say, actually centered about us."[27]

John Strong Newberry was also a member of the club and naturally a leader of it. He served as the club's president throughout the 1880s, during which time its modest bulletin was built into a proper scientific journal. Both Elizabeth and Nathaniel Britton edited the journal for a time. The *Bulletin* became a vehicle for the reform efforts of Britton and his botanical circle in the 1880s and 1890s. Encouraged by Newberry, the young naturalists who joined the club in the 1870s and 1880s transformed it into an active, forward-looking society.

Henry Rusby's career path illustrates particularly well the rapid change in American science, driven by the economic growth of the country, that his generation experienced. Rusby began as a naturalist intrigued by the medical uses of plants, knocking about the country as a collector but with no clear sense of how to develop his scientific interests. By the late 1880s, he was set on what would be a long and distinguished career in medicine and pharmacy. By teaming up with the Brittons, he supported the creation of new institutions while rising in his chosen medical career to a position of leadership. In 1888 he became professor of botany and materia medica at the New York College of Pharmacy (later the College of Pharmacy of Columbia University) and served as dean of faculty from 1905 to 1933. He was also professor of botany at the American Veterinary College in New York (1889–96) and professor of materia medica at Bellevue Medical College (1897–1902). Rusby was active in the development of pharmacognosy, the scientific study of medical plants, and had an important role in the efforts to control drug quality after passage of the Food and Drug Act in 1906.

Joining the Torrey Botanical Club as a young man not only connected Rusby to more senior and influential scientists, but also linked him to people like the Brittons who had great ambition and drive. It is worth a brief digression to consider what young naturalists like Rusby were doing in the 1880s and how, by dint of visionary thinking, hard work, and tireless networking, they groomed themselves to be leaders of a new scientific elite.

Rusby, raised on a New Jersey farm, was physically strong and adventurous. He had a scrappy nature, and his father, a devout Methodist storekeeper and a Prohibitionist, moved the family to a farm to escape urban wickedness. Eventually Rusby trained for medicine and graduated from the New York Medical College in 1884. His love of botany, originally stimulated by his mother, was expressed through his interest in the economic uses of plants, especially in medicine.

During the early 1880s he traveled into New Mexico and Arizona as an agent for the Smithsonian Institution, his first expedition landing him in the middle of the great Apache War. George Thurber, a botanist of some eminence who lived a few miles from Rusby and knew him, described him as someone who grasped at problems but lacked discipline: "He is just one of those unpractical and impracticable chaps that hang around." Thurber hoped that roughing it in the Southwest might "knock some of the nonsense out of him."[28] Rusby studied the natural history of the Southwest and collected aboriginal relics, while linking up with western botanist Edward Lee Greene, who was developing a reputation for his opposition to Asa Gray at Harvard. Rusby collected for Greene while in New Mexico, as well as for the National Museum. To help pay expenses, he set up a little side business selling plants from New Mexico. The Detroit pharmaceutical firm of Parke, Davis & Company supported some of his Arizona explorations, which focused on medicinal plants.[29] The specimen sales also paid for his medical education, and after he graduated in 1884, he continued working for the company.[30]

Parke-Davis dispatched Rusby to South America on a collecting trip that expanded into a two-year jaunt ending in 1887. The main purpose was to locate plants of medicinal value. Parke-Davis wanted him to investigate the uses of the coca leaf, from which cocaine was derived, and to establish connections enabling the company to get supplies of the drug. Cocaine was known as a stimulant, and its value as an anesthetic in surgery, particularly in eye surgery, had just been discovered. Apart from securing supplies of cocaine, Rusby was to study the sources of supply of cinchona bark, from which quinine and related alkaloids for the treatment of malaria were derived; to locate supplies of the medicinal shrub *Eugenia chequan*, used as a remedy for winter catarrh; and in general to collect supplies of other drugs and study native remedies for diseases in order to identify possible new drugs. Locating wild cinchona trees proved difficult, because decades of overharvesting had wiped out the supply except in the most inaccessible regions, but Rusby was able to study the plant and its hybrids in cultivation. As for the coca leaves, Rusby collected twenty thousand pounds, worth a quarter of a million dollars, but a revolution broke out and the supply was lost in the turmoil. Parke-Davis called Rusby home.[31]

Not wanting to return to the United States via Panama on completion of his assigned job, which had taken him into Peru, Chile, and Bolivia in search of plants, Rusby hoped to extend his trip with a transcontinental journey from La Paz, Bolivia, through the Amazonian jungle. Anything that turned up new crops or drugs would have guaranteed interest back home. In 1851–52 William Lewis Herndon and Lardner Gibbon had led explorations of the Amazon Valley for the U.S. Navy Department.[32] They investigated the navigability of the Amazon River in connection with silver mining, looked into possibilities for trade with Peru and Bolivia, and collected information on climate, soil, crops, and drugs, with the idea of bringing home seeds or plants that might be introduced into the United States. Rusby was also following the lead of three great British explorers who had collected in the region in the 1840s and 1850s, Alfred Russel Wallace, Henry Walter Bates, and Richard Spruce. Spruce later published part of Rusby's collection.[33] Rusby had brought Bates's *Notes of a Naturalist on the Amazon* (1863) with him for guidance, along with Tennyson's poems and a travelogue on the Andes and the Amazon. He was well aware of the publicity value of such a jaunt, especially if he discovered new drugs along the way.

He enlisted the help of a local entrepreneur, Cirilo F. Kiernan, a British subject who was supervising his family's copper interests in Bolivia. Kiernan, an adventurer who had traveled widely in the interior, wanted to establish a new commercial route from the Atlantic Ocean to Bolivia via the rivers of Bolivia and the tributaries of the Amazon. He believed that a strategically placed canal could unite the two river systems and open an important trade route. The two men set out to explore the route from La Paz, Bolivia, to the mouth of the Amazon, aided by Indian workers and guides.

Rusby later recounted his adventures in a popular book, *Jungle Memories*, published in 1933, which owed something to the travel-writing style of Theodore Roosevelt, whom Rusby admired.[34] He was a naturalist very much in the Rooseveltian style: adventurous, curious about all aspects of natural history, and concerned to depict his explorations without glamorizing the hardships encountered but with a matter-of-fact tone that gave the impression that Rusby was fully up to coping with them. Like Roosevelt, he was also confident in the superiority of his race. His observations on the native "savages," although often sympathetic, never failed to convey his impression of their backwardness and their complete inability to absorb modern scientific explanations of such things as the tides. Although he was impressed with their knowledge of medicinal remedies, he viewed their medicine as largely based on superstition and outright fakery. He was, however, repelled by the brutal treatment meted out to the natives both by the mixed-race people and by the whites, especially the proprietors of the rubber stations whom he encountered along the Amazonian rivers. Rusby was witnessing the beginning of the genocide of these peoples, as he came to realize. By the time he returned to the same region in 1921, their annihilation was nearly complete.

Aside from the excitement of adventure, the economic value of plants was Rusby's chief interest, and he was ever on the lookout for the profit-making potential of the South American flora. He brought back with him a rich collection of plant specimens, estimated at about forty-five thousand in all, including some new plants with medicinal properties. After he recovered from the rigors of the trip, it fell to him and the Brittons to identify and describe the species in his collection, which he had given to Columbia. This task proved impossible. The reference herbarium at Columbia and the library did not provide enough information to identify the species. The Brittons decided to go to England in 1888 and seek help from the botanists at the Royal Botanic Gardens at Kew, outside of London. The trip was a turning point: it stimulated the New Yorkers to act on ideas they had tossed about in conversation many times and to form a plan that would lessen their dependency on European institutions.

The problems encountered by Rusby and the Brittons were typical of the difficulties that American botanists faced. Taxonomists, to do their work, needed to consult the "type specimens" of known species, which were held by museums. These were the individual specimens of plants and animals that were the basis for naming and describing a new species and were kept by museums to serve as representative examples of each species. When in doubt about the identity of a particular find, the taxonomist needed to compare it to the type specimen. Because Kew had a large collection of type specimens, it could provide the help that the Brittons needed.

The Brittons were treated handsomely at Kew but found its leadership pessimistic about the future of systematics in Britain despite the great success of the Royal Botanic Gardens.[35] Few systematists were being trained in England; Joseph Hooker, Kew Garden's former director, feared Kew's preeminence would suffer. Kew was the center of a large botanical network extending throughout the British Empire, but its relationship with the curators of the colonial gardens was changing at this time. During Hooker's directorship plants had to be sent to Kew for identification, because colonial workers were not considered competent enough to identify their own plants.[36] Kew housed the type specimens of the colonial floras and determined the standards used to name plants. But after being inundated with colonial specimens year after year, the staff of Kew became burdened with identification of the same plants over and over. The result was that the colonials were encouraged to do more, specifically to identify their local plants and to send to Kew only interesting finds. Colonial gardens that had not built up large collections because they had been sending specimens to Britain were now encouraged to establish their own herbaria.[37]

The loosening of Kew's grip on the imperial botanical network must have reinforced the Brittons' sense that American science could develop along an independent path as well. Although it would not be possible to become fully independent of European museums, for future collections and discoveries it made sense to establish taxonomic centers at home. American scientists wanted better control over the further exploration of their own land and over the still unexplored regions of Central and South America. Viewing Kew Gardens, the Brittons realized that what had been created outside London could be emulated in New York. Elizabeth Britton discussed their impressions of Kew at the Torrey Botanical Club soon after their return from England in the fall of 1888.

The center for botany at this time was the Gray Herbarium at Harvard. Asa Gray, America's "captain of botanical industry," had developed his herbarium, library, and garden with his personal money but gave them to Harvard in the 1860s.[38] Wealthy donors put up the money for a new fireproof building, into which Gray and his two hundred thousand herbarium specimens moved in late 1864. Gray's plans to build botany at Harvard depended on philanthropic support.[39] Only as he neared retirement in 1873 was he able to complete his establishment by adding a lecture room and laboratory for students to round out the existing herbarium, library, and gardens. The chronically underfunded Botanic Garden in Cambridge, which he had struggled to support for many years, was never developed. Instead attention shifted in 1872 to the newly formed Arnold Arboretum in suburban Boston and to its director, Charles Sprague Sargent, who had ambitions to develop the arboretum. These improved resources were a great boon to botany at Harvard but fell short of the idea of creating "an American Kew" there. Britton's plan for developing something along the lines of the Royal Botanic Gardens also signaled his desire to challenge the authority that Gray had held in American botany. With Gray's death in January 1888, a few months before the Brittons went to Kew, a more direct attack on Harvard's dominant position in botany became possible. Britton set about creating a coalition that would accomplish these goals.

Britton always claimed that the idea for a botanical garden in New York originated with Elizabeth Britton during their visit to Kew in 1888. Probably she did express the idea, and Britton, in giving his wife credit as a catalyst of this movement, was doing justice to her selfless devotion to the Garden over the years. But Henry Rusby's recollections suggest that the idea had a different origin. Rusby claimed that as growing settlement destroyed the wild areas where botanists had explored, the need for a protected garden occurred to nearly every botanist, whether professional or amateur.[40] In 1874 the Torrey Botanical Club had appointed a committee to act with the New York Pharmaceutical Association in asking the city to establish a botanical garden in Central Park, but nothing had come of this effort. Rusby remembered the subject coming up often in the club in the 1880s "at the little meetings which gathered around the old pot-stove in Professor Newberry's room." As he recalled, "these conversations would invariably turn to the subject of a possible botanical garden in this city, the old Hosack Garden serving as our text."[41]

The Hosack Garden, formally known as the Elgin Botanical Garden, began as a private garden created around 1801 on twenty acres of land owned by David Hosack on Manhattan Island (at present the site of Rockefeller Center).[42] Hosack was a leading physician in New York and at that time was a professor of botany and materia medica at Columbia College. He found it too expensive to maintain the garden and sold it to the state in 1811. The land was given to Columbia College in 1814, but it deteriorated over the years. As the value of the land increased, Columbia profited from the sale or lease of what was now prime land for development. By 1875 the property was fully developed and the garden had disappeared, but people remembered its connection to Columbia.

It was not until after the Brittons' visit to Kew in the late summer of 1888 that any real plan was formulated, and in this respect Elizabeth Britton's discussion of Kew on their return may have helped the club to visualize the kind of garden they needed. But others had similar ideas. In November 1888 the *New York Herald* published two appeals, supported by enthusiastic statements from a variety of botanical experts, for a garden comparable to the gardens of Europe.[43] This kind of enterprise was considered a sign of American cultural maturity: the *Herald* quoted a foreign visitor as remarking that in artistic matters Americans were mere children. The anonymous writer of the first article touted the merits of a garden to further the "artistic, softening and civilizing education of both rich and poor."

Because the emphasis was on art rather than science, prominent florists were canvassed for their opinions. Florists considered the botanical gardens in Washington, D.C., the best, but the *Herald* claimed that these gardens were thought to be a swindle, since they were used to supply senators, congressmen, and other politicians with flowers at the expense of the taxpayers. Others thought Washington just too miserly when it came to supporting botanical science. Most agreed that above all, any garden in New York should be free of political influences: "A garden of this kind should be managed by a gentleman, who should be a 'pope.' He should not be dictated to by governor, Mayor, Aldermen or anybody else, except by the love he has for the things he cultivates."[44] Another expert favored having a connection between the botanical garden and New York's colleges. Yet another idea was to have a local incorporated society manage the garden and run its educational program.

Arthur Hollick directed the club's attention to these articles and wrote a letter to the *Herald* endorsing them, pointing out that the club had long been interested in the idea and had encouraged government support of botanical research and exploration. The *Herald* reported on the club's response and sought the authoritative opinion of the assistant secretary of the British Museum (Natural History), Thomas Nichols, who happened to be in the United States. He was quoted at great length as being highly favorable to the plan, not only because botany lagged behind zoology, but also because botany was an easy way to educate women in science: "In the old country the number of lady students of botany is surprisingly great." Nichols advised getting the men who held the purse strings involved, even if it meant including politicians: "On our Board in England we have the First Lord of the Treasury [the prime minister] (who can be badgered politically in Parliament on matters of museum principle), the Chancellor of the Exchequer (who can be dunned for more cash), the First Lord of the Admiralty (who can use his ships and officers for bringing home specimens from foreign parts gratis), and so on."[45] This proved to be shrewd advice.

The Torrey Botanical Club immediately appointed a committee to see what action they might take. Emerson Sterns, a young man with literary pretensions as well as botanical interests who worked with Britton on a survey of the region's flora, threw himself into the task of promoting the garden. He obtained the support of the city newspapers and got the park commissioners to agree to appropriate land if enough money could be raised within two years.[46] Because he was the editor of a minor literary magazine, Sterns was able to get a pamphlet printed in early 1889 appealing for a public botanical garden for both scientific and educational purposes and as "a place of agreeable resort for the public at large." The pamphlet left the actual name of the garden blank, calling it simply "The X ____ Botanic Garden," with the idea that if anyone coughed up a million dollars or more, then that person's name could be put in the blank.[47] The public would support this kind of venture, club members thought, on the evidence of the popularity of flower shows. The botanists were undecided about the garden's management and how it might be subsidized. But they agreed that it should be devoted to American plants as well as others. They projected a size of about fifty acres as suitable for these purposes, more than double the size of the Elgin Garden but within the size range of most botanical gardens.[48]

There were at this time about two hundred botanical gardens worldwide, of which only a small number were large in area and well equipped as scientific institutions.[49] The largest botanical garden, at 1,100 acres, was at Buitenzorg (Bogor), Java; it was founded by the Dutch government in 1817 and was well funded. Kew Gardens covered about 260 acres and in the 1890s employed about 150 people.[50] Kew had started the world's first museum of economic botany in 1848. In addition it served as the nucleus of the botanical gardens and experiment stations in the British colonies, most of which were manned by staff trained at Kew. There were smaller botanical gardens in Edinburgh (about 60 acres) and Dublin (40 acres). On the Continent the chief centers were the Royal Botanical

Garden of Berlin, the Jardin des Plantes in Paris, and the Imperial Botanical Garden at St. Petersburg, with several smaller gardens located in Austria, Germany, Sweden, Denmark, and Switzerland.

In the United States botanical gardens were generally on a much smaller scale, but there were a few notable exceptions. The Arnold Arboretum of Harvard University, maintained by a cooperative agreement between Harvard College and the city of Boston, had about 160 acres with plans for significant expansion. The Botanic Gardens of the U.S. Department of Agriculture, in Washington, had a large herbarium, extensive greenhouses, and a large tract of land under cultivation, but their laboratories were poorly located and poorly equipped. The 670-acre Missouri Botanical Garden in St. Louis had just been established in 1889 through a bequest by philanthropist Henry Shaw, who had previously maintained the garden privately. There were also much smaller botanical gardens at the Michigan Agricultural College, the University of California at Berkeley, and the University of Pennsylvania.

The club members, though enthusiastic, had no clear plan for raising the money for a garden. The committee was expanded to include Rusby and Addison Brown, president of the club.[51] Brown was a district court judge and a specialist in bankruptcy and admiralty law. He also had serious scientific interests, especially in botany and astronomy, and he lent his energies and his legal expertise to the campaign. The enterprise also acquired two important supporters: Charles Patrick Daly, a former chief justice in the Court of Common Pleas in New York and president of the American Geographical Society, and his wife, Maria Lydig Daly. Charles Daly had long been interested in public parks and scientific exploration.[52] He had married Maria Lydig, from a prominent and wealthy New York family. Maria also developed an interest in the botanical garden, making it one of her pet projects. She first discussed the project with Newberry and Addison Brown in March 1890 and was promptly elected as a member of the Torrey Botanical Club. She and her husband held meetings and dinners at their house, enlisting the support of other prominent New Yorkers.

Brown and the Dalys helped to make connections to the wealthy elite of New York, as did Charles Finney Cox and William Earl Dodge, both of whom became interested in the garden. Cox was a railway officer on the New York Central system and a member of the New York Academy of Sciences. Dodge was a partner in the mining and metallurgy firm of Phelps, Dodge, cofounded by his father in 1833. The family also had extensive landholdings in the northwestern section of the Bronx. Britton knew Cornelius Vanderbilt, whose father owned a farm on Staten Island, and Vanderbilt was soon involved in planning the campaign for the garden.

Soon after the club's pamphlet was circulated, the park commissioners assured the club that space would be found in one of the new parks in the city. Bronx Park and a property known as the Lydig estate, lying on both sides of the Bronx River, were sites well adapted for a garden, the club's committee decided. This estate had been part of the extensive landholdings of Maria Daly's family. Charles Daly had in 1886 been appointed one of the commissioners of the areas that were to become public parks, and he and his wife directly profited when the Bronx site was chosen for a park. These purchases occurred just before the campaign for the botanical garden got under way. Charles and Maria Daly, both heavily involved in the organization of the garden, favored locating it on the Lydig estate.[53]

The park commissioners had stipulated certain conditions for the establishment of the garden, which included raising a million dollars by subscription to ensure the success of the plan. The minimum individual subscription was fixed at thirty dollars. The amount was laughably low in comparison to the total needed, but the campaigners wanted to convey the idea that the drive to launch the garden was one in which ordinary people, not just the wealthy, would participate.[54] In fact the public subscription raised hardly any money. Emerson Sterns, who had worked so hard to promote the garden, succumbed to severe depression and had to be committed to an asylum.

Brown and Daly persuaded the park commissioners to extend the deadline for fund-raising by two more years. The commissioners, impressed by the beauty of the Bronx site, responded well to the idea that there might be an "American Kew" erected there. The botanists stressed that the Bronx site was in every way better than Kew: more spacious, more fertile, more diverse, and more beautiful. The city would acquire not only a beautiful pleasure garden, but also a great scientific institution, comparable to the important botanical gardens in Europe. In the United States there was almost nothing like it, apart from the Shaw Gardens in St. Louis.[55]

By 1891 the fund-raising requirements were scaled back to allow for a more gradual pace of development. Addison Brown and Charles Daly drafted a charter that was designed to ensure the city's financial support, while enabling the scientists to operate unimpeded by political influence. President Low of Columbia drew up the charter on the basis of their suggestions. Rusby lobbied the New York legislature in Albany to help get the act of incorporation passed. It was secured in April 1891 by special enactment of the legislature. Among the fifty men who made up the corporation were the wealthiest and most prominent members of New York society. *Harper's Weekly* gave due credit to the women who worked behind the scenes.[56] An editorial lauded their role as part of a larger contingent of "intelligent women" who were becoming actively involved in various enterprises, notably reform of poorhouses, cleaning up the city, and raising funds for cultural institutions.

Brown presented a new plan to the incorporators in May 1891: instead of raising the money by small subscriptions, he proposed raising a quarter of a million dollars by persuading ten gentlemen to contribute twenty-five thousand dollars each. Brown made the first pledge, then used his promise as leverage to find other donors. He knew that the success of the entire enterprise depended on the support of Columbia College, and he lobbied Seth Low and the other Columbia trustees. By December 1891 the college voted to subscribe twenty-five thousand dollars. The Columbia subscription raised confidence; within six months Vanderbilt, J. Pierpont Morgan, Darius Ogden Mills, Andrew Carnegie, and John D. Rockefeller made matching donations.[57]

While all this fund-raising was going on, Britton was also involved in a parallel enterprise to raise support for the work done by the major scientific societies of New York. The vehicle for this effort was the Scientific Alliance of New York, an affiliation formally established in 1892 between the New York Academy of Sciences and six other scientific societies, including the Torrey Botanical Club.[58] The union of these groups was meant to provide lobbying strength to raise money for scientific publications and for a new building that could house the societies' meetings, provide lecture halls for educational programs, and contain a grand central library of scientific works. The Alliance lasted for sixteen years and never achieved its main goal of acquiring a building. But its members, through their constant lobbying efforts, were able to keep scientific matters well in the public eye and in the newspapers.

The speeches given to inaugurate the Alliance illustrate how the importance of science, and by extension the benefits of enterprises like the New York Botanical Garden, was being sold to New Yorkers at this time. Charles Cox, representing the New York Academy of Sciences, stressed the need for "pure science," meaning science without obvious moneymaking applications but possessed of the virtue of permanent truth, a commodity "which moth and rust do not corrupt," something more valuable than riches in an age of "business and bustle" that was too inclined to value only wealth.[59]

Cox, speaking to an audience with active interest in science, reminded them that their most powerful enemies were the average mass of men engaged in money-making enterprises, who valued science only to the extent that it could be translated into dynamos, telephones, electric lights, dye-stuffs, mining machinery, and other commercial wares. These were the people who spoke sarcastically of "theorists and doctrinaires" and had no use for such subjects as natural history. Cox cautioned the members of his audience to fortify themselves against the "contemners of the impractical." By bringing the scientific groups together to work toward a common cause, he hoped to "stem the discouraging current of utilitarianism which is ever dragging capable men away from the firm anchorage they might find in meditation, study and investigation."[60]

Seth Low, president of Columbia College, argued that a scientific alliance would function something like a chamber of commerce, by representing all scientific interests rather than the special interests of one group; it would thereby contribute to the overall growth of science in New York.[61] In this way the parochialism of the societies could be overcome, and New York could produce science on the highest level, competing with the best in the world. Addison Brown picked up the theme of America's low standing relative to the European states.[62] He held up as a model John Tyndall, the Irish physicist and mathematician who had donated the proceeds of his lectures in the United States to American universities in order to advance the cause of scientific research.

The problem, as Brown saw it, was that research in pure science, except when carried out by people of independent means, could rarely be self-supporting, for money went to those who patented and marketed inventions, not those who made the discoveries on which they were based. Brown believed these represented fundamentally different classes of people with quite different frames of mind. Those who discovered could not be expected to transform themselves into mechanics and inventors. Money was needed to endow their work. Germany, with lavish government support of science, had risen to the top position in the world. Although the United States had not lacked scientific men of eminence, on the whole the level of support for research was well behind that of Germany and Britain; scientists had to struggle to find the time and money to pursue their research. Brown believed that private money, rather than government funding, would best solve the problem; his appeal was meant to catch the ear of wealthy businessmen who might be persuaded that the cause of science was "in the deepest sense their cause" too.[63]

Britton was secretary and treasurer of the Alliance. Less given to rhetorical stances, he merely reminded his listeners that when a calamitous fire in 1866 destroyed the collection of the Lyceum of Natural History, the precursor to the New York Academy of Sciences, the Academy had been left homeless and scientific work in New York had begun to lag behind that done in Philadelphia and Boston. With a new building to support the crucial work of scientific societies, New York could achieve scientific preeminence again. His talk laid out exactly what was needed: a substantial four-story building in the central part of the city made possible by the "generous liberality of New York's wealth."[64]

This talk about the need for more support of "pure science" echoed similar calls from some of the leaders of American science going back to the 1870s and 1880s. Henry Rowland, physics professor at Johns Hopkins University, made this theme the subject of an address given in 1883 to the American Association for the Advancement of Science. Yet, as Simon Baatz has astutely observed, Rowland's argument for raising scientific research beyond a level of mediocrity to a status comparable to that of European science was not in the same spirit as Addison Brown's speech nearly a decade later.[65] Rowland had no time for the avocational scientist, the sort of person who depended on the existence of scientific societies for the presentation of his or her results. He was contemptuous of such work.

Brown, who was an avocational scientist, did approve of those societies and the people who supported them. The spirit behind the Alliance was more inclusive. The call for "pure science" was not an attempt to exclude the avocational scientist from participating in research, but rather to give those with a serious interest the resources to enable them to do better work, to transcend the parochialism that certainly could beset local societies. The Scientific Alliance was meant to give a stronger voice to people who might not be connected with a university but who, through their participation in scientific societies, expressed a positive desire to further scientific research. They needed help to obtain better facilities for research and to acquire meeting places and lecture halls for scientific discussion and educational activities.

Although the Alliance in the end did not achieve its goal of procuring a building for the societies' work, this kind of coalition-building was an important aspect of the campaign for the garden. Britton must have understood that the ability of the scientific societies to present a unified front would not only help to promote science in a general sense, but would also aid specific enterprises, particularly those promising to create lasting monuments to the greatness of New York. These appeals to civic and national pride, and to a broad philosophical spirit that transcended mere moneymaking, also helped to promote the botanical garden as a fitting companion to such grand institutions as the American Museum of Natural History and the Metropolitan Museum of Art, which together anchored beautiful Central Park. Britton's decision to divert energy into pulling the Alliance together, a huge and time-consuming task, right at the time that he was involved in a major fund-raising effort for the garden, seems puzzling. But it shows that Britton understood the value of his connection to the Torrey Botanical Club and recognized the potential power that could be wielded if all of the societies could speak with one voice.

Even with a list of incorporators that included Andrew Carnegie, J. Pierpont Morgan, John D. Rockefeller, and Cornelius Vanderbilt, and with the enthusiastic support of the Dalys and a group of women dedicated to furthering the cause, the fund-raising for the garden grew sluggish after 1891. Financial depression brought new problems, but the plans for the garden were also under attack. The chancellor of the University of New York (now New York University) wanted to have a share in the management of the garden and maneuvered in the New York Legislature in 1892 to have the charter amended so that the management would be divided between his university and three colleges, including Columbia. The garden incorporators compromised by agreeing to include the professor of biology of New York University, W. Gilman Thompson, on the board of scientific directors of the garden. Another attack concerned the garden's location. A movement was started to give the Park Board of Commissioners power to choose the garden's site. The board appointed a committee that recommended another location, Pelham Park, which was also supported by a vigorous campaign in the *Sun* newspaper. This opposition hampered fund-raising efforts.[66]

Maria Daly died in 1894, but Judge Daly pressed on with the campaign in his wife's memory, stressing how important it had been to her. Several smaller subscriptions were secured, and by June 1895 the corporation had gathered the full $250,000. The act of incorporation had been amended in 1894 to authorize the park commissioners to set aside up to 250 acres and to construct the necessary buildings in Bronx Park, provided that the corporation could raise another $250,000 within seven years. Now the idea of raising an additional quarter million by small subscriptions was abandoned, and plans moved forward to secure the garden land. In August 1895 the Bronx Park location that Britton had wanted all along was approved, and the New York Botanical Garden was ready to be developed. The officers of the board of managers were Cornelius Vanderbilt, president; Andrew Carnegie, vice president; J. P. Morgan, treasurer; and Nathaniel Britton, secretary.

An editorial in the *New York Times* signed "Botanicus" extolled once again the great beauty of the proposed site, which was esteemed because its beauty was natural, not artificial (implying a comparison with Central Park). Botanicus was undoubtedly one of Britton's allies, if not Britton himself. Certainly the ideas presented were close to Britton's. Botanicus emphasized the importance of undertaking scientific work at the Garden and placing it in the control of scientists.[67] This idea—that land of exceptional beauty should be turned over to scientists for management and for scientific purposes, in addition to general educational purposes—was by no means taken for granted by New Yorkers. In fact, it had to be constantly sold to the public, because it was not obvious that a scientific takeover of the land would have any special public benefit or offer what the public wanted, namely a recreational park and not a botanical garden as such. Even after seven years of discussion and campaigning, there was opposition by people who were critical of the Garden and perhaps suspicious of the motives of those who supported it. Botanicus responded to these critics tactfully, arguing that scientific use of the land did not mean sequestering it from public use and that the managers of the Garden were not asking for the land for their own benefit but were acting selflessly on behalf of the public.[68]

Negative feelings toward Britton and the Garden supporters as land-grabbing members of a wealthy elite were inflamed by plans to locate the director's house inside the Garden, as though the Garden were private land and not a public park. Bitter opposition emerged as discussion of the design of the Garden continued after development began in 1896. Until that time it was not clear that Britton would in fact become the scientific director of the Garden; others were competing for the post. It was ultimately decided in Britton's favor by the mayor of New York, William L. Strong, who made the appointment in June 1896. Britton's initial term as director was three years.

Charles Sprague Sargent, director of the Arnold Arboretum in Boston and one of the foremost botanists of the time, was appointed chairman of a committee of experts whose job was to scrutinize the detailed plans for the Garden. The committee's report of 1897 opposed certain details of Britton's plans, in particular the locations proposed for the buildings, which included setting the scientific director's house inside the park. The park directors were reminded that they were not designing a suburban neighborhood, that the land was not theirs,

Fig. 1.1. Nathaniel Lord Britton. He fought for a world-class botanical garden in New York, and he won. Courtesy of the LuEsther T. Mertz Library of the New York Botanical Garden, Bronx, New York.

and that the park was "essentially a reservation of romantic and beautiful scenery," where buildings should not intrude on the landscape.[69]

The committee's comments also implied that there was an inherent incompatibility between scientific needs and recreational needs and that the scientific functions of the park should be nearly invisible throughout most of the park. Getting away from the city and back to nature meant being able to escape to a landscape that had no intrusive reminders of city life. Sargent's committee recommended separating the artificial and the natural, grouping the buildings together in a formal arrangement in a section of the park that was not important for its landscape features, while leaving the main portion of the park relatively untouched, to preserve its rustic charm: "If a beautiful landscape prevails in the wilder portions of the park, and good architecture and formal gardening occupies some tame and characterless part near its border, no one will fail to recognize the extraordinary natural charms of the one and the art of the other."[70]

Articles and editorials in the *Sun* focused on these negative comments and voiced opposition through the summer of 1897, stressing the expertise of Sargent's committee and inveighing against the "outrage in Bronx Park." The newspaper relentlessly disparaged the incompetence of those who had drawn up the botanical garden's plans, expressed disbelief that such plans were supported by "well-reputed gentlemen" who were using the park as though it were a personal toy, and lamented the undue influence of such men on Mayor William Strong, who backed Britton's plans even though he had no expertise in the area.[71] Britton and his supporters were forced to make some compromises, in particular moving some residence buildings outside of the park, but in general they kept to their original plans and obtained the necessary approval from the Park Board.

In 1896 a second charter had enabled the Garden to expand its functions so that it effectively served as the botany department of Columbia. In order to devote his full energies to the Garden, however, Britton himself had to resign his professorship and was appointed professor emeritus, at the age of thirty-seven. When Britton's initial three-year term as director-in-chief ended, his position was made permanent, and he remained at the helm of the Garden for the next thirty years.

His comrades from the Torrey Botanical Club likewise took on central roles in the new institution, sometimes while holding other positions elsewhere. Henry Rusby assumed charge of the Garden's economic collections as honorary curator, over and above his work at the Columbia College of Pharmacy. Arthur Hollick had been working as a sanitary engineer for the Board of Health of the city of New York but in 1901 became assistant curator of fossil plants at the Garden, resigning in 1913 to curate and then direct the museum of the Staten Island Association of Arts and Sciences. He returned to the Botanical Garden in 1921 and remained there until his death in 1933. Judge Daly died in 1899 but left an important bequest that subsidized the Garden's scientific publications and enabled its scientists to pursue their increasingly ambitious plans for exploration and cataloging of new species. Addison Brown served on the Garden's Board of Managers from the beginning, chaired its Committee on Plans and its Executive Committee, and was president of its Board of Managers from 1910 until his death in 1913. His bequest to the Garden provided funds for a "high-class magazine" bearing his name (*Addisonia*) and devoted to colored illustrations of American plants, with descriptions in popular language.

The Garden also partly absorbed the Torrey Botanical Club. In 1901 the club started holding one of its twice-monthly meetings there, with convenient daytime hours that would enable more people, especially women, to attend. With this loss of independence, the club effectively lost the fervor, and the power to foment revolution, that it had had when the Brittons first joined. Britton himself had gone from revolutionary to czar; his main concern would be to consolidate power and ensure that the Garden would become the center of research that he envisioned and a serious competitor on the world stage of science (fig. 1.1).

CHAPTER TWO

A Botanical Revolution

The work that Britton, Rusby, Hollick, and their allies were doing to establish the New York Botanical Garden was only one facet of a broader effort to raise the standards of American science and make it more independent of European institutions. The way science was done needed to be changed also—or at least this was the view of some Americans, Nathaniel Britton foremost among them. Although they were building on the European example, they felt the need to reinvent the method and style of science and to adopt a pragmatic approach that was better suited to their immediate goals.

Americans sought not only new American institutions, but also a new set of criteria for how species were to be named. This boiled down to a revised code of taxonomic practice that they called the "American Code." The term *code* refers here to rules of nomenclature, the set of rules that taxonomists used to govern how species were named and how credit was allocated for the discovery and naming of species. What these rules should be and how strictly they should govern practice were matters of perennial discussion within the world community of taxonomists. As Americans entered into these discussions, they tried to reform the rules in a way that would favor them, a way that would enable them to get the maximum credit for their own work in taxonomy. We now turn to this parallel campaign for an American code conducted during the 1890s; it reached its zenith as the Botanical Garden came into existence.

Britton perceived that reforming the rules governing the naming of species was one way to ensure that, as he and his colleagues went out and named new species, their names and descriptions would be accepted as authoritative. Quite simply, he wanted American scientists to receive full credit for their work, especially the work done on "their" flora. At first "their flora" meant the flora of North America, but soon it came to include species found in the Caribbean, Central America, and South America, regions of growing interest to Americans. Redefining the rules of nomenclature was for Britton essential to maintaining an authoritative scientific institution like the New York Botanical Garden. American botany had to advance quickly in its race with European institutions established decades earlier. The introduction of a new code—the American Code—coupled with fast publication of new species, was one way of moving to the head of the pack. The campaign to establish the use of an American code was also fought on the home front. A challenge had to be extended to the existing botanical aristocracy in the United States, namely the botanists at Harvard University, where Asa Gray had presided, where his name was revered, and where the methods used at Kew Gardens prevailed.

The furious controversy over the criteria used to name and describe species, the subject of this chapter, also illustrates one of the major themes of this book, which is the relationship between how American botanists practiced science and how they thought about American society and the way society was developing. Political metaphors abounded as reformers pressed for a democratic, anti-authoritarian method of decision making. In these colorful bursts we can see American scientists expressing their ideas about what kind of scientific culture would allow them to fulfill their aspirations as they explored new lands and claimed new species for science. We see them engaging in their own struggle for place in the world of science.

Declaring scientific independence from British and European institutions began with these efforts of Britton and others to reform how species were named and described, one of the basic tasks of the natural historian. Normally species were given two names, the first identifying the genus and the second the species within that genus, as in *Homo sapiens*, where *sapiens* is the species epithet within the genus *Homo*. But naming a species in this way did not guarantee that its original binomial would be fixed for all time. Often name changes accompanied taxonomic revisions. For example, a species might be moved into a different genus, and in the course of that move the species epithet (also known as the "trivial" name) might also be changed. Alternatively, a single species might be split into two or three species, each requiring naming. Or two species might be lumped together, with the new name departing from both of the original specific names.

When revisions of this kind occurred, the taxonomist responsible signaled his authority by appending his name, or an abbreviation of it, after the revised name. When authoritative texts were revised, for instance Asa Gray's *Manual of the Botany of the Northern United States*, which went through several editions, there might be many name changes that botanists would have to memorize. Sometimes species acquired different names accidentally because two botanists independently discovered and named the same species. Even when species were known to have been named previously, botanists sometimes renamed them and then claimed credit for having done so.

The problem for botanists was to decide when these name changes were reasonable and legitimate. Over the years botanists had developed codes of conduct in the form of rules of nomenclature that would help the community determine what names should be selected for species. The impetus for developing rules of naming came from a leading Swiss botanist of the early nineteenth century, Augustin Pyrame de Candolle. His son Alphonse de Candolle, also a botanist, estimated that by 1845 there were on average two names in existence for each known species.[1] Botanists often preferred the older names if given a choice, but practices varied widely and sometimes reflected patriotic sentiments. The movement to devise codes of naming that would govern all botanists did not develop until a couple of decades later. British naturalists had in 1842 recommended a code known as the Stricklandian code, or "Rules of the British Association," which was revised slightly and reenacted in 1860 and again in 1865. This code applied to zoology rather than to botany. It granted highest authority to the first person to define a new genus or describe a new species. If at all possible, the first name given to a species should be preserved.

In 1867 Alphonse de Candolle proposed a set of rules of nomenclature to the International Botanical Congress at Paris. His code was adopted at the congress and, like the Stricklandian code, was based on the "rule of priority," the idea that the first name should be retained whenever possible.[2] The English botanists did not attend this congress, however, and the Kew botanists followed what was known as the "Kew rule," which allowed a botanist to ignore a known early name under some circumstances when a plant was being moved from one genus to another. The Kew rule granted priority not to the very first specific name, but to the name given to a plant when it was placed in what was considered to be its "true genus," even if the botanist making the revision did not retain the original specific name. Under the Kew rule it was this binomial that should be preserved, rather than the first specific name. Asa Gray at Harvard followed the Kew rule. It gave botanists leeway to use their own judgment in deciding whether to retain an original name.

The complaint of Britton's generation was that lack of standardized practice was producing anarchy, because botanists were altogether too free to alter names. As Emerson Sterns, a colleague of Britton's, complained in 1888: "Ingenious botanists have species at their mercy, and are free to combine, separate and recombine, and to christen the successive groups very much at their own sweet will."[3] Zoologists also were aware of these problems, and in the United States they had already (in 1876) voted to adopt a stricter priority law. Ornithologists were leading the way in this campaign for reform, although other zoologists got into the act too.[4] David Starr Jordan, ichthyologist, put the problem succinctly: "There are only two ways of naming plants or animals, either to give them their oldest names or to give them any names you please."[5]

To botanists like Britton, it seemed that the most logical course of action was to reaffirm the priority rule, which was the basis of the Paris Code of 1867, and try to make it the basis for a universal code of nomenclature. Botanists would still revise their descriptions from time to time, and they might in the process split a species into two or combine two species into one, or they might move a species into another genus. But throughout these necessary revisions, every effort had to be made to preserve the original specific name, even though the generic name was changed. Arbitrary deviations from the original name should not be allowed. Reformers began to call loudly for adherence to the "law of priority" to end the chaos into which botanists were plunged. Britton orchestrated the reform effort, ably supported by his colleagues in the Torrey Botanical Club and by botanists in the western United States who were eager to challenge the eastern hegemony of Asa Gray, with his affiliations to Kew.[6]

What fostered resistance to the priority rule was not so much the basic idea of preserving names, which in general was lauded, but the rigid application of the rule. The advocates of the priority rule wanted to apply it retroactively, so as to fix once and for all those species whose names had been improperly changed over the years. They did not always agree about how far back in time one should go, although in general they thought a sensible starting point would be Linnaeus, who first popularized the binomial system of nomenclature. But going back to the original Linnaean names (where feasible) would mean changing the names of many species immediately, creating great confusion in the short term. Was it really worth throwing botany into this much immediate confusion just for the sake of some ideal of permanent stability? Could that ideal ever truly be realized in any case? The reformers appeared to be trying to fix something that was not really broken.

Sweeping application of the priority rule also meant throwing out many post-Linnaean changes that had become established through long use. In the eyes of those following the Kew rule, including Asa Gray, the younger botanists were adopting a rigid, mechanical approach to taxonomy and were forgetting that it was an art as much as a science. Why throw out familiar names that were sanctioned by the highest scientific authorities, just for the sake of a theoretical principle? In one of the last letters Gray wrote, shortly before his death in January 1888, he reprimanded Britton for revisions that seemed to exemplify this mechanical approach. Gray took it as a sign of Britton's scientific immaturity.[7] But Britton was not afraid to challenge the highest scientific authorities. He bridled at this criticism. Behind it, he might have sensed, was some resentment at the way the younger botanists were claiming species as their own, perhaps a little too quickly.

In the debate that followed, American reformers appeared to be very deliberate in their use of the word *law* in the context of these taxonomic disputes. The "rule of law" was specifically invoked to counter the idea of arbitrary rule by a botanical aristocracy. That aristocracy included Americans, in particular Asa Gray and his followers, who were cozying up too closely to Kew. The language of this debate was chock full of political metaphors that revealed how self-consciously Americans thought of their efforts not simply as scientific reforms, done for the good of all, but as tactical moves in a larger strategic plan that would increase the authority of American scientists, especially when it came to the study of American species.

In his quest for reform, Britton was egged on by Edward Lee Greene, an itinerant naturalist and Episcopal clergyman in the American West. Greene was one of the most colorful, enigmatic, and controversial figures in American botany, a man who elicited strong emotions of either affection or intense dislike. Some colleagues regarded him as an outstanding western botanist; others thought him insane. Transferred by his church to California in 1881, he created a storm when he introduced Roman Catholic doctrines into the ultra-Protestant congregation. As he leaned more toward Catholicism, his congregation rebelled, forcing the bishop to insist on his resignation in 1883. A year later Greene became a lay Roman Catholic. In 1885 he was appointed instructor in botany at the University of California at Berkeley, his first job as a professional botanist. He remained there for six years and became chairman of the Botany Department when it was established in 1890.

Greene had been an active collector in the West for many years, and he normally sent his specimens to Gray in Cambridge to be described and named; he was following accepted procedure. But around the time he moved to California in 1881, Greene began to rebel against this practice. He stopped sending his specimens to Gray and instead began to publish his own collections, as well as those of other botanists. The conflict between Greene and Gray grew intensely personal, and toward the end of Gray's life, he simply stopped corresponding with Greene.

Greene objected to the way field collectors like himself did the work but the museum men in the East got all the scientific credit. This struck Greene as unjust. He took it personally, feeling that Gray's methods were specifically designed to rob him of scientific credit.[8] He also felt that Gray's descriptions were flawed because Gray was a "closet botanist," that is, a museum man who spent too much time studying dried specimens and had no understanding of how these plants actually looked.

Greene was setting Gray up as a straw man with this criticism, since the importance of field observation was generally accepted by this time. Gray certainly valued direct observation, but it was far more efficient to have specimens sent to him for analysis and organization than to spend months in expensive and arduous travels, without being certain of getting many new species. As Gray's biographer remarks, Gray's exceptional skills and experience made him better suited to be a "headquarters man," and from the mid-1840s to the 1870s he was indeed more tied to the working table than occupied in the field.[9] Greene's comments, although true up to a point, were not entirely fair, but then fairness was not among his virtues. Greene continued to quarrel with Gray's successors and became persona non grata at the Gray Herbarium. When Merritt Fernald, later a professor of botany at Harvard, came to the Gray Herbarium as a teenager, he was instructed that if Greene were ever to appear, he should slam the door in his face—and lock it.[10]

Part of Greene's pitch for independence was to press for adherence to the rule of priority in naming. But when he thought about priority, he was not content to start with Linnaeus, as most botanists thought sensible. He wanted to extend priority back to the very beginnings of natural history. He seemed to resent the authority granted to Linnaeus and worked to recognize the priority of earlier botanists, including writers of Greek and Roman antiquity. Even if it were impractical to extend priority to the beginnings of botanical history, he believed credit should at least be accorded to those earlier botanists whom Linnaeus himself had recognized. As he wrote to Sereno Watson at Harvard, one of his targets, he was "not well disposed toward revolutionists like Luther and Linné."[11] Just as he was concerned to maintain Catholic tradition in the face of Protestant sectarianism, Greene held to botanical traditions that were endangered by Enlightenment reforms. In this respect Greene broke with his otherwise like-minded colleagues, who were generally in favor of starting with Linnaeus.

One of Greene's students, Frederic Bioletti, recalled the zestful spirit of Greene and his student group during these years in California. Abhorring any idea of utility in botany, they approached the subject in a romantic rather than a scientific spirit. The field of botany was not so much a workplace as a playground, "a substitute for football games, track events, writing poems for the *Occident* and similar campus activities." The core of the "sport" of botany was the "art and mystery" of collection and the search for a perfect specimen. They made liberal use of the senses to distinguish plants from each other: "We became experts in detecting families, genera and species by smell and taste."[12] Such methods were the antithesis of the Linnaean approach because they lacked objectivity; Linnaeus's criteria had emphasized features of plants, such as the number of stamens, that did not depend on subjective judgments of this kind.

Because of their emphasis on a sensual approach to botany, Greene and his students disdained the "closet botanist" and "his paraphernalia of microscopes and of desiccated and fragmentary type specimens." In the field, Bioletti recalled, they were like hunters pursuing game in competition with fellow collectors, "interlopers and trespassers" who sometimes invaded their territories. Greene, the "Great Chief," was the only one allowed to roam unrestricted: "We had too much to gain from his friendship to object to his hunting on our grounds." What they gained was the privilege of having a genus or species named after them, and "for this we were deeply grateful."[13]

The reference to the "Great Chief" not only conveys a sense of Greene's charismatic personality but also raises the question of how Greene viewed his science in relation to aboriginal knowledge of plants and their properties. He certainly encountered Indians on his various travels in the West. Greene is circumspect on the question of what he may have learned from those encounters, although it seems likely that the Indians would have used a similar multisensory approach to the identification of plants. However, Greene's historical writing located the origins of scientific botany in the first treatise to adopt a nonutilitarian, philosophical approach to the subject, namely that of Theophrastus, student of Aristotle. Theophrastus was for Greene the true father of botany instead of the farmers, gardeners, and herbalists who had come before him.[14]

But Greene did admit that the utilitarian interests of ordinary medical men had influenced philosophers like Theophrastus. The practical men who most interested him were the "rhizotomi," the class of men in ancient Greece who gathered, prepared, and sold roots and herbs for medical uses. Some of them wrote books and approached their subject in a broader philosophical spirit. Their example taught Theophrastus and other botanists the importance of describing the form, texture, color, odor, flavor and active properties of plants; it taught them, in short, to think in those "ecological" terms that Greene would later commend as superior to the study of dried specimens. In this way the "half-literate rhizotomi" made "their peculiar impress upon the character of descriptive botany."[15] As for the American Indians, Greene did observe that knowledge of North American plants, and the popular names themselves, owed much to aboriginal knowledge, reflecting the continuing role of the "herb doctor" or "root doctor" in modern times. But his historical writings leave no doubt that he traced the origins of his ecological approach to the philosophical texts of ancient Greece, not to the practices of the people he encountered on his travels.

As noted, Greene could elicit strong positive and negative feelings from others. He was considered by the young men who knew him to be physically striking, a perfect specimen of manhood, and Bioletti's description has homoerotic overtones. Robert McIntosh points out that the charge of homosexuality was widely circulated about Greene but remains unsubstantiated.[16] One writer thought that the harsh criticism leveled at Greene by Katherine Brandegee of the California Academy of Sciences was motivated by unrequited love for Greene.[17] Another botanist, writing in a private publication many years after Greene's death, dismissed this opinion with the comment that she would have despised his homosexuality.[18] Scientific criticisms may have nothing to do with emotional attachments or moral judgments, of course, and whether Greene's personal style affected his standing among scientists can only be a matter of conjecture.

There were other reasons to react deeply to Greene, for he was an extremist in scientific matters. His style of reform was in fact intensely reactionary. He never accepted evolution and seemed most at home in the intellectual milieu of the Middle Ages. He might have held a particular grudge against Asa Gray, who was one of the first champions of Darwinism in America. Greene's rejection of evolution was taken as a mark against him by some critics, although it seems not to have disturbed others who supported his push for taxonomic reform. Britton regarded Greene as an ally and was always cordial, even when they disagreed. But Greene's style of criticism could often be uncompromising. Described as having courtly manners and a genial style, in his writings he threw caution to the winds, attacking his colleagues with open sarcasm. As he wrote to Britton, "I know I must repress all sarcasm and ridicule, but it is hard for me."[19]

With all of these idiosyncrasies, Greene acted as a catalyst for change. The arguments Greene was making about having better control over one's local flora fell on sympathetic ears in the West. George Engelmann, a prominent botanist of the midcentury, wrote to Gray in 1881, "It is amusing to see State Pride cropping out in these Western botanists, they seem to want their flora for themselves!"[20] Engelmann thought the emerging western identity was fine as long as the botanists in the West did their work well. But inasmuch as Greene was asserting the right of botanists to publish their results without first submitting specimens or manuscripts to Asa Gray, Gray was less amused. He regarded Greene's idiosyncratic methods and dismissal of British traditions as unscientific.

The movement for local autonomy affected easterners as much as it did westerners. Britton's zeal was evident as he urged reform of the rules of nomenclature. The rational basis of reform was the need to standardize the practice of naming new species and reduce the confusion in the field. But beneath this appeal to rationality, he shared Greene's dislike of the Cambridge group and their tendency to revere Asa Gray.

After Gray's death in January 1888, discussion of the priority law became even more intense, in part, it was suggested, because botanists were no longer restrained by fear of offending Gray, for whom they felt respect and affection. Greene wrote to Britton in 1888: "Cambridge must lose prestige gradually and it is our religious duty to pull it down as fast as we discreetly can." Britton replied, "Keep up the fire my glorious colleague, we will bring them all down after a little."[21] Another probable stimulus was annoyance at the adulation heaped on Gray after his death and the fact that the botanists of the Gray Herbarium stuck loyally to Gray's methods.

One of the first published declarations of the New Yorkers' reform intentions was in a preliminary catalog of the plant species in the New York region, compiled by six members of the Torrey Botanical Club, including Britton, and published in 1888. The catalog explicitly stated their desire to follow the priority rule, that is, to maintain as far as possible the earliest specific or varietal name (the "trivial name") for each plant, even when transferring plants into a new genus. The reaction from England was swift. James Britten, an English botanist and the editor of the *Journal of Botany*, reviewed the catalog and complained that the "revolutionary tendencies" of these Americans would result in "chaotic confusion."[22] The English botanists resisted adopting the priority rules that their Continental colleagues followed and that the Americans now seemed to be pushing to extremes.

Britten's review suggested general resentment at the very impertinence of these young Americans, for his complaints bordered on the petty. His reason for criticizing the American revisions was that each new name would be followed by the names of the revisionists, in this case, by "Britton, Sterns, & Poggenburg," the first three authors of the catalog in question. To the Englishman this cumbersome appendage was unallowable. Britton had a ready answer to this gratuitous remark: they would simply tack on the initials "B.S.P," as was common in multiauthored European works.[23] He pointed out that an American committee had considered the matter and believed that it had adopted a better method than the old system.

Not all American reformers targeted Gray's work in the spirit of Greene and Britton, however. Charles Sprague Sargent, director of the Arnold Arboretum at Harvard, held Gray in the highest esteem. He edited a collection of Gray's scientific papers that came out during the nomenclature dispute and made sure to include Gray's views on nomenclature in the selections. But he too believed that the priority rule was needed in order to create a stable nomenclature. His monumental work *The Silva of North America*, the first volume of which appeared in 1890, was one of the first major taxonomic works that adopted the priority rule. "The rigid application of this rule leads to the change of many familiar names and considerable temporary confusion," he admitted. "But unless it is adopted, anything like stability of nomenclature is hopeless, and the sooner changes which are inevitable in the future are made, the more easily students will become accustomed to them and acquire a knowledge of the correct names of our trees."[24] Sargent dedicated his volume to Gray, his "friend and master."

Outside Cambridge the reverence felt for Gray was less strong and the reform movement gained momentum, with Gray as its frequent target. The ideas behind the reform movement were not exclusively American; European scientists were also pressing for similar reforms, and these issues were hotly debated at international congresses. In Germany, botanist Otto Kuntze was leading the way in calling for reform. Zoologists had similar problems and were working on their own reforms. But the American reform movement in botany was not piggybacking on European initiatives and was not slavishly following the zoological lead. It was the expression of a growing feeling of independence and professional competence among American botanists.

The Torrey Botanical Club was buzzing with discussion of the reforms, its members largely in favor of them. At the U.S. Department of Agriculture, Frederick V. Coville, chief of the Division of Botany, pushed the cause of reform, as did the botanists at the National Museum in Washington. The reform contingent, those who supported the "law of priority" and put it into practice, included many western botanists at state universities. They hailed from such states as Missouri, Nebraska, Indiana, Minnesota, Wisconsin, Ohio, and Kansas. These were the people exhibiting the new "western pride" to which George Engelmann referred. Charles Bessey in Nebraska wrote to Elizabeth Britton in 1896, "I do not believe at all in the policy of sending American plants abroad for naming or for first distribution. I believe with you fully that the types should be kept on this side of the water. . . . In the Botanical Seminar we have been following such a policy and we have gone so far that we object to the sending of specimens that can be worked up here, outside of the state."[25] Bessey was a solid ally, arguing for reform through the 1890s.

The issues involved questions of ownership and credit. Reforming the rules of nomenclature was part of a larger goal of repatriating American natural history, that is, keeping the type specimens of North American plants at home in institutions such as the New York Botanical Garden. But in addition, these botanists wanted to ensure that American botanists got the credit for discovering and naming "their" species. As discoverers and describers of new species, their names would have priority and would need to be preserved. In the words of one reformer, Emerson Sterns, the idea of a fixed nomenclature was to do "strict justice" in the case of "every North American plant" to "the man who first gave it a published name accompanied by an adequate description."[26] The use of the priority rule, the centerpiece of the American Code, coupled with the practice of rapid publication of new discoveries, would quickly accumulate unquestioned scientific credit for these young botanists.

John Merle Coulter, editor of the *Botanical Gazette* and always the voice of moderation, hoped in 1888 that differences of opinion could be settled at the next meeting of the American Association for the Advancement of Science. Coulter had been an ally of Gray's and had leaned more toward the conservative camp until this time. After Gray's death he seemed to be willing to listen to the rebels, for the sake of achieving consensus. Rusby tweaked him a little in the Torrey Botanical Club's *Bulletin*, "It is just possible that when it is found that the mountain will not come to Mohammed, Mohammed may conclude to move to the mountain; not at all because there is any principle involved, but just for the sake of 'uniformity.'"[27] Elizabeth Britton, opening her copy of the *Bulletin* in England, was amused at Rusby's "delicious" hit on Coulter: "The rascals have done real well."[28]

Others were not convinced that agreement would come so easily. Emerson Sterns, Britton's collaborator on the catalog of flora that had prompted the ire of James Britten, noted the confusion in botany created in part by "our great leader," Asa Gray, who had sometimes obeyed the law of priority and at other times had increased the number of names for no clear reason. Reform, Sterns allowed, would encounter "the opposition of inertia, perhaps of jealousy, and certainly of honest difference of opinion." The answer, he thought, was a good, honest fight to the finish. He urged botanists to form two leagues of supporters and opponents of the law of priority. Each side was enjoined to formulate its doctrines precisely, apply them strictly, and defend them boldly. Then let nature take its course: "In due time the right side will convert and absorb all its antagonists and hold undisputed possession of the field."[29]

An editorial in the *Botanical Gazette* in 1892 pointed to the new spirit that was taking hold. Combining the idea of science as a political contest with the Darwinian metaphor of survival of the fittest, the editorial's language revealed how these botanists thought of social change in broad Darwinian terms and with an emphasis on progress or improvement.[30] The point was that the progress of science was guaranteed by the same process of competition and selection that governed the evolutionary process as a whole. The editorial described how botany had passed through an "aristocratic" period, when the favored few reigned supreme over the less favored, and had emerged finally into a new era, in which "the spirit of democracy is prevalent, one that proposes to fight not only its own battles but also those of all ancient neglected worthies." The modern period of American science was one "in which the voice of authority is to come from 'the people.'" The botanical landscape would look more like a "uniform forest" than a "cluster of sequoias towering in the midst of their lowly neighbors." (The giant sequoias, discovered by white men in the 1850s, had by the 1860s come to symbolize American greatness and manifest destiny.[31] Here, of course, the metaphor is turned around so that the sequoias represent oppressive authority, as opposed to the American spirit of democracy.)

The analogy to the uniform forest was not meant to suggest a state of consensus or stability, but the opposite, for in a uniform forest there would be intense competition and, by analogy, a Darwinian evolution toward an improved scientific system: "Everything wrought out will have to run the gauntlet of the many instead of the few," the editorial explained. The result would be improvement, a more rapid development of science and production of more efficient workers. There was a disadvantage to this democratic impulse, for it permitted "follies," "erratic movements," and "fads" that a central power would have "repressed." But extremists posed no lasting dangers: "When there is nothing left to rebel against they usually settle down into staid and comfortable citizens." Subsequent events showed that the editorial's views on progress, efficiency, and the virtues of democracy were overly optimistic, but for the moment the comments signaled that the reform movement was gaining crucial support from more conservative quarters.

Britton assumed the editorial had been written by the senior editor of the journal, Coulter, who had been associated with the camp of Asa Gray and the other Harvard botanists. Coulter had been highly critical of Greene, viewing his work as lacking any sense of propriety and verging on craziness. That Coulter should be coming around to the reform position, even expressing criticism of the "aristocratic" domination of men like Gray, emboldened Britton.

He seized on the editorial's rhetoric to reply to a pointed article by Sereno Watson, Asa Gray's successor at Harvard, that appeared shortly after Watson's death in 1892. Watson had attacked the reform movement. Britton in response derided the "aristocracy" that had acted too much on its own, without considering the opinions of others. The choice to be made was between the "Kew rule" of the "botanical aristocracy" and the Paris Code of 1867. When the time came, Britton said, that "two or three American botanists not controlled by the 'aristocracy' were by nature impelled to think for themselves," the best course seemed to be to adopt the Paris rule. The wisdom of the choice, Britton suggested, was shown by the fact that more botanists were leaning in the same direction. To cement this sense of agreement, more formal discussions at botanical meetings were needed, with platforms expressed in writing and votes properly taken. The way to do science would be decided democratically.[32]

At the time his article appeared, Britton was preparing for the upcoming meeting of the Botanical Club. The club consisted of a group of botanists from around the country who met annually with the American Association for the Advancement of Science (AAAS), which did not then have a separate section for botany. The August 1892 meeting was held in Rochester, New York. Just before the meeting, news reached the Americans of a manifesto circulated by the Berlin botanists protesting the methods used by Otto Kuntze in his monumental *Revisio Generum Plantarum*, the first volume of which was published in 1891 and where the priority law was sweepingly applied. The bone of contention concerned the starting date of nomenclature, with the Berlin group preferring later Linnaean texts than Kuntze had used.

Spurred by these European developments, Britton moved for the appointment of a committee to consider the nomenclature question; he had been agitating on this question for five years. Henry Gleason, a botanist who later worked with Britton, believed in retrospect that Britton probably came to the Rochester meeting with the rules already written and then orchestrated the committee's activities so that it would make the recommendations he favored.[33] Britton's friend Henry Rusby was appointed chairman of the meeting, thereby ensuring that the right committee appointments would be made. Not only was Rusby close to Britton, but he also knew Greene from his stint in the mid-1880s as a Smithsonian agent in the Southwest. He had collected for Greene, and in return Greene had named the genus *Rusbya* after him.[34] The committee included Rusby himself, as well as Britton and other men sympathetic to reform.[35] The committee's recommendations were submitted and adopted at the close of that very meeting.

The two main tenets of the "Rochester Code," as it was first called, were to recognize priority of publication as a guide for both species and genus names and to use as a starting publication the first edition of Linnaeus's *Species Plantarum* of 1753. The choice of this later publication brought the Americans into agreement with the Berlin botanists but into disagreement with Otto Kuntze. Further rules were given to determine whether a name was legitimately published or not. Rules to establish "legitimate authorship" were needed because sometimes naturalists would discover and name species but would not adequately describe them. This had often happened on earlier expeditions, when European collectors had named American plants in the field and sent the specimens to their home herbaria, but failed to describe the species properly in the first published account of their journey. Did this kind of discovery really deserve scientific credit? Britton and his supporters thought not; the person who published a proper description was the one to whom credit belonged.[36]

The Botanical Club of the AAAS also arranged for publication of a list of plants of the northeastern United States, to provide an example of what the new rules would accomplish. Charles Bessey commented that the book was "the sign that the day of 'authority' as such, is ended, and the day of 'law' has begun."[37] The "humblest botanist," he continued, could now lawfully correct the greatest. He added a good-natured and prescient warning that making the process of scientific judgment more transparent to the ordinary citizen could have untoward consequences. Those who now challenged the authorities would one day be vulnerable to similar attacks:

> When the world learns that the pronouncements of "Science" are after all only the judgments of, say, Professor Britton, Professor Coulter, Professor Scribner, or some other mortal, it may not stand in such ignorant, open-mouthed wonder as it formerly was wont to do. It may cry out against them, and demand that the golden calf be set up again. But if these professors set forth plainly that their work is plain work, the plain and straightforward statement of facts, the world will eventually cease to be the blind idolaters of that which they do not understand.[38]

An editorial in the *Botanical Gazette* described the Rochester meeting as unusually large and therefore representative of American botanists. Those who could not attend sent letters of opinion on the nomenclature question. Coulter probably wrote the editorial, which expressed approval of the frank and thorough but largely amicable discussions: "Everyone was ready to make concessions for the sake of agreement, and the principles finally adopted represent a resultant of various concessions." The Rochester Code was a "thoroughly wise compromise, . . . honorable to all concerned." The meeting marked "an epoch" for American botany in that the botanists had finally managed to get their own section at the AAAS, marking the divorce from the zoologists that they had long sought.[39]

In the fall of 1892 the American delegation attending the International Botanical Congress at Genoa came equipped with several copies of the Rochester resolutions for distribution and discussion. The American delegate from the AAAS, Lucien Underwood of DePauw University, Indiana, allowed that there were differences of opinion in the United States and that these differences largely emanated from the new curator of the Gray Herbarium at Harvard, Benjamin Lincoln Robinson, who succeeded Sereno Watson. These differences stemmed, Underwood said, from dissent against the system used at Kew, which was followed by "our late lamented Professor Asa Gray."[40] But Underwood pointed out that, quite apart from the American reform movement, Otto Kuntze had also opened the path to reform in Europe. At the Genoa meeting, an International Standing Committee of thirty botanists was formed and asked to report to a future congress. The American contingent consisted of Britton, Greene, and Coulter, all of whom were in favor of the Rochester reforms, although Coulter was more moderate than the other two. Britton, Coulter, and Underwood were also members of a committee of the AAAS that was to report on the nomenclature question at the Botanical Congress at Madison, Wisconsin, in 1893.

Botanists had high hopes for the 1893 congress, for 1893 was the year of the Chicago World's Fair, which they thought would act as a magnet to attract Europeans who might not otherwise attend American scientific meetings. The original plan was to meet in Chicago, but when this proved impractical, the venue was changed to Madison, chosen for its proximity to Chicago. Americans still hoped that the congress would attract an international gathering, one that would give discussions the "sanction of authority" and thereby help to settle these disputes.[41]

In the end, even with the World's Fair going on, the Europeans did not come. The attendance of foreigners was so low that the word *international* was dropped from the title of the congress. But about one hundred American botanists did show up, and they affirmed the Rochester Code with some minor changes. The reforms were again adopted at a meeting in Brooklyn, New York, in 1894. The reformers could now argue that the reform agenda had been sanctioned repeatedly, and in a democratic manner, by groups that fairly represented a cross-section of American botany. The Rochester Code was becoming the American Code.

At the meeting in Madison botanists organized the Botanical Society of America, itself a controversial move because not all botanists thought a new society was necessary, and some were suspicious of Britton's growing influence.[42] The nucleus of the society consisted of ten botanists selected by the group; these were the leaders recognized for their high standards of science, and Britton was among those selected. Although the Botanical Society remained fairly small for several years and did not serve as the main organization for American botanists, it was one step up the ladder for Britton, into the new elite of American science.

These American actions were a bit too independent, and a bit too democratic, for Otto Kuntze to stomach. Although the Americans had presented their reforms at Genoa as being part of a larger European reform initiative, Kuntze resented the news that the Madison congress had been used to claim consensus on a new code. Writing to Coulter in 1894, he expressed dismay that Americans were working on their own code. He argued that the Paris Code could never be set aside, as the Americans appeared to be doing, but could only be amended. Such amendments had to be worked out at international congresses—and then only if the congress was "competent." The reforms discussed at the Genoa congress, he felt, had been poorly handled and were a mistake. Incompetent congresses carried no weight. He believed nothing decisive could be done before the international congress scheduled for 1900 in Paris.[43] (As it turned out, the Paris Congress was unable to settle the question and deferred it to the Vienna Congress of 1905.)

His grievance concerned the starting date for setting priority, the same issue that had prompted the "Berlin protest." Whereas the American Code recommended starting with Linnaeus's *Species Plantarum* of 1753 for both species and genus names, Kuntze held that the starting point for genera had to be 1737, the date of Linnaeus's *Genera Plantarum*. He branded American reform efforts as revolutionary in intent, because they were not meant to amend the existing code but actually to overthrow the Paris Code and replace it with a new American code. He bridled at the way the American reform efforts threatened to sap his own authority: "A new code . . . would deprive me of my rights as emendator of the Paris Code." Such revolutionary actions would only create a schism in botany, he warned.[44]

Greene in America actually agreed with the spirit of Kuntze's criticism. Although named to the standing committee at the Genoa congress, Greene was not at the meeting in Italy and did not take an active role in formulating the new code. In fact he did not favor the idea of having a fixed starting date, arguing that "history has none."[45] Greene was president of the botanical congress at Madison but did not take an active role in the nomenclature debates, feeling, much as Kuntze did, that decisions made at these congresses need not be followed. Greene continued to nitpick about the rules and to go about his business in his own way, without much concern for the opinions of others. His colleagues grew exasperated at his overly critical stance. Coulter took issue with Greene's habit of citing pre-Linnaean authors, using this obstinacy as a reason to break with Greene: "We ourselves have participated in revolutions of nomenclature in the interest of peace, and not that one revolution may simply be the prelude to another. . . . Highly as we esteem Professor Greene we cannot just now follow him any further in this ghoulish business, and we trust that he will understand that we have deserted him, not for his own sake, but on account of the company he keeps."[46] Greene appeared to want the revolution to continue indefinitely.

The controversy came to its peak in 1895, by which time the Rochester Code had been affirmed enough times that the opposition was compelled to act. The strongest defenders of Gray's methods were those who succeeded him at Harvard, first his colleague Sereno Watson, who took over Gray's work but died soon after Gray died, and then Benjamin Robinson, who inherited all of Gray's and Watson's unfinished work. He upheld traditional methods and resisted the calls for a new code of nomenclature. The Harvard botanists had held back from the debates at Rochester and Madison, but in 1895 Robinson circulated a petition signed by seventy-four botanists recommending (1) that names that had been established by long usage should not be revised on theoretical grounds and (2) that the question of retaining such names be left to the discretion of individual writers. The petition, in short, reaffirmed the "Kew rule" and the methods used by Asa Gray.

The petition prompted impassioned responses from reformers, with Lester Ward and Frederick Coville taking the strongest stance in favor of reform. Coville lashed out at Robinson, denying his insinuation that Britton alone was responsible for the reformed nomenclature and pointing out that although Britton was the most "active and influential in *hastening* the reform," many other botanists agreed with his judgment. In fact the reforms had been "adopted by overwhelming majorities in democratic botanical assemblages," leaving Robinson to defend a "now discredited system" under which botanists had "been chafing more and more for past fifteen years."[47] Robinson replied that trying to quash opposing views on the new rules was both undemocratic and unscientific.[48]

Lester Frank Ward, as noted in the introduction, was fascinated by the general problem of adaptation, a theme running through his botanical studies and extending into his later sociological theories. The struggle for existence, Ward perceived, was also a struggle for place, that is, a struggle to extend one's territorial reach and enjoy sole possession of resources in the face of competitive pressures. How fitting that he should join the debate to defend American botanists' efforts to gain control over their flora. He penned a brilliant rhetorical defense of the reform movement in 1895, which pulled out all the stops and depicted the debate as a struggle of idealistic reformers thwarted by conservative sentiment.[49]

Ward spurned the crude suggestion that the reform efforts were all about grabbing credit by tacking one's name onto the end of a freshly named species. The reformers were not concerned with credit, he claimed; they were taking a higher moral stance. He conjured up an image of a future botany in which scrupulous quoting of authority, branded as "one of the worst evils of botanical writing," would simply not be required. If botanists could use the priority rule to agree on the names of all known plants, then it would not be necessary to append after each plant's name the abbreviations of the various people who were involved in naming it. The custom of tacking on one or two abbreviations after a name added up to "an ugly cacophony," and the object of the reform, declared Ward, was to eliminate this kind of clutter. Furthermore, the reforms had to be made retroactive, no matter what immediate confusion this caused, so as to "remove the old weeds" from the garden.[50]

In any case, Ward declared, the American movement was not revolutionary at all. It was but a continuation of the reforms begun by the greatest naturalists of the century. Here he invoked the names of Augustin and Alphonse de Candolle, Darwin, Alfred Russel Wallace, Thomas Henry Huxley, and other leading men who had been involved in the writing of the earlier codes. The steps taken by the Americans simply carried on an evolutionary process to which the great men of science had contributed. The Harvard petition was the real revolutionary document, a call for a "laissez faire system" that was capable of "producing most chaotic results."[51]

Ward's defense of the reformers fitted in with the sociological theories that he was advancing at the time.[52] He saw progress as dependent on individual initiative and rational thought, which enabled people to build on past accomplishments in a cumulative way. In the debate over the relative weight of nature and nurture in human development, Ward leaned heavily toward nurture, opposing the views of Francis Galton. Here he took aim at Herbert Spencer, the English philosopher who had championed the idea that progress was an automatic unfolding of an evolutionary process. Ward denied that progress could be achieved by letting nature take its course. He also challenged the laissez-faire philosophy that for Spencer was a necessary component of his evolutionary worldview. Achieving progress meant taking charge of one's life and building a better world; it was a self-conscious act, combining the power of reason and initiative. For Ward, the botanical reform movement was just such an example of the combined effect of reason and initiative that together enabled science to advance.

For the supporters of the reforms, the issue, as they chose to characterize it, was one of democracy and justice, in contrast to blind submission to authority. Those opposed to reform were depicted as at best unduly influenced by sentiment and at worst following blind prejudice. Ward idealized the reformers as a group who were striving selflessly to advance science as a "labor of love," without any expectation of personal gain. The opposition, he argued, were trying to thwart the democratic processes of the society of botanists.[53]

The reform agenda in botany fitted well with a liberal, progressive view of society. In such a society progress depended on the spread of a scientific method, specifically a method derived by discussion and consensus politics, not one imposed by any single higher authority. Reform, in other words, entailed not just fixing some rules for naming species, but also creating an environment that might nurture scientists, that might make it possible for more people to do science effectively. This debate was also about broadening the constituency for science, giving more people the vote, as it were, and allowing the previously disenfranchised to have their views count.

Britton and his colleagues continued to agitate for nomenclature reform, becoming even more radical as time passed. The discussions continued for several years, and Britton never wavered in his views even as he watched fellow reformers drop their radical stance. When the International Botanical Congress at Vienna in 1905 decided to base discussion on the 1867 code, the American nomenclature committee countered that it was better to abandon the code altogether and substitute a simpler set of rules, which they laid out in the Torrey Club's *Bulletin*.[54] They especially objected to the Vienna Congress's decision to affirm that Latin would continue to be required for taxonomic diagnoses. They regarded this decision as arbitrary and unwise at a time when modern languages were replacing Latin in school curricula. In New York the language of botany would be English, and floristic catalogs would be written not for the erudite specialist but for ordinary people. These people were self-educated: they might have been introduced to botany at their mother's knee and learned their science as best they could in the absence of adequate school and college courses.

As Britton and the other reformers knew, the argument was not to be won by rhetorical flourishes. What mattered was not so much achieving complete consensus as achieving enough agreement—a critical mass, as it were—to justify implementing the new rules in taxonomic work. What mattered was what people actually did, not how loudly they shouted or how cleverly they disparaged the opposition. Although the shouting continued, and the schism that Kuntze had feared did indeed open, creating a rift between the New York Botanical Garden and the Gray Herbarium, the crucial thing for the reformers was that they had enough of a mandate to implement the American Code. At least they felt justified in claiming such a mandate in their own work. And despite Ward's claim that the question of credit was beneath their dignity, it must have been very much in their minds as they discovered new species, named them, and housed them, and in so doing claimed their rights of ownership over the natural history of the Americas.

At the Department of Agriculture's Divisions of Botany and Forestry, the National Herbarium in Washington, the Arnold Arboretum, and the Missouri Botanical Garden; in a number of midwestern state surveys and in western universities; at Columbia College (soon to be Columbia University); in the Torrey Botanical Club; and in the revisions of the U.S. Pharmacopeia (on which Rusby worked), botanists began to implement the new system in their publications. And most significantly, Britton made the new code the core of his own work and the work of the botanists he would bring together at the new institution he envisioned in New York City. The botanists at the New York Botanical Garden were veterans of the nomenclature wars, many of them fighting together in the Torrey Botanical Club regiment. They shared with Britton an enthusiasm for the American Code, and they continued to push for its acceptance, voting as a block when they could at scientific meetings in the United States.

One of the first taxonomic works published by the Garden that built on the new reform principles was the *Illustrated Flora of the Northern United States, Canada, and the British Possessions*, by Britton and Addison Brown, the first volume of which appeared in 1896 and the third and last in 1898. Britton must have done the lion's share of the work, for he was much more experienced in botany than Brown. Brown proofread and worked on the indexes, which comprised not only a scientific index of Latin names but also an exhaustive list of popular plant names in English. Brown had a particular interest in botanical folklore and dug up much of the information about regional variations in popular names. Even more importantly for the amateur botanist, Brown invested thirty-three thousand dollars in the publication of the *Illustrated Flora*, thereby ensuring that it could be sold at moderate price and be widely used.[55]

Not only did the *Illustrated Flora* inaugurate the Garden's scientific work, but it epitomized the style that Britton preferred as the most efficient way to educate people in botany: short, easily understood descriptions accompanied by illustrations. These volumes were not only for the professional botanist but also for the self-educated amateur, young people just like the adolescent enthusiast that Britton had been. The copious illustrations in the work were very important in broadening its audience by making it easy to use. It was indeed a huge success.

Henry Gleason quoted Britton as saying, some thirty years later, "If I can write a better flora than anyone else, my names will be used, no matter what code of nomenclature I follow."[56] Gleason conjectured that Britton must have had exactly this idea in mind as he began work on the flora around 1891, about the time of the Rochester meeting. If he could write a reference work that people would prefer to Gray's *Manual*, then his work would prevail over that of the opposition. More than a century later the revised Britton and Brown, as it is known, is still in use.

CHAPTER THREE

Big Science

We turn now to the details of institution building, to the scientific and educational work of the New York Botanical Garden, and to the ecological work that it engendered. Botanical gardens, as Janet Browne has remarked, represented the "Big Science" of the nineteenth century.[1] The New York Botanical Garden merited the descriptor Big Science: it was expensive and expansive and had a great impact on the further development of American science. From the start Britton adopted an expansionist strategy, quickly building up the holdings in the museum's herbarium, expanding the greenhouses, mounting expeditions, founding new journals and magazines, creating vigorous programs in experimental botany, and exploiting opportunities to develop field stations as satellite operations.

This expansion depended crucially on patronage from wealthy New York businessmen and industrialists, whose money was especially needed to fund scientific explorations. New York businessmen had been deeply involved in the industrialization and agricultural development of the West during the great railway boom that followed the Civil War.[2] To the extent that they were persuaded to invest in research, the biological sciences in the United States were able to advance rapidly, soon achieving the recognition that Americans so anxiously sought. Experimental and field sciences grew apace as the tentacles of the Botanical Garden reached outward from New York to the world.

Exploration and taxonomic work were the primary activities of the Garden. However, the Garden was an unusual institution in that it also included laboratories for experimental science, where many of the latest ideas emanating from European laboratories could be developed in the American context. The research done on experimental evolution by the laboratory's director, Daniel Trembly MacDougal, dealt with some of the most exciting and controversial topics of the day. Most crucially, these areas of research were capable of attracting the attention of patrons of science. In this mix of field research, taxonomy, and experimental science, ecological work could flourish alongside other fields of study.

The commitment to expansion of research in all directions that characterized the first few years of the Garden's activity did not last. After the departure of the first two directors of the laboratory, Daniel MacDougal (who left in 1906) and Charles Stuart Gager (who left in 1908), the experimental program became narrower, focusing on problems of plant pathology and genetics. By the 1920s the taxonomic work of the Garden was predominant, and broader ecological studies that were not directly related to the Garden's main botanical surveys were pushed to the margins. Explorations also suffered during the Great Depression: Britton, who remained director of the Garden until 1929 and died in 1934, lost much of his personal wealth after the stock market crash of 1929, which meant less money for exploration. But in the Garden's first decade or so, the combination of scientific methods and problems ranging across the spectrum of botanical subjects helped to boost interest in ecology. In addition, the Garden's educational activities encouraged an ecological approach to nature study and fostered interest in conservation. The embryonic field of ecology thus benefited from everything that the Garden undertook.

In energetically pursuing three facets of Garden work—exploration and collecting, experimental science, and public education—the Garden's scientists acquired great authority during this crucial phase in the expansion of American science. Their work and their influence nurtured the science of ecology, which at this time was not an obvious candidate for development into a new field of research. It could in fact be easily dismissed as a passing fad. The momentum created by the Garden helped to tip the balance toward respectability. For the consequences of this influence, see chapter 4. Here, I focus exclusively on the Garden to emphasize the importance of institution-building for the foundation of new disciplines.

Creating an aesthetically pleasing recreation ground was important, but the scientific and educational work of the Botanical Garden was paramount.[3] The park was to be open and free to the public daily, and the educational and scientific privileges were available to all, male and female. Development of the Garden grounds included planting theme gardens; excavating a boggy area, which was made into two lakes; building roads; and lowering the Bronx River by changing the height of a mill dam. A system of drains had to be laid to channel water and improve drainage. Britton was active in all the planning stages.

By the end of 1898, there were more than two thousand species and varieties under cultivation, up from seven hundred species in 1897. The head gardener, Samuel Henshaw, toured British botanical gardens as well as commercial nurseries and private gardens to obtain seeds and shrubs. Eight botanists were at work on the different museum collections, temporarily housed at Columbia University. Britton arranged specimen exchanges with various institutions, including agricultural experiment stations. In 1897 Columbia had moved from its midtown Manhattan location to Morningside Heights, closer to the Garden. In 1898 the bulk of the herbarium of Columbia College, nearly a half million specimens, and its botanical library of five thousand bound volumes, were turned over to the Garden, in trust and for its use.

In spring 1900 the museum building opened to the public (fig. 3.1). It was mostly completed, with two wings left for future expansion. Now the botanical work and educational programs began in earnest. Britton considered his building to be the "best adapted of any similar edifice in the world"; he had chosen the Italian Renaissance style for its "frank and dignified" look, and the construction and furnishing of the building were accomplished at a cost of three hundred thousand dollars. What Britton did not emphasize was how his extreme frugality affected his staff: the need to economize on heating and gas lighting left them working in a cold and dark building for most of the year.[4]

The museum contained a lecture theater; exhibits on the economic uses of plants, on the relations of plants to each other, and on fossil plants; a library; laboratories for instruction and research; and an herbarium to which additions were made at the rate of about fifty thousand specimens per year.[5] The economic collections grew as a result of donations by several companies, especially pharma-

Fig. 3.1. Big Science. Britton hoped that his grand museum building would be the largest in the world. Courtesy of the LuEsther T. Mertz Library of the New York Botanical Garden, Bronx, New York.

ceutical, tobacco, rubber, and paper and wood manufacturing companies. By 1906 the herbarium had more than doubled and the library had increased to eighteen thousand volumes. Plants that were growing at the Columbia greenhouse were moved to the new conservatories at the Garden in the summer of 1900.

Elizabeth Britton worked alongside her husband at the Garden until her death in 1934; Nathaniel died four months later. Elizabeth took charge of the moss collections as a "voluntary assistant" and from 1912 was designated honorary curator of mosses. She taught taxonomy in her areas of expertise, continued her research on mosses, and was active in campaigns to preserve American wildflowers. Her conservation work dovetailed with her concern to preserve the Garden itself, of which she was fiercely protective (fig. 3.2). She was equally protective of her husband's standing and supported him wholeheartedly. It is remarkable that she was not even granted the title of honorary curator until 1912; the delay was indicative of her willingness to remain in the background, despite being fully devoted to the Garden and her botanical studies from the start.

Nathaniel Britton was reserved, frugal, and obsessed with the smallest detail. Those obsessions underlay the Garden's success. Henry Gleason, a student and later curator at the Garden, emphasized Britton's exceptional qualities as a

Fig. 3.2. Elizabeth Britton. She fiercely protected the New York Botanical Garden from people who wanted to pick its flowers. Courtesy of the LuEsther T. Mertz Library of the New York Botanical Garden, Bronx, New York.

manager and his single-minded devotion to the Garden.[6] Britton planned every action carefully and left nothing to chance. His managing skill included knowing when to get expert advice from the two political appointees connected with the Garden. Arthur J. Corbett, the superintendent of buildings and grounds, was a Tammany ward boss with influence among New York city politicians. In addition there was a former Tammany politician known as "Old Chief" who was on the payroll for fifty years as membership clerk, custodian, and gatekeeper. He would tell Britton which politicians he needed to see and how to talk to them. Gleason remarked that although Britton was known for his "Blitzkrieg tactics," which were much in evidence during the nomenclature battles, he tended to get involved only in campaigns that could be won. His Board of Managers supported his decisions consistently. Gleason attributed this record to the care with which Britton prepared every proposition.[7]

This kind of personality requires the softening influence of more outgoing types. The individuals who had been Britton's comrades in the Torrey Botanical Club and the nomenclature debate provided not only enthusiasm, labor, and money, but at times offered the warmth needed to balance his controlling nature. One such balancing personality was Lucien Underwood, one of the most radical of the botanical reformers during the nomenclature debate and Britton's successor as professor of botany at Columbia in 1896.[8] He served on the Board of Scientific Directors of the Botanical Garden and replaced Seth Low as chairman of the board in 1901, when Low was elected mayor of Greater New York.[9] At the Garden Underwood had charge of the fern collections. Described as "big and bearded, bluff and hearty," he was affable and never at a loss for words, qualities that made him a good orator and teacher.[10] With his outgoing personality he provided an important counterweight to Britton, who was as reserved in emotion as he was driven by ambition. Able to rib "Lord Britton" for his imperious ways, Underwood helped resolve tensions between Britton and the staff and students.[11] His life took a tragic turn in 1907 when, at age fifty-four, he suffered a mental breakdown and committed suicide. Henry Rusby succeeded Underwood as chairman of the Board of Scientific Directors.

Colleagues such as Underwood, along with Henry Rusby, Arthur Hollick, and Addison Brown, were Britton's peers, often giving their labor freely to the Garden as honorary curators and not as his employees. His relationships with his paid staff were different: they were his serfs, and Lord Britton ruled his fiefdom sternly. Starting in 1900 no salaried officer was allowed to engage in work for pay other than for the Garden, without Britton's written permission, which was seldom granted. Even outside professional activities required approval, as did any writing submitted for publication. This degree of planning and the strict enforcement of the Protestant work ethic helped to ensure rapid growth and relatively high productivity. But there was friction on a personal level, for Britton was highly demanding even of his friends, could not easily delegate authority, and was unwilling to compromise on fundamental issues.

The close connection between the Garden and Columbia University ensured the success of the Garden's research-oriented educational program. The scientific directors of the Garden included the president of Columbia and its professors of botany, geology, and chemistry. Britton was still professor emeritus at Columbia. In collaboration with Columbia, twenty-one research courses were approved for advanced degrees. University courses at the Garden were free for Columbia students. For others there was a fee, but students could pay by working for the Garden or by donating specimens or books. The success of the Garden in providing research training not available in many other places is indicated by the fact that the cohort of 43 students in 1902 included graduates of thirty-one different colleges and universities. From 1900 through 1909, the average number of students was about 33 per year. The minimum number was 18, in 1900; and the maximum was 46, in 1904.[12] Roughly one-third of these students were women, several from Columbia's affiliated women's college, Barnard.

The best facilities for botany in the East were at Harvard, Cornell, and the University of Pennsylvania; most schools had not yet developed much laboratory work, as Rodgers notes in his history of the "new botany."[13] We can compare New York's facilities with those of another East Coast institution, Johns Hopkins University, which was the first graduate school in the country and a leading school for the training of zoologists. Botany was included in the Johns Hopkins advanced biology program but was a minor part of it. The biology courses in 1899–1900, for example, enrolled a total of about one hundred students (including medical students), but only five were enrolled in the two graduate botany courses offered each semester by Duncan S. Johnson.[14] For access to an herbarium and a botanical library, students had to rely on a private citizen in Baltimore, who agreed to let them use his collections and books. The New York Botanical Garden offered a far more comprehensive program of study in botany.

The Garden also accepted people who wanted to pursue special research problems. It was willing to consider almost any subject in botany for investigation. The laboratories were never closed. In 1900 the Garden started convening a weekly research discussion group, which all the active botanists in the city were invited to join. Soon after its opening, the Garden became a leading center of botanical research in the United States. In 1916 the Garden acquired another 140 acres, bringing its size to 400 acres. By that time it stood as one of the top botanical gardens in the world.

Scientific subjects at the Garden included morphology, physiology, cell physiology, studies of special taxonomic groups, and ecology and evolution. Ecology had been newly identified as a modern way of approaching the study of biology, but its roots went back several decades at least. The word *ecology*, coined in the 1860s by Ernst Haeckel, denoted the study of the processes that made up the struggle for existence that Darwin had described. Haeckel did not promote ecology as a separate field of biology, but in the late nineteenth century there was increased interest in biogeography, the study of adaptation, and plant physiology, all of which helped to define the emerging discipline that would in time be called ecology. A few European biogeographers, such as the Danish botanists Eugen Warming and Andreas Schimper, were developing Alexander von Humboldt's ideas into the science of ecological plant geography; and at the Berlin botanical garden scientists were using ecological principles to organize their collections.[15] Americans took note of these new ideas.

In the United States the term *ecology* came into vogue in the mid-1890s, following the Madison botanical meeting of 1893 at which Britton's American Code of nomenclature was adopted. A committee of botanists got together after this meeting to discuss nomenclature issues in plant physiology. They decided to adopt *ecology* to distinguish their interest in plant adaptation from the narrower kind of physiological study normally undertaken in the laboratory. Ecology, as Gene Cittadino has aptly described it in his survey of the "new botany" in America, was a form of "outdoor physiology," a science devoted to investigating, through both experiment and field observation, the relations between organisms and their environment.[16] As Cittadino also points out, American botany owed a lot to European models: the physiological work of Julius Sachs and his students was especially influential in the United States. Ecology included the study of adaptation, the effect of the environment on plants, the form and function of plants, and the geographic distribution of plants.

The new ecology was not just the laboratory method of experiment extended to the field. It also sought to broaden museum-based taxonomic work, so that the intellectually barren practice of collecting plants without paying attention to their surroundings might be superseded by research into the living organism and its environmental context.[17] The word *survey* implied just this modern approach: a rounded analysis that included an understanding of the life history of the plant and its ecological context. Britton, although not himself engaged in ecological research, agreed with modern ideas about expanding and reforming botany. In 1894 he had urged a reinvigoration of systematic botany.[18] Darwin's theory of evolution had posed new problems for systematists, he noted. The concept of "species" now had to be thought of in dynamic terms, and past conditions as well as present had to be considered when classifying species. The modern systematist needed broad training that included the study of many different organisms.[19] Britton continued with illustrations of how taxonomic decisions might take into account knowledge of the laws of variation and hypotheses about geographic distribution and evolutionary history.

The idea was to create a broader, contextualized botany, one that spurned the limitations and idiosyncrasies of the "collectors" of the past and looked forward to a future full of promise as new research problems opened up. As Lucien Underwood remarked in 1900, the new systematics would bring morphology, embryology, and laboratory and field work to bear on taxonomic problems.[20] It would not be the kind of work that one could leave to amateur triflers; it would require intelligent study by trained botanists.

The articulation of this modern vision was one facet of the broader reform-oriented movement that ushered in the American Code and spurred the formation of the Botanical Society of America. The New York Botanical Garden was the place where all the threads of the various reform movements came together, and the science of ecology was part of this mix. The research-oriented educational program was broad, embracing traditional studies in natural history and systematics, with their emphasis on museum work, but also bringing in newer subjects that involved laboratory work and experiments. Ecology and physiological ecology, including field and laboratory work, were supplemented by experimental morphology, the study of what caused variation in form and structure.

Although the Brittons were not themselves experimentalists, they brought into the Garden scientists whose chief aim was the development of experimental botany. The Garden aimed for improvement and modernity on all fronts, combining systematics, ecological surveys, and the latest experimental science into the research program. At the intersection of these diverse ways of envisioning an expanding scientific enterprise lay the new ecological science that was emerging as a distinct field of study. The Garden's leadership did not self-consciously try to create a new discipline of ecology, but the combination of these scientific interests in a major institution provided crucial support for ecological research. The building of this institution and the creation of a professional infrastructure—graduate courses, new journals, and a vigorous research program—were critically important to the birth of ecology as a new discipline.

Among those touting the advances to be made by experimental science was Daniel Trembly MacDougal, who came to the Garden as Britton's "first assistant" in 1899 and largely directed the laboratory operations. A midwesterner of Scottish ancestry, MacDougal was every bit as energetic and entrepreneurial as Britton, and a good deal more jovial. While an undergraduate at DePauw University in Indiana in 1890, he had taught himself the rudiments of the new physiological botany from the laboratory exercises developed by Joseph Charles Arthur, a professor at Purdue University. MacDougal worked through the exercises on his own, setting up in a windowsill laboratory his homemade apparatus fabricated from lamp chimneys, remodeled alarm clocks, kitchen appliances, and tinware. From 1890 to 1893 he worked as an instructor in Arthur's laboratory and eventually received a doctoral degree in plant physiology from Purdue in 1897.

Joseph Arthur was prominent among the group promoting botanical research in the mid-1890s. He was very interested in broadening American botany, which he described in 1895 as "more and more exhibiting virility by its adaptability to the needs of the times."[21] Botany could aid the economic growth of the country, he pointed out, but it required capital for its necessary development. Research into fundamental questions was important, but the support for that research might be found in the practical world. Arthur hoped in particular that the newly created agricultural experiment stations would be the loci for advances in research; his own scientific reputation rested on his studies of plant diseases in the agricultural context.

These new places of scientific study were, he thought, especially important for the pursuit of ecology, which awaited definition and development. Ecology was a branch of science that was both "rich in possibilities for the philosophical and speculative mind, and bristling with queries demanding experimental solution." And although ecology was conceived as a branch of physiology, he thought it would also engage the highest intellectual powers of the naturalist. Darwin was his inspiration, but he had to admit that the science was still in a "rather chaotic condition," badly in need of a textbook so that it could be taught and, through teaching, be defined as a subject.[22] It was an embryonic science, full of promise but with its boundaries not yet delineated and its purpose in American society not yet clear.

MacDougal did not have quite the same philosophical leanings as his teacher, but he did have the same interest in investigating the potential of cutting-edge research in physiology and evolutionary biology. The work that earned his doctoral degree was completed under Wilhelm Pfeffer at Leipzig, where MacDougal had gone to get more experience in advanced research. MacDougal also did research at the botanical garden at Tübingen under Hermann Vöchting, who succeeded Pfeffer as director after Pfeffer moved to Leipzig. MacDougal's account of the Tübingen laboratory showed that his European sojourn left a deep impression on him. The well-furnished laboratories, intelligent assistants skilled in designing apparatus, superb library facilities, well-stocked gardens, and most importantly the encouragement of men like Vöchting, gave him exposure to an ideal research environment.[23]

MacDougal was as much a man of the field as of the laboratory. An enthusiastic traveler, he had worked in 1891–92 as an agent for the U.S. Department of Agriculture. Equipped with Navajo blanket, tin cup, and a jar of beef extract, he roamed and collected through Arizona and Idaho. In 1898 he studied the desert and alpine flora of Arizona for the Department of Agriculture and experimented at the western edge of the lava desert near the Little Colorado River. With this combination of field experience in the American West and an interest in experimental science, he was an energetic up-and-coming young botanist who would have appeared ideal for the Garden.

He quickly set about furnishing the laboratory after assuming his duties in July 1899. His space encompassed thirteen rooms, with an embryological laboratory, a morphological laboratory, a physiological laboratory (sky-lighted), a physiological darkroom, a chemical laboratory, and photographic rooms. Instruments included microscopes and accessory tools, dissecting tools, and photography equipment. The chemical laboratories were arranged for practical work, especially work on vegetable dyes and pathology. MacDougal himself, students, and visiting researchers investigated a variety of problems in physiology, ecology, and morphology. By 1901 every available table was occupied.

MacDougal was hard at work on one of the hottest topics of the day, the experimental investigation of the mechanisms underlying evolution. Although he considered himself a physiologist, his approach to the study of evolution bore a natural affinity to ecological problems involving the relationship between organisms and the environment. MacDougal's entrepreneurship in advancing these studies and drawing the attention of wealthy patrons of science helped to advance the nascent field of ecology, as we shall see in chapter 4. Here, let us look briefly at what MacDougal's experimental program entailed.

Darwin had left many thorny problems for his followers to work out, and one of the most central was to discover the mechanisms that produced novel variations and explained how traits were passed from generation to generation.[24] One influential new theory was the "mutation theory" advanced by Dutch botanist Hugo de Vries in a two-volume study, *The Mutation Theory*, published in 1901 and 1903.[25] MacDougal was immediately attracted to de Vries's ideas and experimental method, and he became the chief promoter of the mutation theory in the United States.

De Vries argued that evolution occurred by discrete jumps, or mutations, not slowly and continuously as Darwin had supposed. His theory was based on many years of breeding experiments on several species of plants. Of these experiments the most promising were on the evening primrose of the genus *Oenothera*. This genus, which was believed to have originated in the southern United States, had been cultivated and hybridized for over two centuries. In 1886 de Vries thought that he had discovered an entirely new species of primrose originating in the wild and living alongside the parent form in apparent harmony. The plants appeared to be examples of saltatory change or "mutation." He collected seeds from the plants and began cultivating them to determine the laws governing the appearance of these mutations, which he called "elementary species." His interpretation of these results in the light of Darwinian selection theory was the most thorough treatment of the origin of species by mutation up to this time.

Although he offered a critique of natural selection, de Vries in fact revered Darwin and was trying to bring Darwin's ideas up to date, not to destroy them.[26] He did believe that these mutations or elementary species were subject to natural selection. Some elementary species would survive, and others would become extinct. However, this kind of selection operated like a sieve: it eliminated the unfit but did not create anything new. The actual *origin* of new species was located in the unknown physiological processes that induced the mutation, not in selection. In a nutshell, natural selection explained the survival, but not the arrival, of the fittest. However, the "arrival" of new species was not an automatically progressive process. Mutations could occur in any direction, and the daughter species might well be less fit than the parent form.

Biologists understood the term *natural selection* to mean "constant and progressive variation in one or many directions," with the most improved forms, those designated the "fittest," surviving and reproducing.[27] After a long period of adaptive change in a given direction, a new species, fitter than the old, would gradually be created. According to the mutation theory, the idea of gradual, adaptive change leading to a new species was false. Of course, whether the elementary species ultimately lived or died still depended on natural selection. But it was the entire "elementary species" that was selected, not the individuals within the species. The mutation theory suggested that the origin of new species was a fairly fast process and was also potentially directly observable.

For MacDougal, de Vries exemplified the European school of physiological botany to which he himself had been most attracted. Led by plant physiologist Julius Sachs at Würzburg, botanists were investigating problems that were aimed at the unification of plant and animal physiology. Sachs's ideas reached the United States under the influence of people like J. C. Arthur, MacDougal's teacher. The European mentors with whom MacDougal had worked, Pfeffer and Vöchting, were both followers of Sachs.

As far as evolutionary ideas went, Sachs and his school focused on the general Darwinian problem of adaptation: how did organisms acquire their adaptations? In his later life, Darwin had become very interested in the power of movement in plants, which was most intriguingly demonstrated in insectivorous plants.[28] The German physiological school picked up the investigation of plant "behavior" and tried to analyze the mechanisms underlying such behavior in a way analogous to the study of behavior in lower animals. This experimental program was to some degree in opposition to the botanical studies undertaken by Darwin, whom Sachs thought was not experimentally rigorous.[29]

Pfeffer's approach to these questions was highly mechanistic. He saw organisms as machines that were set free by a spring or trigger, which released the energy of the machine.[30] Vöchting was interested in plant behavior and how plants responded to stimuli. In addition to studying how the organism's "machinery" operated, this line of research led to questions about how plants might change their form permanently in response to external stimuli. This topic was related to the broader problem of how evolution occurred. How were morphological changes related to physiological processes? This was a central question in a number of research laboratories in Europe. De Vries had also been a student of Sachs and was part of this physiological school of botany. Of all the approaches to evolution that were current around 1900, de Vries's mutation theory, with its dynamic, physiological viewpoint, came closest to exemplifying what MacDougal considered to be the most analytical, precise, and direct way to tackle the subject.

For de Vries, and for MacDougal, the key point of interest was that mutations could be easily observed, which meant that speciation also was observable. One had only to seek out mutating species. De Vries considered any form that remained "constant and distinct from its allies in the garden" to be an elementary species.[31] More importantly, evolution could potentially be controlled if the cause of mutation was discovered; this possibility awakened MacDougal's keen interest in de Vries. As he said, there was no more profitable subject for research in all of natural history than the causes that produced new species. If characters appeared suddenly and did not need thousands of years for their "infinitely slow realization," then evolution could be studied in the controlled environment of the laboratory. Acquiring such power over life would "rank well with that of any biological achievement of the last half century."[32] De Vries, in MacDougal's eyes, had made the most important discovery since Darwin's time.

MacDougal was campaigning on behalf of the mutation theory even before he had begun any research to duplicate de Vries's results. As early as April 1902, as soon as he had finished reading the first volume of *The Mutation Theory*, he lectured on the subject at the weekly colloquium of the Botanical Garden. His long and laudatory review of the book appeared that summer in *Torreya:* this was the first American review.[33] He obtained seeds from de Vries and began growing them in the Botanical Garden in May 1902. In 1903, when Anna Murray Vail, the Garden's librarian, had to visit Europe to buy books, MacDougal sent her on a side trip to examine de Vries's collections at Amsterdam and the type specimens at the Muséum d'Histoire Naturelle in Paris. Without waiting to obtain any new mutations in his cultures, MacDougal began proselytizing for the theory, discussing it and exhibiting his seedlings to his scientific colleagues.[34]

He was convinced that the question of the origin of species had to be resolved not by observing plants within a single generation, but by careful breeding over several generations of plants. The control necessary to eliminate all variables such as hybridism, disease, and parasitism could not be achieved in a field study, but only under an experimental breeding program. By 1904 MacDougal had found what he was looking for: mutations among his primroses. De Vries assured him that these were the first authenticated mutations seen in America.[35] With the help of Anna Vail and an assistant, J. K. Small, he expanded his research into the life histories of the parent and mutant forms and looked for mutants in other species of primrose. A species named Lamarck's evening primrose turned out to be a most obliging source of mutations. Between de Vries and MacDougal, it yielded fourteen mutants by the end of 1904.

The mutation theory lent itself to an ecological perspective because investigation of the theory involved trying to produce mutations, or new species, by subjecting plants to various external stimuli. These might include chemical treatments, changes in temperature or light, or even radiation treatments. MacDougal's research focused on the relation between organism and environment and the possibility of controlling variation by application of an external agent.[36] In this way he hoped to uncover the physiological processes governing the appearance of new variations.

His experiments were designed to cause permanent changes in his plants. He found that when various substances were injected into plant ovaries, the seeds were atypical and the new characters persisted in the second and third generations. This research showed that environmental agents could act directly on the germ plasm. One of his colleagues at the Garden, Charles Stuart Gager, took up the problem of radiation treatments and mutation. Gager thought the radiation did affect the germ cells and occasionally the next generation, but he did not believe he had produced new species or that the variations were the same as de Vries's mutations. MacDougal, less cautious, asserted that if radiation produced chromosome abnormalities, and if chromosomes carried the specific characters (as biologists were beginning to think), then radiation treatment would afford a "ready means of suppression or substitution of characters."[37]

Research on the mutation theory was related to the neo-Lamarckian theory of evolution, named for Jean-Baptiste Lamarck, the French biologist of the early nineteenth century who had advanced a controversial theory of evolutionary change in 1809. Lamarck had postulated that organisms could respond to environmental conditions in an adaptive way and that these responses could become hereditary. The neo-Lamarckian school of thought was at its height in America in the 1890s and had many adherents at the time de Vries's book was published.[38]

The difference between Lamarckian theory and the mutation theory lay in the idea of an adaptive response: in the mutation theory, the mutation that originated the "elementary species" need not be adaptive at all, and the elementary species might quickly die out. The Lamarckian theory emphasized the ability of organisms to respond adaptively to environmental stresses. Because both schools of thought focused on the impact of environmental conditions on organisms, they can be characterized as ecological theories of the evolutionary process. The importance of these ecological theories lay in their suggestion that the evolutionary process might be brought under control by simple experimental manipulations.

At the time, such researches could be provocative, for Darwin's ideas were still controversial; even small results were seized upon as potentially significant. In December 1905 MacDougal went public, announcing the results of the evening primrose studies with panache at a lecture to the Barnard Botanical Club of Columbia University.[39] The *New York Times* reported on the lecture prominently in the Sunday paper on Christmas Eve, under the headline "Dr. MacDougals [sic] Botanical Feat Threatens Evolution Theories," emphasizing the challenge that this research posed to Darwinism.[40] The piece began with a disparaging reference to Jacques Loeb, who had not long before prompted spectacular headlines of his own when his discovery of artificial parthenogenesis was announced.[41] Artificial parthenogenesis referred to the ability to cause an unfertilized egg to develop by stimulating it mechanically or chemically. Immediately the popular press linked his work to the creation of life and the explanation of immaculate conception. By 1902 he was on the front page, with appreciative profiles appearing in popular magazines. Loeb himself was deeply embarrassed by the exaggerated publicity, which was depicting him increasingly as a quack. The *Times* piece profiling MacDougal contrasted Loeb's "inconclusive experiments" with MacDougal's apparent demonstration that injecting strong osmotic agents and mineral salts into the ovaries of plants actually transformed one species into another.

MacDougal saw the mutation theory as a vehicle for reorienting research away from natural history, comparative anatomy, and embryology, the traditional grounds for evolutionary speculation, and toward the direct observation of the origin of species.[42] Experimental botany suggested a way to resolve the theoretical controversies of nineteenth-century evolutionary science by substituting hard data for speculation. However, MacDougal pressed his conclusions much further than the data warranted. Like Britton, he was engaged in a broader campaign to establish the primacy of a home-grown research program in experimental botany that could compete with European science, even while it drew inspiration from Europe.[43] As we will see in chapter 4, the mutation theory was particularly attractive to patrons of science who were interested in the idea of controlling evolution. Because of this interest, the experimental study of evolution served as a stimulus for the development of research programs in the life sciences beyond the confines of the Garden.

Garden scientists placed themselves on the frontier of scientific research, taking their lead from European science but adapting the research to the American context. The breadth and level of training in botany at the Garden was an enormous improvement over what the Garden scientists had themselves experienced two or three decades earlier, when they were just starting out. Lucien Underwood, surveying the educational benefits to Columbia, noted that compared to his earlier experiences at colleges where instruction in botany meant a two-hour course lasting ten weeks, the affiliation of Columbia with the Garden seemed "like the realization of an impossible dream."[44]

To Underwood, and no doubt to Britton and the rest of the group of reformers, this was also an American dream, a forward-looking expression of American energy and possibility. Although the Garden had modeled itself partly on Kew, Underwood pointed out that Bronx Park was superior in beauty and even in climate to Kew Gardens. Moreover, Kew was somewhat backward-looking, he thought, a conservative institution slow to adapt to modern ideas. Underwood criticized Kew for failure to illustrate "the modern principles of ecology" in its arrangement of floras.[45] The methods of the botanical garden at Berlin, under the leadership of Adolf Engler, were preferred, because the principles of plant distribution underlay the arrangements. Engler was interested in interpreting the present-day distribution of plants and also the history of the world's vegetation, going back to the Tertiary period.[46] Underwood lauded the careful and philosophical floristic work done at Berlin, which he thought exceeded what was being done at Kew.

On top of that, the attitudes toward botany and botanists displayed at Kew left something to be desired. The Kew museums were too crowded and not well designed, and the herbarium was not in a fireproof building, which he thought amounted to a "national disgrace."[47] The English botanists, so well endowed with their collections, were faulted for their neglect of other herbaria and their habit of belittling work done elsewhere. Underwood admitted that for all these conservative tendencies, there were positive signs that the Kew botanists were starting to recognize the need to modernize.

The point of this criticism was to stress that American botany need not be hampered by the traditions that were thought to be holding back botany at Kew. If Americans concentrated on their own flora, using the most modern ideas that were available from European institutions, they would surpass Kew Gardens. This enterprise was the destiny of citizens of the United States, whom Underwood imagined to be specially equipped for this kind of study by virtue of their racial qualities, in particular their "practicality, vitality, and energy," traits he felt were lacking in many Europeans.[48] Despite his admiration for German research, Underwood thought the future lay with the Anglo-Saxon people of America, who had the right racial qualities but were not bound by tradition. "The whole American continent, from Alaska to Cape Horn, with all that immense dark continent of South America, must be the working field of the American botanist," he proclaimed.[49] "The Anglo-Saxon is the only race that can enter a country, hold it firmly and elevate it in the scale of civilization by making it more productive. . . . The Anglo-Saxon blood, English or American, is destined to be the leading colonizing and civilizing spirit throughout the world in the future, as it has been in the past." Hence, he concluded, "the Anglo-Saxon blood in the New World, as in the Old, must originate and direct all exploration and development, and this will form one portion of the work of American botanical gardens."[50]

His patriotic fervor aroused in the wake of the Spanish-American War, Underwood proclaimed his compatriots' ownership of the natural history of the Americas. Americans did not need to compete with Europeans in the Eastern Hemisphere but should concentrate their attention on what was closest to them: "We can show our friends across the sea that in botany at least the Monroe doctrine is still a living and practical issue." Nor did he think it worth waiting for Europeans to settle questions of nomenclature or any other policy, mired as they were in conservatism. As he told his fellow botanists, "Americans must take the initiative in all these matters if they are ever to be settled, and the Old World must follow the lead of America in every field where progress is involved." The time had come for American botany to assert itself "in the modest way that becomes Americans" and adopt a position of leadership in science.[51]

Britton seized any opportunity to establish a foothold in regions that were botanically interesting and within reach of the Garden. Money for the Garden's work came from a special exploration fund built up from private donations and coming in piecemeal year by year. The city paid for the physical maintenance of the Garden and its buildings but not for all of its scientific work. The Garden had not raised the million dollars that was the goal of its fund-raising campaign, and it fell well short of the half-million-dollar goal that the botanists hoped to raise in their initial campaign, prior to development of the Garden. This failure to find support left it chronically short of funds for scientific work. For exploration in particular it depended crucially on its patrons.

The Spanish-American War turned that exploration decisively toward the Caribbean, as the United States became an imperial power. By the Treaty of Paris, signed on December 10, 1898, Spain renounced its claim to Cuba and ceded Guam and Puerto Rico to the United States. Immediately in 1899 the Garden sent private collectors to Puerto Rico to collect plants; in June they returned with eight thousand specimens. Cornelius Vanderbilt, the first president of the Garden's Board of Managers, paid for the trip. Vanderbilt asked Theodore Roosevelt, assistant secretary of the navy, to intercede with the secretary of the navy and obtain help. Darius Ogden Mills, a wealthy banker and patron of the Garden, arranged transportation to and from San Juan and got the War Office to request aid from the general commanding the U.S. forces in Puerto Rico. Mills took over as president of the Garden after Vanderbilt's death in 1899. The naval station provided all necessary help to the botanists, who obtained a complete outfit for collecting and photography, as well as steam launches, small boats, horses, and laborers when needed. This began a long association between the Garden and Puerto Rico.

Britton did not visit Puerto Rico until 1906, but later he was instrumental in organizing a scientific survey of Puerto Rico and the Virgin Islands, which began in 1914 under the direction of the New York Academy of Sciences and with the cooperation of the American Museum of Natural History and other institutions. He was in charge of the Puerto Rican survey until his death in 1934, and the survey itself, originally intended to last four years, continued until the mid-1940s.[52] Britton generally took his winter vacation in Puerto Rico, where he stayed at the expensive Condado-Vanderbilt Hotel, entertained lavishly (in contrast to his relatively modest lifestyle when in New York), and enjoyed a reputation as the "millionaire botanist."[53]

From 1898 to 1916 the Garden sent out seventy-three expeditions, involving 131 people in all, to the West Indies. Britton made twenty-five trips himself, usually accompanied by Elizabeth, who continued her studies of mosses. Most of the expeditions went to Cuba (where Elizabeth had grown up), Puerto Rico, Jamaica, and the Bahamas. Britton raised money for expeditions and other activities by pressuring the wealthy members of the Board of Managers. He had considerable inherited wealth, so he would write to the members of the board, explaining that he was personally donating a sum of money for some purpose and then requesting that the board member make an appropriate donation. This proved to be an effective tactic.[54] An estimated 150,000 specimens were added to the Garden's herbarium as a result of these expeditions. Hundreds of new species were discovered.[55]

Britton supplemented his collections with specimens obtained from other museums, an important source being the Berlin Botanical Garden, which housed Ignaz Urban's important collection of Caribbean plants. Urban, a specialist in West Indian flora, had never visited the region himself, but he obtained material from resident botanical collectors and distributors. Britton, by contrast, never considered building up his Garden's collections through purchases and exchanges alone. He always thought it important to have his own collectors in the field. His own workers not only could become experts on certain regions; they could gather information about plant distribution and related ecological problems while they were collecting. Henry Gleason, who took part in the ecological survey of Puerto Rico, noted that Britton chose men who were "young, strong, and zealous."[56] Rather than being sent to a hotel base in the main towns, they were posted far afield in small villages, remote plantations, or peasant houses. The same men went out repeatedly, honing their collecting skills with each trip. The "collector" of the past was evolving into the "ecologist" of the future.[57]

Britton promoted intensive exploration of these regions as not just desirable from a scientific standpoint, but also relevant to the commercial interests of the United States in the countries to the south.[58] Henry Rusby continued his work on the plants of South America, searching for important drugs and other economic products.[59] After his first expedition in the region in 1885, he maintained his relationship with Parke-Davis for two decades. He kept a collector in Bolivia for several years and later arranged to get collections from the Bolivian Department of Agriculture. In 1896 he collected in Venezuela for the Orinoco Exploration and Colonization Company of Minneapolis; and after a new source of rubber was discovered in Mexico in 1909, he spent several seasons surveying Mexican forests for a New York company. In 1917 he was asked to go to Colombia to investigate certain of its drug supplies and used the opportunity to make extensive general collections, with the help of an assistant from the Garden.[60] His final explorations to Colombia were in 1921 for H. K. Mulford Company, a Philadelphia pharmaceutical firm, although ill health forced him to leave the expedition early. These expeditions also afforded opportunities for general collecting for the Botanical Garden.

Collectors roamed across the United States as well, often in cooperation with state geologic surveys, their work made easier by the construction of railways. By 1900 there were nearly two hundred thousand miles of railway in the country, enabling scientists to reach remote areas and transport crates of specimens from the West back to New York. The railways also benefited the Garden by creating wealth and increasing the number of rich patrons of science who might have an interest in funding exploration work. William Dodge, a member of the Garden's Board of Managers, promoted the botanical study of Montana, and in gratitude the scientists named Dodge Mountain after him. Dodge, who died in 1903, made special gifts for explorations to the West Indies, Central America, and the Philippine Islands, as well as the Rocky Mountain districts.

Per Axel Rydberg, a Scandinavian botanist, devoted several summers to collecting in the Rocky Mountains for the Garden. Expeditions also went to the southeastern states and across the western territories. Arthur Hollick collected fossil plants in the Chesapeake Bay region. He also spent four months in Alaska in 1903 with the U.S. Geological Survey. The Garden helped support the work of botanists at other institutions as well. Joseph Nelson Rose of the U.S. National Museum made several expeditions in the United States that enriched the Garden's collections; he and Britton collaborated in the study of cacti, Britton's particular interest. Because the National Museum did not have enough funding at this time to finance any extensive explorations, the cooperative relationship with New York was very important, enabling Rose to take leaves of absence and pursue his botanical studies under the auspices of the Garden.

Establishing new research stations in the field was an important aid to these collecting enterprises. Daniel MacDougal was quickly dispatched to Jamaica to secure a tropical research station for the Garden. His efforts on behalf of the Garden followed several years' work to try to establish a government-run tropical station for American botanists. In 1897 a committee of American botanists, MacDougal among them, had been commissioned to select a site in the American tropics. They knew that their European rivals had been keenly interested in tropical American plants for decades, and the American group used that interest to promote their own cause, while suggesting that their approach would be more successful on account of the greater willingness of Americans to cooperate. As an editorial in the *Botanical Gazette* noted, in the United States there was less of the bickering and animosity that characterized scientific life in Germany, "where no one is really satisfied until he has a *Feind* [enemy]."[61] The key to success was better organization of government work in science and more cooperation between government divisions.

Eugen Warming, one of the fathers of European ecology, had himself made two trips to the tropics, once as a student to Brazil and thirty years later, in 1891–92, to the Antilles and Venezuela. In a speech to his Scandinavian colleagues, reprinted in the *Botanical Gazette* in 1899, he stressed the great scientific interest of the tropics, which showed unusual species diversity, although many species were sparse in abundance.[62] The tropics provided rich ground for the study of adaptation and the many interesting problems relating to coevolution and the competitive struggles between species that Darwin's work had raised. Warming believed that a historical perspective aimed at understanding the long process of development through geologic time was the key to understanding species diversity. The physical environment might explain the general appearance of vegetation, whether it took the form of forest or savanna or marsh, but the full explanation of a region's vegetation required both short-term and long-term scales of reference. The science of ecology that Warming promoted incorporated both immediate physical causes and historical causes as explanations of diversity and abundance.

American editorials supported the concept of a tropical laboratory as part of a plan to organize and develop American science. They had in mind the kind of wide-ranging ecological analysis that Warming described to his colleagues. Major funding was not necessary, just enough to provide a place adequate for research. The Spanish-American War interrupted their efforts, but botanists returned to the theme in 1899, focusing on Jamaica as a possible site. Jamaica had had a private botanical garden as far back as the 1740s; by the late eighteenth century, plans to develop a government botanical garden were moving forward. The breadfruit plants that Captain William Bligh (of *Bounty* fame) brought from Tahiti in 1793 to help feed the slave population were sent to two gardens in Jamaica. By 1797 Jamaica had an official island botanist to collect and describe native plants and investigate their medicinal and other uses.

In the 1860s a new garden located nineteen miles from Kingston was created for the cultivation of cinchona, from which quinine was derived. That marked the start of serious efforts to develop a botanical garden in cooperation with the Royal Botanic Gardens at Kew. By the 1890s the Jamaicans were looking toward the United States and its agricultural experiment stations as models.[63] The United Fruit Company also had plantations in Jamaica.

Jamaica had a pleasant climate, good roads, and an established British colony sympathetic toward American science and American business. It was easily accessible by steamer from Boston, New York, Philadelphia, and Baltimore. There was some concern about the isolation of the gardens from Kingston and the inadequacy of the Kingston hotel by American standards.[64] Health was a consideration as well. Johns Hopkins University already had a laboratory in Jamaica, consisting of several converted rooms in an American hotel, but the botanist there had contracted a fever and died. Trinidad and Puerto Rico were plausible alternative sites.

As botanists debated the merits of the different locations, the Jamaicans found in 1902 that they had to move their cinchona plantation to a lower elevation. With this move the laboratory and surrounding grounds at the original experiment station site, known simply as Cinchona, suddenly became available. Lucien Underwood, who went to Jamaica in 1903 to collect ferns, immediately saw that the location might be just what the botanists needed.[65] It had a furnished six-room house, with stable and servants' quarters, three low buildings suitable for laboratories, a storehouse, a small building for lodging visitors, and two greenhouses. He thought it a good site for American students and suggested optimistically that it could even rival the impressive Dutch gardens in Java.

MacDougal, by then employed at the Garden, was ordered on one day's notice to go to Jamaica, and, with wife and daughter in tow, he rushed to secure the laboratory. In July 1903 he negotiated a lease of the buildings and about ten acres of land at Cinchona and arranged to obtain herbarium specimens for New York. Having secured the Jamaican site, the New York Botanical Garden controlled a research station that would facilitate not only collecting but also the pursuit of ecological research. At an altitude of five thousand feet, the climate at the station was not strictly tropical, so that scientists could work there without experiencing the lassitude common to tropical regions. It was also a region of exceptional diversity, with many endemic species (species native to the area and possibly existing only there), which increased its ecological interest.[66] Botanists there could also study at the Hope Garden, located at sea level about five miles from Kingston. Britton considered the acquisition of the station to be the most important recent step taken by the Garden to further its research.[67]

The rapid expansion of the research and collecting activities was supported by aggressive publication. From the start Britton appeared to recognize that the production of both scientific and popular publications would help the Garden gain ascendancy over its rivals. Gleason noted that Britton tried to complete every project with the least delay. This meant focusing on feasible projects and publishing quickly. As Gleason remarked, Britton's motto was "Get it into print."[68] He pushed forward the publication of his *Flora of Puerto Rico* as soon as he had enough material at hand, without waiting for the survey to be completed. Gleason interpreted this haste as an indication of Britton's desire to make scientific work available to others, so that people could begin to build on that foundation. His work was a beginning, not the final or definitive statement.

But it seems evident also that Britton was acutely conscious of the need to publish quickly to establish priority, to claim ownership of those new American species that were being discovered daily. He was in a tight race with other museums, and he had to win. Gleason commented, "If Britton could have done so, he would have extended the Monroe doctrine to botanical research."[69] Unable to do that, he worked hard to establish his priority in the naming of species by getting into print before his competitors. His style of science was therefore quite different from that of his chief European rival, Ignaz Urban of Berlin. The German work consisted of detailed descriptions, full citations of collectors and localities, and extensive discussions of species relationships, all in Latin. Britton's descriptions were short, his citations meager, and little attention was paid to comparisons and kinship. And they were in English. When duplicate specimens were sent to New York and Berlin at the same time, "before Urban could complete his minute dissections, his ponderous descriptions, and his erudite discussions, Britton had his brief diagnoses actually in print and the type specimens, christened with a Brittonian name, safely deposited in the herbarium at the New York Botanical Garden."[70] This is what the nomenclature war had been about: ostensibly its purpose was to bring order and uniform standards to botany, but an equally important goal was to grab the scientific credit that went with priority. Ironically, the obsessive attention to detail in planning the Garden's growth, combined with workaholic habits and a deep need for control, did not produce work at a level of erudition comparable to the best European taxonomic work. But it did produce a flood of work, some of it flawed and incomplete, which by its sheer mass put the Garden on the world map.

Britton's American Code did not win general acceptance; most botanists followed the new International Code formulated at the International Botanical Congress in 1905 and amended at the next congress in Brussels in 1910. The Harvard botanists followed the International Code, which was used for the revision of Asa Gray's *Manual* in 1908, the same year that the international community affirmed Latin as the language of taxonomy. Botanists became more vocal in their opposition to the American Code as its supporters dropped away. At the conference of the Botanical Society of America held in Chicago on January 1, 1908, to discuss "Aspects of the Species Question," it must have been particularly painful to Britton to find himself attacked by former ally Charles Bessey of the University of Nebraska, who now firmly sided with the European botanists and condemned the American Code for its effect on scientific quality.

Bessey complained about the lack of authority in botany and the license that this had given to unqualified people to think they could practice taxonomy: "The rule which goes into effect today, requiring diagnoses of new species to be in Latin, will prove a deterrent to the tyros who would rush into print with their diffuse English descriptions."[71] Britton's strategy had made it too easy to do botany. Designed as a way to boost the move toward professionalism in American science, it was now threatening that very professionalism. Bessey continued to rail against the anarchism that now reigned in American botany: "When we had masters in botany, who were kings to whose authority all must bow, we complained bitterly. Now that the kings are dead the democracy of botany is suffering from the misrule of anarchy. If democracy will not control its subjects we shall have to return to a botanical oligarchy, or even to a dictatorship, for anarchy cannot be endured."[72] Henry Cowles, an ecologist at the University of Chicago, spoke in similar terms, urging Americans to join their European colleagues in adopting the International Code, putting an end to "taxonomic chaos," checking the "voluminous contributions of amateurs," and bringing scientific standards back to taxonomy: "The recent ebullitions of the taxonomic radicals have evoked in botanists in general successively dissatisfaction, contempt and rage. These things will not be endured much longer; a little more and the sinning taxonomists will be 'cast out into the outer darkness where there shall be wailing and gnashing of teeth.'"[73]

So much for the spirit of friendly cooperation that was said to distinguish

American from European botany. Both Britton and MacDougal presented papers at the same conference, and in both cases their focus was far removed from concern over the legitimacy of the American Code. Instead they discussed the intellectual problem of how to define a species and what impact the new experimental sciences, such as the mutation theory, might have on the concept of a species. One can only imagine how these colleagues must have felt as they were chastised, given the impressive accomplishments that Britton could look back on in creating the Botanical Garden. But however much he may have gnashed his teeth, Britton was not persuaded to change his ways by the fear of being cast out. In 1919, in a review of a catalog of plants from the Washington, D.C., area, he praised the authors' use of the American Code and disparaged the International Code, saying that it had been "forced down the throats of the Vienna Botanical Congress by a German majority." "Internationalism," he concluded, "is proving a dangerous principle to play with, and in many aspects has much to condemn it."[74]

As Britton's former allies appealed for accommodation with European practices, foreign botanists dependent on American institutions complained bitterly about being forced to adopt an outmoded code. Henri Pittier, a Venezuelan botanist who had regularly sent specimens to North America, both to New York and to the National Herbarium in Washington, was frustrated by their insistence on using the American Code when its popularity had waned. Writing to Smithsonian curator W. R. Maxon in 1920, he remarked that this form of "tyranny in the so-called free America is discriminating, not to say more."[75] Pittier's problem was that he had to send his findings to the United States in order to get them published. Britton had well understood that gaining freedom meant being able to publish your own work. With the help of his wealthy friends, he achieved the control over scientific publication that ensured the success of his enterprise.

With a critical mass of loyal supporters, and with control over scientific publication, Britton was able to use his code without any thought of conforming to what the rest of the world was doing. He and his colleagues published scores of species without the Latin diagnosis. Not until 1930, close to the end of Britton's life, was the International Code revised in a way that brought it closer to the principles of the American Code. At that time too, the names published by Britton and his colleagues in defiance of the rule were validated, although the requirement for a Latin diagnosis was upheld.

But the point is not really who "won" or "lost" this debate, whether the Kew rule was really a better means of achieving stability than the American Code, or whether the erudition of Ignaz Urban revealed his higher standard of science. The point rather is that Britton's American Code served an important purpose for him. It enabled him to launch a taxonomic Blitzkrieg that routed the opposition and allowed the New York Botanical Garden to dominate the study of American flora. American botany had to advance quickly in its race with European institutions that were far better established. Using the American Code, along with fast publication of new species, was one way of moving to the head of the pack. Britton's colleagues eventually disagreed that this strategy could work: it was democracy run amok. They felt it was too inclusive, too likely to encourage amateurs with low standards, and they rejected the strategy in favor of conformity to international authority. Moreover, Britton's position at the Botanical Garden was becoming czarlike, and his intolerance of other views was hardly democratic. But at first the strategy worked remarkably well at channeling resources into science.

Lucien Underwood's description of the Garden, quoted earlier, expressed both the nationalist and racist attitudes underlying these advances in science. The corollary of these attitudes could be seen in the educational programs of the Garden. Not all Americans, of course, were Anglo-Saxon, and by the eugenic logic of the day it followed that not all shared the racial characteristics most conducive to the pursuit of science. Ordinary people had to be taught the value of science, which involved teaching them how to behave in a botanical garden as well as how to behave toward plants when they were left to their own devices on public lands. One of the purposes of the Garden was to improve people's character, not just by giving them a beautiful landscape to help relieve the stresses of the city, but more deliberately by reshaping the way they approached their environment.

The Garden had a mission for public education, starting with the labeling of the trees and public lectures to promote interest in science. In 1901 the Torrey Botanical Club started meeting regularly at the Garden. The meetings were held during the day to enable more people, especially women, to attend. The membership of the club was nearly half women. Most members did not attend the meetings, though; generally only around twenty people showed up. A new magazine, *Torreya*, began publication with the purpose of offering general botanical information to lay people. Education meant not just providing magazines, lectures, and exhibits, but in a more general sense teaching people how to act in a botanical garden, how to move through it and get not only information but the right aesthetic experience from it. As early as 1897 Britton was giving thought to how to control the flow of people, imagining that one day the New York garden would be as hugely popular as Kew Gardens. The buildings of most interest had to be separated in order to divide the crowds and prevent people from becoming dangerous to each other and destructive of the Garden.[76]

One of the goals of the Garden was to improve and refine the people of New York. New York's Central Park had been transformed from a genteel and exclusive park to an eclectic recreational site for working-class New Yorkers by the 1890s; Britton envisioned a more restrained atmosphere for his Garden, one in keeping with his lofty scientific and educational aims.[77] Alcoholic drinks were strictly forbidden within the Garden, but in the absence of a wall or fence it was impossible to control what people brought in. The approaches to the Garden had to be kept free of "demoralizing influences." Border nuisances that visitors complained about included the drivers of cabs and hacks soliciting patronage at the entrances and other points along the driveways.

Some visitors—an estimated 10 percent—were destructive in various ways. Littering was a common problem, but vandalism was also rampant. People broke branches off shrubs and picked flowers by the dozen. Fine plants were stolen, flower beds were sometimes stripped, and tops were cut from coniferous trees to be sold as Christmas trees. Careless smokers tossing matches from the paths caused serious fires. Even the labels were stolen to be sold for the metal they contained. School children were sent by their teachers to collect specimens for the next day's science class. Barrelfuls of flowers were confiscated by Garden staff from visitors as they left the gates. Signs were posted to try to prevent littering, a misdemeanor carrying a fine of one to five dollars, and Britton hoped that when more paths were built, people would stick to them and thereby preserve the forests. A police patrol helped to enforce neatness, and on Sundays and holidays the park staff itself kept guard. Stuart Gager, who was at the Garden in the early years before leaving to direct the Brooklyn Botanic Garden, remembered the despair they felt that a botanical garden open to the public could ever be maintained under these conditions.[78]

The nearby New York Zoological Park, established in 1899 in the southern part of Bronx Park, was experiencing similar problems. Its director, William Hornaday, attributed the bad behavior to "lower class aliens" and in his annual reports made scathing remarks about "low-lived beasts who appreciate nothing and love filth and disorder."[79] The Brittons were more restrained, refraining from disparaging remarks of this kind. But they felt the need to educate constantly, and Elizabeth in particular took keen interest in reforming people's manners. Within the Garden she kept a sharp watch for destructive behavior and politely but firmly instructed the guilty parties to mend their ways. Gager recalled Elizabeth's "rapid, somewhat nervous step" as she hastened to reach a visitor who, "imbued with the obsession that this is a 'free country,'" was helping himself to some flowers from the Garden. The encounter was one "that the overzealous lover of some one else's flowers would not soon forget."[80]

The problem, as Elizabeth pointed out, was that people behaved within the Garden exactly as they did when in the open country, picking flowers as they liked.[81] It was necessary to train students and teachers not to view the garden as a type of common land to be used by anyone, whether it be for a goat pasture or a picking ground for flowers, but as controlled land that was like private property. Having learned to behave properly in the garden, people would then be expected to carry that learning back into the wild and refrain from destruction. They were also to treat unprotected open land as though it were private property, no different from a garden. And developing loan collections for teachers would make it unnecessary for them to pick plants each year, whether from the Garden or from open land.

Stuart Gager's later reminiscences also focused on this theme, which was an early expression of the argument popularized in the 1960s as the "tragedy of the commons." The notion of a common land was seen as antithetical to preservation of species, for common lands could only be misused. When the automobile increased people's access to the countryside, this need became even more acute, for the car carried "a more miscellaneous type of person, including those whose idea of property rights was primitive, who had no more appreciation of the importance of conserving natural beauty and esthetic natural resources, such as wild flowers, than the early lumbermen had (or should I say have?) of forest conservation."[82] He too shared the idea that the central task was to make people believe that all land had to be treated as private.

In 1901 several newspaper articles appeared with this theme, and Elizabeth drew attention to them in an issue of *Torreya.* Spring had brought its annual "systematic destruction of a large proportion of all wild flowers within reach," claimed the *New York Tribune.* Ignorant people failed to understand that the "vegetable world has rights which the lord of creation himself is bound to respect." The arbutus and mountain laurel were almost exterminated in the Bronx region as a result of overenthusiastic picnickers and nature lovers. *House Beautiful* noted the tendency of people to rush upon every growing thing, uprooting it and then tossing it aside: "Nothing of any value remains out of all the lives butchered to make an East Side holiday." No thought was given to the flower's "right to existence" or the "harm done to the children by allowing them to think that they may destroy life as they choose." The preservationist's sermon stressed nature's inviolable right to exist, an idea that ultimately depended on a religious viewpoint. The human role was to study and appreciate nature, not destroy it.[83]

Elizabeth noted that various areas had their favorite plants, and these were often endangered there: the fringed gentian in the Berkshires, rhododendron and azalea in parts of Pennsylvania, and the holly and *Prinos* berry in New York. Fashion was the great enemy. Filling jardinieres with wild ferns was causing much destruction in Bronx Park. In Connecticut a law had to be passed to protect the climbing fern. Casual botanizers were also a problem. She expressed amazement that the journal of the New England Botanical Club actually advertised the locations where rare plants might be found along the Bangor and Aroostook Railroad. "What botanist sold his birthright for a few railroad passes?" she asked.[84] Even members of the Torrey Botanical Club had their knuckles rapped for wanton destruction of plants. The solution was to cultivate, not pick from the wild, those plants that fashion pushed to center stage, to educate children and teachers to study nature but leave it alone, and to encourage botanists to be more responsible in their collecting activities. As the language of these articles suggests, campaigners like Elizabeth Britton approached their task with religious zeal.

In 1901 two women, Olivia Phelps Stokes and Caroline Phelps Stokes, prompted by Elizabeth Britton's concerns, gave three thousand dollars to the Garden for the preservation and protection of native flora. The gift established the Garden's first fund for conservation. The Brittons used the money to fund prizes for essays on conservation. The essays selected often had as their subject the New York Botanical Garden itself, proposing various methods for protecting the garden's collections from ignorant New Yorkers.

The frequency with which the Garden appeared as a subject of these essays reflects the Brittons' preoccupation with their creation, but it also points toward a deeper assumption that was not questioned by these preservationists: that the problem was to be located in the selfish and uncivilized behavior of new Americans. The solution had two components: first remove land from public use by privatizing it and controlling how it might be used, for moral suasion by itself would never work, then reeducate people so that they would accept new standards of behavior. An aesthetic appreciation for plants had to be cultivated, in part through scientific education about nature but also through nature poetry and nicely illustrated popular magazines. But in addition public attitudes toward common land had to be reversed.

In this respect the Garden embodied a set of attitudes common to the American scientific elite at this time that set scientists in opposition to the values and views of ordinary people, who did not see themselves as culpable in the way that scientists saw them.[85] The prevailing assumption was that ordinary people required management. They simply could not be trusted to behave responsibly if left to their own devices. Gardens provided the environment in which this transformation of public attitudes could be effected.

Directly following from discussions like those reviewed in Elizabeth Britton's piece in *Torreya,* the Wild Flower Preservation Society of America was formed in 1902. Its president was Frederick Coville of the Department of Agriculture, who had also been a loyal Britton ally in the nomenclature wars. The vice president was Daniel MacDougal of the Garden, and the secretary was Charles Louis Pollard of the U.S. National Museum (the Smithsonian). Elizabeth Britton and eleven other botanists from around the country (ten men and one other woman) were named as managers of the society. The dues were set at one dollar annually, for which members received the *Plant World,* a magazine of popular botany. In the first year 264 people joined, the largest contingents being from New York and Baltimore. Elizabeth's role within the society eventually grew more prominent, and she became a leading advocate of plant protection. She led a national boycott of the American holly as a Christmas green, wrote regularly about conservation, and helped to arouse public sentiment for the passage of conservation laws.[86]

Botanists took their cue from zoologists, who had become concerned about extinction of species two or three decades before. Ornithologists, alarmed at the destruction of bird populations, were leading conservation efforts in the United States but at the same time were careful to protect their own scientific activities.[87] They did not favor too stringent legal restrictions on collecting because scientific ornithology required killing birds in order to learn about them. Observing through glasses was not adequate.

The wildflower preservation movement likewise targeted certain endangered species and then mounted campaigns to discourage picking those species, but it did not interfere with the collecting practices necessary for the Garden's scientific functions. Essays advocating conservation measures could be followed by announcements that the Garden had acquired collections of tens of thousands of specimens, without any hint that a contradiction was perceived. And although preservation themes were pushed with religious zeal by people like Elizabeth Britton, the preservation society did not always agree with the more extreme conservation views of the time. Daniel MacDougal, collecting for the Garden in the Southwest, noted that the daily press was circulating sensational reports about the danger of extinction of the giant saguaro cactus. Since the cactus was growing over hundreds of square miles and had no commercial importance, he pointed out, its destruction in certain irrigated areas was of little importance.[88]

From the perspective of the Garden scientists, the goal of preserving species was an extension of the immediate goal of preserving the Garden and its botanical mission. The Garden was a microcosm, a repository of so many of the continent's species. Each individual's experience of nature had to be mediated by institutions like the Garden, run by professional scientists who would scour the earth to gather collections, describe and house them, and make them available on loan to teachers and students of nature. The staff of the institution would place the plants in attractively arranged and labeled gardens and make it possible for people to experience these plants from the safe distance of the pathway. Just as the Garden was to be treated as private land, so all of nature was to be taken as private property and likewise respected. Teaching people to accept the Garden and to behave properly in it was part of a larger scientific enterprise. Civilizing people, preparing them for a modern, more crowded world, meant preparing them also to accept the authority of science.

The emphasis on public reeducation and blaming the ordinary citizen for the ill effects of "unsocial" behavior was characteristic of urban reform movements of this time, whether aimed at public health, eugenics, or crime and delinquency problems. The focus of reform was often the individual, whose behavior had to be altered in the cause of greater health, efficiency, social stability, or racial improvement. Conservation was akin to missionary work, aimed at converting the individual and setting him or her on the path of righteousness. But what sort of individual might all this training produce? One prize-winning essayist contrasted New Yorkers unfavorably with the Japanese, who had learned to revere and protect nature. Significantly, the author, a man, pointed out that the Japanese were not made effeminate and weaker as a race because of this love of nature; they could be as warlike as anyone![89] People needed to be restrained, yet they needed to retain their virility, and with it their racial superiority. It was not obvious that these qualities fitted together.

Preservationists took a dim view of the average person. The approach adopted by the Wild Flower Preservation Society and by Elizabeth Britton echoed the arguments of the first winner of the Stokes prize in 1902, Frank Hall Knowlton, an eminent botanist and paleontologist and the editor of the *Plant World*. Knowlton, thinking of the Audubon Society as a model, wanted to emphasize public education, focusing not on the practical uses of plants but on their aesthetic and scientific uses. In fact, he warned against drawing attention to the idea that plants might have real value: "The cupidity of the average human being is so great, that, if it were simply rumored that these plants could command a money price, their doom would be fixed."[90]

Another theme, often advanced by women, was to devise methods of educating children about nature, so that the next generation would be better informed. The essays on the education theme helped to publicize the innovations of the trendy nature-study movement that was being advanced in some New York schools.[91] As applied in nature study, the endeavor consisted of showing children how plants and animals were related to their own lives, to their needs and activities. This approach was intrinsically ecological. No organism was to be studied in isolation; rather the connections between species, and ultimately the connections to humans, were stressed. In awarding first prizes to these essays, the Garden promoted an ecologically aware form of education.

This style of conservation rhetoric drew harsh words in 1940, at a celebration to commemorate the conservation work of Elizabeth Britton. Percy L. Ricker, president of the Wild Flower Preservation Society, pointedly criticized the conservation literature's emphasis on sentimental and aesthetic values. The movement focused too much on children and the general public, he believed, and too little on the problems of unregulated land development that presented the real threat to species. As he pointed out, thousands of acres of land were being "burned, grazed, or cleared every year for agricultural, manufacturing, residential, recreational, and highway purposes, destroying many times the number of flowers that are picked by wild flower lovers, roadside and market dealers, gardeners, nurserymen and drug collectors."[92] Such rapaciousness had of course been noted long before; it was addressed nearly a century earlier by people like George Perkins Marsh, as well as by conservationists after him. But Ricker felt the public face of the conservation movement had overemphasized sentiment and not adequately attacked the prevailing ethos of growth.

With hindsight one can see how the bias that Ricker identified would arise, for the Garden drew strength from the economic development of the country and depended on the wealth created by industrial expansion and territorial conquest. True, the Garden was a way to protect land from development and destruction, and its authority might persuade others of the values of conservation. But the Garden's objectives, even in conservation work, were not to subvert the ethos of economic growth, but to mould the American people into an orderly, forward-looking population, supportive of Britton's vision of American science and its continued progress.

One consequence of Britton's aggressive growth strategy was to boost the kind of research that was considered new and modern. In botany, the buzzword of the 1890s was *ecology*, a new approach to the study of plants in their environmental context. Britton and his colleagues at the Garden did not set out in a self-conscious way to create or promote a new discipline called ecology. Britton himself was primarily a taxonomist, and his chief concern was to build his collections and, by imposing the American Code, to gain credit for the Garden's explorations and discoveries. The Garden's scientists did not make programmatic statements about the need for a new science of ecology. But the Garden's emphasis on modern research and on fieldwork, and its inclusion of experimental botany as well as taxonomy and morphology, had the effect of boosting the kind of research that we now identify as ecological, at a time when ecology was emerging as a distinct field of inquiry but had not yet achieved definition as a discipline. The public educational programs also helped to spread an ecological point of view by trying to teach people how to appreciate both the Garden and nature in the wild.

As we continue with the story of the development of ecology in the chapter 4, I will emphasize the relationship between field investigation and experimental biology. It is important to understand this connection because experimental work generally has an objective that distinguishes it from descriptive field sciences: it seeks to manipulate and control nature. My argument is that ecology attracted patronage in part because ecological work fitted into a larger quest for control over the evolutionary process, which in turn was related to the broader development of the country's resources as the population expanded. Controlling evolution was the scientific Holy Grail that Americans avidly pursued along several different paths in the early twentieth century. The scientific context of ecology includes not just the expansion of taxonomy and biological surveys, but also deepening interest in the experimental study of evolutionary mechanisms.

The new science in various ways expressed the aspirations of a particular group of Americans who were white, middle-class, Protestant, of British or northern European descent, and imbued with a strong sense of the superiority of their race as America expanded its empire. Ecology, broadly speaking, is the study of the relation of organisms to the environment and to other organisms. But humans are one of the species that must relate to and fit into the landscape, and in the early twentieth century Americans had to reflect on their own relationship to the land and the transformations they were effecting as they settled the West and looked toward the Caribbean and Central and South America. As we unfold the story of botanical science, we will explore the relationship between the growth of ecological understanding as a scientific development and the concurrent discussions that were taking place, in myriad contexts, about American identity and the future of American society.

CHAPTER FOUR

Science in a Changing Land

The success of the New York Botanical Garden elevated Britton to a leadership position in the community of American botanists very quickly, even though he lost some allies along the way. His prominence had far-ranging impact on the development of botany as philanthropists came to have a more important role in American science. This chapter examines how the authority of the Garden was translated into new research programs of particular significance for the growth of ecology. We will look closely at the Desert Botanical Laboratory near Tucson, Arizona, founded by the Carnegie Institution of Washington in 1903. It was the first laboratory in the United States devoted to basic research in botanical science. Its first director was Daniel MacDougal, who brought to the job the same energy and drive he had exhibited while running the Botanical Garden's laboratories. The Desert Laboratory operated somewhat as a satellite of the Garden, although it was not controlled by the Garden. MacDougal's leadership style was quite different from Britton's: he seemed content to allow scientists freedom to pursue research as they saw fit. He envisioned a mecca for botanists, and that was what he achieved.

In establishing a research tradition that embraced a wide range of approaches to botany, from physiological experiments to descriptive studies of vegetation and climate, the laboratory's work helped to define the subject matter of ecology. The cumulative effect of this work was to give substance to the new science by creating a solid body of research and inaugurating a tradition of long-term inquiry in various locales. The Desert Laboratory, as a research community, was a nucleus around which the discipline of ecology could form. It was not the only such nucleus in the country: others formed at the University of Chicago, the University of Michigan, the University of Illinois, the University of Wisconsin, and the University of Nebraska, schools that produced many of the leading first-generation ecologists in the United States. The agricultural experiment stations operating in conjunction with land-grant colleges also dealt with ecological subjects, particularly in relation to soil science, but scientists were often overwhelmed by the need to work on practical problems.[1] The Desert Laboratory was the only purely research organization among this group.

One historian, despairing at how a science can have such breadth, being defined on the one hand as a branch of physiology and on the other hand as a branch of biogeography, has referred to ecology as "schizoid" and concludes that ecologists must have been caught in an intolerable "double bind" as they "lurched" between the cultures of laboratory and field.[2] Certainly ecology is an extremely challenging subject and is therefore often frustrating and taxing. Furthermore, descriptive field sciences can be in competition with experimental science; at the end of this chapter we will see evidence of this competition. But the Desert Laboratory created an environment where, at first, these interests were not competing but were complementary, just as the Botanical Garden created an environment, at first, that brought together all areas of botany. These conjunctions may not be sustainable, but the important thing is to note their significance when they exist.

The scientists at the Desert Laboratory did not so much "lurch" between laboratory and field as move easily between different scientific enterprises, different places, and different communities. They felt no conflict between any disparate or antithetical cultures of laboratory and field. Their mixture of field research and experimental interests did not always yield conclusive answers or fulfill their starting ambitions. Sometimes they underestimated the difficulty of understanding ecological and evolutionary processes. But the research did help to define a subject that otherwise might never have achieved coherence as a discipline, although it might have remained important as a way of viewing biological problems.

My goal here and in the next chapter is not to survey all of the concurrent developments that affected the growth of ecology, but to treat the Desert Laboratory as a microcosm that illustrates the range and variety of ecological work in the early century, as regards botany, and reveals why such work should have found support at that time. Focusing on this institution, a new type of research community, enables me to place ecology in the context of the economic expansion of the country and the need for more science to support that expansion. Much like its applied counterparts, conservation and range management, ecology was pursued in aid of a broader quest to expand human dominion over the land, and to do it in a deliberate, rational way rather than by trial and error, which could bring disaster.

What values and motives underlay early ecological work? What did people think ecological analysis would achieve? There is no single answer to these questions. Of course it is always possible simply to point to the general desire to acquire new knowledge. Human curiosity is its own motivation. I would like to go beyond such general motives and consider the way ecology served broader social purposes during this period of rapid expansion in the United States, and how ecological work derived momentum from the changes that this expansion entailed. My short answer is that ecology promised to address questions dealing with the control of life and the problems of adaptation in diverse landscapes. These questions were important to a population that was settling new regions and hoping to exploit new resources. These questions demanded both experimental work and fieldwork and justified combining disparate forms of research in a single enterprise.

I am not, however, suggesting that there was a fixed and explicit purpose that unified ecological thought, or that there were well-orchestrated efforts to articulate a unified vision of what ecology was about and what it should accomplish. Ecologists were not handed marching orders and told to create a science that would address specific goals. Ecology is a broad, eclectic science that embraces many kinds of research. It does so now, and it did so then. Moreover, it was advanced as a basic science, that is, one that sought fundamental new knowledge about living processes, without immediate regard for applications but with practical ends in view down the road. The underlying justification for ecological knowledge came from the desire by Americans to understand, predict, and control living processes, so as to improve the ways in which humans were moving into and adapting to new lands.

A stronger statement would be that underlying early ecology is an ideology of domination and control that is similar to that of other modern sciences. It is important to understand that ecology in its infancy included experimental analysis of problems of adaptation, as well as long-term studies of how species were faring in a variety of landscapes. Experimental science is chiefly about prediction and control, and long-term studies help one to pin down the causes of change in a changing world. Ecology, like its sister discipline genetics, was part of an effort to control life and to apply rational methods to a complex set of problems generated by the American desire to migrate into and adapt to new landscapes. Ecology was in this respect a quintessential Progressive Era science emerging in response to rapid growth and transformation of the land. Ecology was a science called forth by change; it sought to render natural processes more predictable so that further change could be better controlled. It was not "schizoid" but in step with American values and interests. The problem was that these values and interests were redefined over time, and ecologists had to adapt to new conditions and expectations as their science slowly took shape.

When Andrew Carnegie founded the Carnegie Institution of Washington in December 1901, the officers of the Institution naturally looked to Britton and his associates for advice about how it should spend its money. Carnegie had been one of the original incorporators and patrons of the Botanical Garden and served as its first vice president, a position he held when the Carnegie Institution was created. The Institution's initial mandate was to seek out exceptional men and promote their research, but soon the directors began to think about setting up new laboratories, in effect small research communities, rather than merely funding individuals. To decide where these laboratories might be located and what kinds of research might be most profitable, the Institution looked to the leaders of American science for advice. Frederick Coville of the U.S. Department of Agriculture chaired the Carnegie Institution's advisory committee on botany and was aided by Britton and two other botanists of note, John Muirhead Macfarlane and Gifford Pinchot. Macfarlane was a professor of botany at the University of Pennsylvania and director of the university's small botanical garden. Pinchot was U.S. chief forester and a prominent conservationist and advocate of scientific forest management. The committee's report, presented in June 1902, reflected the interests of these four men.[3]

The committee urged the Carnegie Institution to direct its resources to two broad problems: the relation of vegetation to environment in the United States and the botanical exploration and study of Central America and the West Indies.

The total cost to the Carnegie Institution was estimated at about forty thousand dollars per year, and it was assumed that these tasks would need an initial commitment of five years of support. The Carnegie Institution was encouraged to complement the work of the Botanical Garden. Indeed, the report specifically noted that further research in the North American tropics would help the Garden to complete the work it had already begun. Britton had not yet secured the Jamaican laboratory and was likely using the report to entice the Carnegie Institution to take the lead in developing tropical research. The report envisioned that such a laboratory might become the foremost tropical research center in the world, drawing European as well as American investigators.

The committee emphasized that a social and industrial revolution was under way in the tropics. The commercial needs of the United States were driving forces for a transformation of tropical agriculture, which had occurred already with sugar cane and was in process with coffee, rice, and bananas. They described this process in explicitly racial terms, as the unfolding of a racial destiny: the "white race," having learned how the "germs of yellow fever and malaria are transmitted," was penetrating into the tropics, searching for healthy places to settle and establish industries.[4] These newcomers had to learn to adapt to the new landscape and therefore needed to draw on the storehouse of information that native people had accumulated over the centuries. For this reason ethnobotany was included in the research plan. The relationship between native peoples and the environments in which they had evolved would yield valuable information about these people's movements, their historical development, and their industries, information needed by those who would colonize the land.

The report's recommendations for studies in the United States also focused on the need for precise scientific knowledge to facilitate economic growth. One pressing need was for study of the relationship between forests and water supply. Seventy thousand square miles of national forest reserves had been created to protect the water supply in the West; yet, the report claimed, scientific data to demonstrate the effect of this measure was lacking. The Carnegie Institution was in the right position to pursue and coordinate a more systematic scientific study of forests and their environment, so that the efficacy of these conservation efforts could be assessed.

Another recommendation was to establish a field laboratory in the western desert. Britton was especially interested in the study of cacti, but the idea was chiefly Coville's. It was a project he had had in mind since 1891, when he made a botanical survey of southern California's desert regions, in particular Death

Valley and the surrounding area. In 1890 Congress had enlarged the scope of government surveys to include botany, and Coville's Death Valley survey was one of the first to focus on botanical questions. This increased funding came right at the time when botanists, people like Britton, Charles Bessey, John Merle Coulter, and Joseph Arthur, were pushing for the expansion and development of botany.

Coville also hoped that the momentum created by these surveys could be used to develop botany, not just for the obvious economic reasons, but also because of the interesting biological problems that the desert regions posed. He had been struck by the observation that the desert contained many perennial shrubs and annual herbs that appeared to have few adaptations for water storage. These discoveries raised questions about the nature of adaptation, questions of broad ecological significance that needed to be studied in the field. A field laboratory in the desert would provide basic knowledge about plant physiology and relations to the environment.

Studies of this kind also made economic sense. As Coville pointed out in the Death Valley report, if the objective of these surveys was to assess the agricultural potential of a region, the easiest way to collect pertinent information was to study the adaptations of the plants already growing there. Agriculturalists, he said, had usually confined their studies to climate and soil composition, or got bogged down in laborious farm experiments testing various crops. It would be cheaper and more efficient to gauge the agricultural capacity of a region by analyzing its natural vegetation.[5]

A decade after the Death Valley expedition, it was becoming more common to conduct ecological studies of adaptation and of the distribution and abundance of species in order to provide a scientific basis for agricultural and other economic development. This research was linked to conservation work, where conservation specifically meant the management and efficient use of resources. As Samuel Hays has aptly termed it in his study of conservation science in the federal government, the ideology underlying the conservation movement was faith in the "gospel of efficiency," a faith characteristic of Progressive Era reform movements.[6] Frederick Coville and Gifford Pinchot, being directly involved in formulating policy to guide western development, were two prominent preachers of the gospel of efficient management.

The National Reclamation Act of 1902 provided for the construction of irrigation projects using funds from the sale of public lands in the West.[7] In connection with this act, Pinchot and Coville helped establish government policy for conservation and resource management. Coville's special area of expertise was range conservation; his studies supplied crucial technical advice for the Public Lands Commission, which turned to the problem of formulating a range policy in 1903.[8] The committee report for the Carnegie Institution drew attention to the need for scientific studies that would help to assess the prospects for population expansion and agriculture in the Southwest:

> The economic ground for the establishment of such a laboratory is the enormous development of population and industries that is bound to take place in our arid region during the next hundred years. The basis of that development is agriculture, both with and without irrigation. At the present time comparatively little is known about the peculiar fundamental processes of plant growth under the unusual conditions surrounding plant life in that region.[9]

The emphasis on "fundamental" knowledge in this report was important. The committee did not intend that the future Carnegie laboratory should be tied exclusively to practical concerns. Such work could be left to the agricultural experiment stations that were being established in the Southwest. The practical subjects pursued at those stations, even when aimed at studies of soil and climate and the effect of irrigation on plant growth, did not constitute a well-rounded research program in ecology, which addressed broader questions about the relationship between plants and their environment with the aim of clarifying the nature of adaptation and the mechanisms of ecological succession and evolution. The committee's plan was to provide basic research that might be used by the agricultural experiment stations in the arid states but would not be confined to the practical needs of these stations. The development of ecological research was therefore linked to the drive for efficient resource management, which in turn was an outgrowth of western expansion.

The combination of an appeal for basic research to advance scientific knowledge with the promise of eventually contributing in a practical way to economic development appealed to the Carnegie officers. Instead of following all the recommendations of the botanical report, however, the officers singled out the desert laboratory as a worthy project. In addition the Institution created a station for experimental evolution at Cold Spring Harbor, New York, and a marine biological laboratory in Dry Tortugas, Florida.

The Board of Trustees approved the desert laboratory plan in 1903 with an appropriation of eight thousand dollars to establish a desert laboratory and pay a resident scientist. To locate the laboratory, the Institution sent Coville and Daniel MacDougal, "a desert specialist of the first magnitude, and a jolly good fellow all the time and everywhere," out west to find a suitable site.[10] They explored a large area along the Mexican boundary, rejecting regions that were too severe in climate or—because they feared the depression that accompanied too much solitude—areas that were too remote for long residence. They settled on the area surrounding a flat-topped hill, Tumamoc Hill, outside Tucson, Arizona. Tucson, with seventy-five hundred people, was described as "a wide-spreading, wide-awake little city on a level, sub-tropical plain that is encircled by granite mountains; a city with a strong Mexican accent, a city neither fast nor slow; a city with wide, clean streets, good buildings, abundant electricity and all the respectable concomitants of a metropolis—this is Tucson, Queen City of cactus-land."[11]

Tucson was originally the site of a Papago Indian village, the Papagos having been accomplished desert farmers well before the arrival of Spanish missionaries in the seventeenth century. The land was acquired from Mexico as part of the Gadsden Purchase of 1853. The Southern Pacific Railway put the town less than four days' travel from New York and just over one day from San Francisco. The laboratory location combined the advantages of a varied desert flora, including mountains and plain, with easy communication access by telephone to Tucson and by telegraph and cable to the greater world. Geographically it was well placed for scientific expeditions to the other deserts of the southwestern United States and northern Mexico.[12]

MacDougal and Coville were gratified to find the "progressive Americans" who ran the town well disposed toward science. The town leaders offered subsidies of land for the building site and a nature preserve, installed a water system, provided telephone and electricity, and built a rough road from Tucson.[13] Tucson was already the home of a Carnegie library, a provision store, and a hospital. The convenience of the location was enhanced by the presence of an agricultural experiment station, created in 1890 in conjunction with the opening of the University of Arizona in 1891.[14] The Carnegie Institution provided the money to erect a one-story laboratory in 1903. William Austin Cannon, who had just completed his Ph.D. in botany at Columbia and had been hired by the New York Botanical Garden, was installed as the resident investigator.

A later account of the origins of the Desert Laboratory by William Hornaday, who accompanied MacDougal on an expedition in the region in 1907, captures what must have been Coville and MacDougal's sense that they had found an idyllic location in which to pursue their scientific destiny. Having landed in Tucson, recounts Hornaday,

> MacDougal intimated to the proletariat of Tucson his belief that the Desert Botanical Laboratory might do worse than settle in their midst. Forthwith, the Tucson Board of Trade carried the botanist to the top of a high mountain close by, and showed the world that lay at his feet. "All this," said the Board, "shall be thine, and more, if thou wilt pitch thy tent herein, and become one of us."
>
> A mountain of many moods and tenses, and a belt of plain around it, both of them covered with weird things with stickers all over them, was offered, as it were, on a silver plate. Inasmuch as the site was the finest bit of real estate for the purpose in all the southwest, Tucson's offer was blithely accepted; and thus was born into the world the Desert Botanical Laboratory.[15]

MacDougal could hardly wait to start exploring, using the Desert Laboratory as a pied-à-terre. He immediately started planning an expedition with a local geographer, Godfrey Sykes. Sykes was an Englishman who had traveled in the Southwest since 1890, indulging a scientific interest in the Colorado River, its delta, and the Gulf of California, regions that were relatively unknown at that time.[16] He and his brother Stanley Sykes were both civil engineers and ran an engineering business, based in Flagstaff, that often involved scientific enterprises. Among other things, they built the telescope dome and installed instruments at the Lowell Observatory in Flagstaff.

Godfrey Sykes and MacDougal became fast friends.[17] With Sykes's help, MacDougal explored extensively in the southwest desert regions, making collections for the Garden. Using an Indian guide, the group traveled on horseback, collecting westward from Arizona toward the Gulf of California. They rode at night to lessen the chance of an encounter with Yaqui Indians. The most difficult task was the collection and shipment of the giant saguaro cactus. MacDougal found six transportable specimens less than fourteen feet tall, the largest weighing about a thousand pounds. Six men dug out the roots and packed the cacti in creosote bushes in long, coffinlike boxes for the ride east. Tons of cacti rode the rails to New York. In 1904 and 1905 Sykes and MacDougal made more ambitious explorations; Sykes mapped the Colorado Delta region for the first time.[18]

The Desert Laboratory, although owned by the Carnegie Institution, also functioned as a satellite of the Botanical Garden. The original plan was to provide funding for five years and then decide whether it should continue. But the botanists who worked there, and MacDougal in particular, had so impressed the

Carnegie officers that in 1905 they determined to continue the laboratory's operation. Robert S. Woodward, a geophysicist, became president of the Carnegie Institution in 1904 and strongly backed MacDougal's bid to become full-time director of the Tucson laboratory. Woodward was enthusiastic about the possibility not only of contributing to the development of economically important work for the government, but also of bringing the process of evolution under control. Both the Tucson laboratory and its sister laboratory at Cold Spring Harbor, under Charles Davenport's direction, could be focused on experimental studies of evolution, the former from an ecological and the latter from a genetic point of view. Woodward was especially impressed when it appeared that MacDougal had produced mutations and perhaps also new species experimentally. In December 1905 the Carnegie Institution created the Department of Botanical Research in order to correlate the work of the Desert Laboratory with the Station for Experimental Evolution at Cold Spring Harbor. MacDougal must have seemed not just a good choice, but the only choice for director.

In January 1906 MacDougal left the New York Botanical Garden to head the new Carnegie department and the Tucson laboratory, returning to New York regularly to complete the mutation theory experiments that he had started there. When he took charge of the laboratory, he had the enviable task of assembling a research community dedicated to the idea of basic research. The scientists would be free to pursue problems without constantly having to justify their results in practical terms. The Desert Laboratory would give scientists freedom to engage in long-term studies that might require decades or even longer for solution. As one observer noted, such freedom to do research marked an important distinction between this laboratory and government-funded agricultural stations, "for legislatures are not yet educated to the point where they can see that abstract science pays."[19] Moreover, the scientists at the desert laboratory would not be burdened by time-consuming administrative or teaching duties, as they would have been in university positions. MacDougal assiduously touted the advantages of his laboratory for research, and it quickly developed, exactly as he had planned, into a mecca for botanists who were lured by the promise of a fascinating landscape and intellectual freedom.[20]

Anticipating the Carnegie's approval of the new plans for its botanical department, in the fall of 1905 MacDougal sent Sykes instructions for expanding the buildings and grounds of the laboratory. Once the new budget was secured, they moved quickly. The size of the laboratory building was doubled, a greenhouse went up, and a pumping plant and a reservoir for well water and rainwater were built. The local contractors figured out ways to transport thirty-foot Oregon pine timbers up steep roads with hairpin turns, and they excavated foundations next to rooms that held delicate glassware. The laboratory was designed to be adapted to the climate. The thick stone walls, made of local volcanic rock, absorbed heat slowly; the overhanging roof was built to form a ventilated air chamber. When the windows were open, the dry breeze through the building made even the summer heat tolerable, or so it was claimed. The intense light was the main source of discomfort.

One of the chronic problems of fieldwork was lack of water. MacDougal estimated that a worker at the laboratory during an ordinary day in May would consume sixteen pints of water, and a horse would drink fifteen or twenty gallons in the same time. Sykes designed a still or condenser that enabled workers to get drinkable water from the sap of cacti or from alkaline waters. The apparatus, which produced several gallons a day, could be set up anywhere, enabling a research party to work as long as they needed to in a place that had undrinkable water.[21]

Most important for scientific study, the tract of land around the laboratory had to be fenced in to keep out animals and people. Ironically, MacDougal's concern to avoid an isolated location had prompted him to pick a site where the desert he intended to study had disappeared. The terrain had been degraded by grazing cattle, mules, and goats; pothunters were defacing the petroglyphs and Indian artifacts, and other people were collecting rocks from the site for building material. The land controlled by the laboratory was expanded to 860 acres, comprising almost the whole of Tumamoc Hill and a large mesalike slope west of the hill, and up went the fence. Only then could the land be restored to its original condition and the full measure of species diversity be taken.

Its value as scientific property increased as the diversity of the former landscape was brought out, revealing a land rich in species, if not in actual numbers of individuals, and as worthy of interest as any tropical locale. MacDougal also got permission from Gifford Pinchot, the U.S. chief forester, to use the forest preserve in the Santa Catalina Mountains for experiments; this meant putting up fences to exclude wildlife. To MacDougal's chagrin, the creation of a nature preserve excited too much interest in the people of Tucson, who visited the laboratory unannounced, disturbed the experimental plots, and expected the staff to entertain them. All this unwelcome attention forced him to keep two guards on Sundays to protect the grounds. He appealed to the Chamber of Commerce to stop advertising the laboratory as a recreational showplace.[22] Just as the New York Botanical Garden had had to establish defenses against visitors, training them to take their recreation in a way that did not disturb the Garden, the Desert Laboratory had to protect itself from intrusive locals.

The laboratory was within easy commuting distance of Tucson, but those who wished to avoid commuting had the option of living in tents on the laboratory grounds. The laboratory quickly grew into a small community of families, "Tumamocville," with some of the wives of the staff sharing in the scientific work. After MacDougal and other staff members built permanent houses at the bottom of the hill, the tents were used to house visitors. Scientific work included a wide range of topics in plant physiology, ecology, and evolutionary studies, anchored by a core staff of scientists and ably assisted by Sykes, the jack-of-all-trades who ensured that everything remained in running order.

From the Carnegie Institution's point of view, part of MacDougal's job at the Desert Laboratory was to help coordinate related botanical research that the Carnegie was funding across the country. On the East Coast, the Cold Spring Harbor Station for Experimental Evolution was involved in a variety of genetic studies, including intensive investigation of the mutation theory recently advanced by Hugo de Vries and so avidly promoted by MacDougal. George Harrison Shull, a botanist working there, was one of MacDougal's chief collaborators on the mutation theory studies, and several other scientists at Cold Spring Harbor were independently investigating the evening primrose to discover the reasons for its high level of mutability.

On the West Coast, the Carnegie Institution had developed an interest in the horticultural work of Luther Burbank, the "wizard" of agriculture who began his climb to fame after publication of his catalog *New Creations in Fruits and Flowers* in 1893.[23] His creations included such things as the spineless cactus, the stoneless plum, and a berry dubbed the "wonderberry" that he promoted as a new fruit. In 1905 the Institution approved a five-year grant to Burbank of ten thousand dollars per year, one of the largest grants given to a research operation outside the Institution's own laboratories and departments.[24] The problem was that Burbank rarely wrote anything down, so a young scientist had to be sent to observe his methods and record them. Woodward consulted MacDougal and Charles Davenport, the director at Cold Spring Harbor, and settled on George Shull.

Woodward was enthusiastic about the Burbank project because of its economic potential: Burbank's new breeds of fruits and vegetables, he thought, would likely generate wealth that would exceed the entire $10 million endowment of the Carnegie Institution. But he also saw that Burbank's experience might help solve the evolutionary problems being studied at the three new Carnegie biological laboratories at Tucson, Cold Spring Harbor, and Dry Tortugas, Florida. In fact, he used the Burbank project to advocate hiring MacDougal as permanent director of the Tucson laboratory, on the grounds that the resources of the three laboratories could be combined to create a scientific account of Burbank's work.

Hugo de Vries had also been interested in Burbank's work, had visited him in 1904 and 1906, and wrote popular accounts of Burbank's achievements.[25] However, Burbank and de Vries differed sharply in their interpretation of Burbank's results. Burbank vehemently rejected the mutation theory, as he did Mendelism, whereas de Vries interpreted Burbank's results as being strongly supportive of the mutation theory. Moreover, many of Burbank's claims about his creations were controversial and generated criticism. MacDougal, as a representative of the Carnegie Institution, had to be supportive of Burbank's reputation in general, but his and Shull's position as champions of the mutation theory virtually guaranteed that they would not see eye to eye with Burbank scientifically.

The plan to have Shull work with Burbank quickly unraveled. Shull's initial report in the summer of 1906 indicated that Burbank's methods were hopelessly unsystematic, his quantitative estimates wildly inaccurate, and his record-keeping virtually nonexistent. These negative assessments did not derail the project. Woodward pressed forward with the plan and blamed Shull for not being flexible enough to work with the illustrious breeder. The Carnegie Institution maintained its support through 1909, when it became clear that no amount of study could bring order to Burbank's methods.

MacDougal lost no time as he built a research program along several different paths that complemented the intellectual goals of the Burbank project but seemed more likely to yield solid scientific results. These goals boiled down to understanding the basic mechanisms of growth and heredity well enough to be able to control evolution, or to shape the future course of evolution. Understanding the mechanisms of evolution from an experimental standpoint linked the work at Tucson and Cold Spring Harbor and offered potentially great rewards, as great as those envisioned in the Burbank project. President Woodward was drawn to this grand vision. Of all the research funded by the Institution, he told the trustees in 1908, MacDougal's work on experimental evolution seemed particularly worth pursuing because it opened the possibility "in the future to apply invention to living things."[26] Ecological research and genetics research combined to move Americans closer to this goal of shaping the future by controlling life.

Learning to control life meant more than unveiling the mysteries of the chromosomes: genetics provided only half the picture. It meant understanding the relationship between organisms and environment, which in turn required a variety of locales for observations and experiments. At first the daily work of the laboratory was limited to a twenty-five-mile radius, representing a single day's travel for the riding and pack animals the scientists used, but MacDougal had his sights on more distant terrain. In 1906 he set up two experimental sites on the slopes of the nearby Santa Catalina Mountains, a distance of about thirty miles. This gave them, in addition to the desert climate of the Tucson region, a dry mountain and a temperate mountain environment.

As the Tucson operation expanded, the more congenial climate of California beckoned. Cannon had gone to Carmel, California, in 1907 and 1908 to investigate Luther Burbank's work, as a substitute for George Shull. While in California, Cannon made contact with the Carmel Development Company, which provided him with laboratory space and expressed an interest in helping to set up a coastal laboratory there. Carmel was an eclectic and relaxed community about 150 miles from San Francisco, heavy on writers and artists and known for its scenery and pleasant climate. MacDougal took the idea to Woodward, who accepted the company's offer. In 1909 the Carnegie Institution established a coastal laboratory at Carmel, which served as a summer retreat from the scorching desert heat for the Tucson staff.[27]

These different sites were used for experiments on how plants adapted to different environments and whether these adaptations were inherited. The experiments, which began in 1906, involved removing plants from their native habitats, growing them in a new environment, then returning them to their original location to see whether any changes in form were preserved. The experiments were meant to test the theory that had been elaborated in some detail by the French evolutionist Jean-Baptiste Lamarck in the early nineteenth century, the theory that organisms could make permanent, adaptive responses to the environment. Lamarck's proposals about the inheritance of acquired characteristics were the subject of debate throughout the nineteenth century, and in the early twentieth century they were still being debated along with Darwin's ideas about natural selection. In fact Darwin, who was not a follower of Lamarck, did accept the idea that the environment might have a direct hereditary effect on organisms, and that repeated use or disuse of a part might have an inherited effect. These concerns were also central to ecological science, where Darwinian ideas about the struggle for existence coexisted easily with Lamarckian ideas about the hereditary impact of the environment on organisms.

MacDougal's work on the mutation theory had led to experiments designed to stimulate the appearance of mutations by direct manipulation. In the laboratory, MacDougal had injected plant ovaries with various salt solutions in an effort to induce mutations. Although most of these experiments were unsuccessful, he believed that he had in a minority of cases succeeded in creating new mutations that were inherited. These effects were not necessarily adaptive and therefore did not confirm the Lamarckian theory, which emphasized adaptive responses to the environment. But they did suggest the value of continuing research along these lines. The long-term field studies of adaptation to different environments attacked the same problem of species mutability.

Determining the physiological mechanisms behind adaptation would enable scientists to answer evolutionary questions about how species developed over time, how they adjusted to changing climate and environment, and how species migrated in response to changes in the climate. These experiments would also help to determine whether species were in fact best adapted to their home habitats, or whether a change in environment might prove more suitable for growth. Although by about 1910 many experimental biologists questioned the validity of Lamarckian views, the possibility of direct environmental action on the organism was still an active area of research and discussion.

Darwin's son Francis Darwin, who was a botanist, drew attention to these discussions about plant responses in his presidential address to the British Association for the Advancement of Science in September 1908, on the eve of the centennial celebrations of Darwin's birth that ran through 1909.[28] Francis Darwin's research dealt with plant movements and how the environment stimulated plants to execute certain movements. This was an extension of his father's work, published toward the end of his life, on the power of movement in plants. Charles Darwin had been struck by the resemblance between movements of plants and reactions in lower animals; European physiologists had developed this idea of the connection between plant and animal "irritability." Francis Darwin drew on this experimental research to argue for the importance of studying plant movements and responses to the environment, suggesting that from a Darwinian perspective these kinds of reactions might be adaptive. He meant that reactions to external stimuli might produce permanent changes in form, a direct adaptation by the plant to the environment.

Darwin begged his audience's indulgence as he embarked on a lecture that he knew would make him appear a champion of what some considered a lost cause—the doctrine of the inheritance of acquired characters. He drew especially on the ideas of Richard Semon, a German physiologist who proposed a theory of heredity, the Mneme theory, that postulated that stimuli left traces on protoplasm, a kind of physiological or morphological impression or "memory" (which Semon called an "engram"). Such memories could then be the basis for the inheritance of acquired characteristics. Darwin thought this theory had merit. He came down firmly in favor of the idea of somatic inheritance and argued that changes in the body cells could be "telegraphed" to the germ cells through a form of nervous transmission. Plants were therefore similar to primitive animals, and the inherited effects of habit, or the influence of the environment, provided a mechanism of adaptive variation. Speaking metaphorically, Darwin described evolution as a type of learning process, in which learning was achieved through repeated drills until the information became part of the system. Organisms that failed to learn were eliminated.

These ideas, which suggested that plants had a kind of primitive consciousness or intelligence, caused widespread debate. MacDougal, quick on the uptake, used Francis Darwin's theories as a springboard to publicize his own findings. Darwin had been drawing on much of the same European literature in experimental botany that had shaped MacDougal's approach. Within a month of Darwin's address, the *New York Times* published a feature article on MacDougal's research, connecting it to Darwin's theories and emphasizing its broad significance.[29] MacDougal's experiments, despite failing to confirm the Lamarckian thesis of adaptive response to the environment, seemed to show that external causes could produce new variations or breaks in a plant's heredity. The *Times* reported that the implications of this work were that humans might control and direct the evolution "of the entire organic world"; the headline of the article claimed, "Man Able to Change Form and Color of Flowers at Will." The article described the work of the Desert Laboratory, including experiments on the saguaro or giant tree cactus, a good plant for experiments because of its large ovaries.[30]

The publicity was likely aimed as much at the Carnegie Institution as at the general public. Nathaniel Britton and his collaborator Joseph Rose had renamed the giant cactus *Carnegiea gigantea* (from *Cereus giganteus*) in honor of Andrew

Carnegie, just in time for Carnegie's visit to the laboratory in March 1909. MacDougal reported to his friend Charles Cox that Carnegie had been under the impression that "Carnegiea" applied only to a few plants in the Bronx greenhouses, not realizing that the plant's range stretched for a thousand miles in the Southwest.[31] MacDougal made sure to convey to Carnegie the great honor intended in the dedication of the new genus.

In December 1910, in his presidential address to the American Society of Naturalists, MacDougal spoke of the need to base evolutionary biology on a foundation of physiological experiments. The study of evolution was advancing rapidly in the light of experimental work, he claimed, and the Lamarckian thesis, having been debated for a century without proof or disproof, was at the critical point of being conclusively put to the test. As he reviewed the literature dealing with environmental responses in a wide variety of organisms, from humans down to bacteria, and discussed the research of the Desert Laboratory in this field, he emphasized that these experiments could be conducted even without understanding the underlying mechanisms of inheritance. But he was convinced of one thing: the experiments on transplanting organisms from one place to another, in conjunction with laboratory studies of mutation, promised results "of the greatest value" and demonstrated above all that evolutionary debates could be settled only by experimental work.[32]

By this time, cracks were already appearing in his cherished mutation theory. A consequence of the concentrated research effort focused on *Oenothera* (the evening primrose) was to show that genetic mechanisms—and evolutionary processes—were a lot more complicated than had been thought. One of the main results of this unfolding research on experimental evolution was to render de Vries's theory of mutation obsolete within a decade and a half of its introduction, although de Vries himself maintained his views until his death in 1935.[33] But American research advanced enough that the mutation theory dropped from favor by about 1915. As the evening primrose became a "model organism," that is, the focus of intensive research by a growing number of scientists, scientists gained better understanding of the genetic mechanisms involved in the apparent mutations.[34] It became clear that what de Vries thought were mutations were the result of unusual chromosomal behavior. The term *mutation* came to mean not abrupt species-producing events, but small inherited variations that followed Mendelian laws. Even as research evolved and theories fell by the wayside, the underlying idea that experimental research should aim for understanding and ultimately control of the evolutionary process remained important. If the claims made for the mutation theory turned out to be a little hasty and optimistic, there were still other investigations in experimental evolution that needed to be done.

Although Carnegie president Robert Woodward supported these goals, he did not fully understand what resources were needed to do the related research and balked at the expenses incurred by the expansions of the laboratory and expeditions of the scientists. As early as 1906 Woodward reprimanded MacDougal for what he believed was an inflated budget. He took a jaundiced view of such expenses as a team of draught horses and mounted assistants to help maintain the laboratory and accompany the expeditions.[35] By 1910 MacDougal was using a secondhand car, which he and Sykes specially outfitted for their expeditions. Car travel extended the radius of operations by sixty to one hundred miles. In 1913 a private telephone system had to be installed, after MacDougal fell out with the Tucson Chamber of Commerce over who should pay for the original public telephone extension to the laboratory. The frequent demands for funds provoked Woodward to remark that MacDougal's efforts to "work the Institution for all it is worth" reminded him of a Tammany politician: "To a greater extent than in any other department you and Mr. Sykes have manifested a disposition to overreach your opportunities instead of seeking to establish terms of generous reciprocity with the Institution."[36]

MacDougal's entrepreneurial drive, however it may have grated on Woodward, was an important factor in the laboratory's success. It was also important for the development of the science of ecology, which was a prominent activity in the first decade of the laboratory's life. By 1916, a decade after assuming the laboratory's directorship, MacDougal could boast that his laboratory's reach stretched over five thousand square miles.[37] With such a large area to work in, its scientists could begin to transform ecology from an activity thought to lack rigor into a more systematic quantitative and experimental science (fig. 4.1).

Scientific explorations afforded the opportunity for detailed, long-term scientific studies of the desert. One near-disaster became a fortuitous experiment in ecology that could not have been duplicated by design. In 1905 an irrigation channel on the Colorado River flooded, and as a result the river itself was diverted so that it flowed not into the Gulf of California but inland into an area called the Salton Sink, a large desert basin below sea level, to the north of the gulf. The Colorado River was unstable, and in geologic history this region had been inundated with water from time to time: marine fossils and marks of a shoreline remained as evidence of an ancient lake. But although there had prob-

Fig. 4.1. A scientific Mecca. Daniel MacDougal welcomed all to participate in the Desert Botanical Laboratory's work. Here he explains his work to western writer Mary Austin, whom he helped with her book *Land of Journeys' Ending* (1924). Courtesy of the Arizona Historical Society / Tucson (AHS 11476).

ably been small inundations of water at fairly frequent intervals, the most recent being in 1891, Americans settlers had never witnessed any major flooding of the basin.

The Salton Sink had already been the subject of scientific interest in connection with the building of the transcontinental railway. In 1853 R. S. Williamson of the U.S. Topographical Engineers led an expedition into the region to find a route for a railway. William Phipps Blake, later a professor of geology at the University of Arizona, accompanied this expedition. As the party moved eastward from the Pacific, they found an unknown break in the mountain range that went through the western Cordillera and led into the interior wilderness; it became known as the San Gorgonio Pass. They saw the potential of this route for a railroad across the Southwest, although it was necessary to purchase a strip of land, the Gadsden Purchase, from Mexico. That acquisition in 1853 fixed the final boundary between Mexico and the United States. Blake made the first geologic examination of the Colorado Desert, determined that the Salton Sink was below sea level, and studied evidence of the ancient sea that had covered the area.[38]

American businessmen in the 1890s revived the idea of developing the region for agriculture, and one of the first steps was to remove ominous words like *desert* and *sink* and give the area a more alluring name, the Imperial Valley. The whole concept of what constituted a desert was redefined: this was no longer a forbidding land to cross as quickly as possible on the way west, but a land that could generate wealth.[39] The Imperial Land Company promoted the potential of the land for farming, hoping to attract colonists, and the California Development Company constructed irrigation canals bringing water from the Colorado River—the American Nile—to the Imperial Valley. By 1902 about four hundred miles of irrigating ditches were completed, making water available for one hundred thousand acres of land. The land was suitable for vegetables and fruits, barley and alfalfa, and Egyptian cotton. By 1904 about ten thousand settlers had moved to the region, and a branch of the Southern Pacific Railroad crossed the Valley.[40]

But the recent building of irrigation channels that ran parallel to the river, supplying water to the valley, destabilized the river. As more and more openings were made between the river and the canal, the river was unable to sustain the storm waters that would rush downstream periodically. Its shift in course caused nearly the whole of the river to flow directly into the Cahuilla Basin, located to the west of the southern part of the Colorado River's main delta. This was the most dramatic inundation of the basin that Americans had witnessed. For two years the river flowed into the basin, creating a large shallow lake, the Salton Sea, and threatening to deprive about twelve thousand people of their property. Edward H. Harriman, owner of the Southern Pacific Railroad, organized and helped pay for efforts to fight the "runaway river." The engineers mobilized all the Indian tribes in the Southwest, who with Mexican laborers and various drifting adventurers began work in August 1906, dumping thousands of carloads of rock, gravel, and clay, trying to fill the break faster than the river could take it away. The crevasse was closed, the river was pushed back into its old bed in February 1907, and a levee was built to bring it under better control. A decade later the agricultural production of the Valley was estimated to be worth $500 million.[41]

For MacDougal this diversion was as close to a large-scale ecological experiment as one could get. Located about three hundred miles west of Tucson, the Salton Sea was well within reach of the laboratory scientists. This was a splendid chance to study the desert while it was going through a series of rapid physiographic changes. What made it an "experiment," a term that MacDougal used to describe it, was that the water effectively sterilized the area, killing the vegetation and changing the condition of the soil. Thus it was comparable to what might occur after a volcanic eruption on a remote island. MacDougal thought this event could provide a way to study evolution in action because plant species might evolve while recolonizing the land. There were already a few endemic species in the area, that is, species that were indigenous to that location, and MacDougal thought it possible that they had arisen after earlier floods in the basin. The extreme conditions of the region, caused not just by the rushing water but also by the harsh climate, would be expected to exert a formidable selective pressure on the new seedlings, like a series of sieves that the species had to pass through. Furthermore, as the lake evaporated and exposed larger portions of drying land, MacDougal imagined that the sieves operating in the "great selecting machine" would constantly change in pattern and mesh. Finally, birds attracted by the fish in the lake had introduced a whole new means of seed dispersal that did not exist before the flood.[42]

This was an "experiment" not in the sense of a carefully controlled manipulation as might occur in a laboratory, but in the sense of a total transformation of the normal environment, which would possibly set in motion evolutionary processes that could be watched. The experiment would help scientists address questions concerning, for example, the agents of seed dispersal, the response of plants to altered soil chemistry and to the rapid shift from wet to dry soil, the processes of ecological succession, invasions of new species not native to the region, and how disturbance in the environment might promote speciation. These were complex problems, involving the interplay of many biological and mechanical agencies. The scientists were only able to begin making inroads into these questions by describing the changes occurring in the region as the water evaporated.

One of the limitations to the exploitation of these unusual events was that often the conditions before the disturbance were not well known. This was a problem for the analysis that MacDougal and his collaborators undertook. Luckily, MacDougal and Sykes had made prior investigations of the region, in particular the expedition of 1904 done for the New York Botanical Garden and the trip by Coville and MacDougal in 1903, so that there was some information available as a benchmark for this study. These were not thorough studies, but fortunately MacDougal had taken photographs of areas that were subsequently submerged, so a rough idea of what had grown there could be obtained. In 1907 Sykes built a sailboat on the shores of the rising lake so that he and MacDougal could survey the 150 miles of shoreline, preparatory to two years of intensive fieldwork and later short expeditions to chart the progress of the vegetation as the lake receded. A steel-hulled power launch and small steel rowboat were built in Tucson and used to cross the lake, and wagons and automobiles were used for land trips, although the deep arroyos and soft sand made travel extremely difficult.

The analysis of the Salton Sea involved the collaborative efforts of ten scientists from the Desert Laboratory and other institutions (including a contribution from William Blake, who had made the original survey of the area and who died in 1910). These men contributed their expertise to the analysis of the geology, the geography, the water chemistry, the soil composition, the microorganism populations, and the plant ecology of the region. This was not an orchestrated study operating under a manager with a focused objective, but a collaborative effort with each person collecting information on field trips at different times and sharing the information with others engaged in their own work. Sykes focused on geographic studies, while MacDougal studied the movements of vegetation as land became wet or dry. The studies of the Salton Sea continued in this way for several years.[43] (Although the original expectation was that the sea would evaporate in eighteen years, the local population wanted it preserved. The sea was maintained by excess irrigation water from the Colorado River used to flush salts out of the agricultural lands from the surrounding area. Being thus replenished, it is heavily polluted by salt and nutrient chemicals but has become a valued nature reserve. Its preservation is the subject of ongoing debate.)[44]

In the light of these kinds of opportunities for the study of desert vegetation, MacDougal needed not only more facilities but also an expanded staff. He kept William Cannon on staff, along with Volney Spalding, a physiologist and ecologist who had recently retired from the University of Michigan. Spalding began a detailed ecological survey of the laboratory hill and surrounding area. Continuous studies were very important because it took a few years to discern the cycles of desert life. Animals and plants went through cycles of higher and lower abundance. Figuring out the causes of these cycles, whether climate or something else, required long-term analysis. Both Cannon and Spalding saw the potential of the laboratory for long-term research and urged its development.

The key question was what the conditions were that enabled some plants to successfully occupy the desert but prevented occupation by others. Spalding argued that careful ecological studies would correct a major misconception about life in the desert: that desert plants were limited only by the physical environment and not by competition with other plants. It was his view that severe competition between plants was the rule in the desert. Ecological studies of the limits of adaptation in the face of this competition would help to determine what plants could adapt to desert life. Studying the behavior of organisms on the edges of the desert, and transplanting species beyond their natural limits to see how well they survived, would show how important the living environment was to plants and what was involved in their adaptation and adjustment to the desert. The problem of adaptation was a vexing one that raised all kinds of questions about heredity and the direct influence of the environment, but Spalding felt that it cut to the heart of the "deep mysteries of life" and was central to the task of ecology.[45]

MacDougal also brought in physiologist Burton E. Livingston, a student of Spalding's, and anatomist Francis E. Lloyd, who had worked with MacDougal in New York. Lloyd confessed to initial skepticism about the choice of the Tucson location for the laboratory but was won over after he spent a summer there.[46] The region got enough rainfall to sustain a diverse vegetation, yet the hilly land was not so desirable for development as to be threatened by irrigation. Lloyd brought with him the magazine *Plant World*, which was founded in 1897 and served mainly as a journal for science teachers. When he took it over in 1904, it was foundering, but under the care of the Desert Laboratory staff it revived and developed into a more professional-looking and even self-supporting scientific journal, with serious contributions aimed at a scientific readership. MacDougal used it often to publicize the work of the laboratory.

Livingston left in 1909 for a post at Johns Hopkins University, in its new Department of Plant Physiology, but, moving between laboratory and field station, he returned to the desert in the summers to study the relation of plants to the environment. Livingston understood ecology to mean *field physiology*, a term he took to be synonymous with *physiological ecology*.[47] He thought one of the big obstacles to ecological work was confusion over the interpretation of such words as *adaptation*, *use*, and *purpose* and that the way forward was greater precision in measurement. The development of better instruments for measuring environmental conditions was, he thought, a good field for mathematically inclined ecologists. He also regretted that agriculture and ecology were too far apart, so that data collected in an agricultural context were often of little value to the ecologist. He believed there was great need to develop physiological plant ecology as an analytical science.

Meanwhile Spalding, suffering from arthritis in 1907, urged MacDougal to find a younger replacement to continue the ecological survey work that he could no longer do. MacDougal hired Forrest Shreve, a promising young ecologist, in 1908, and Herman Augustus Spoehr, a chemist and physiologist, in 1910. Shreve had been working at the New York Botanical Garden's tropical station in Jamaica. MacDougal, impressed by the high quality of his research, offered him a research position that would enable him to quit his teaching post at the Women's College in Baltimore (now Goucher College) and finish his research. Shreve was enticed by the prospects of the Tucson environment and became a specialist in desert ecology; his wife, Edith Bellamy Shreve, did research on plant physiology.[48] Spoehr's research focused on strictly physiological problems, primarily studies of carbohydrate metabolism and photosynthesis, rather than ecological matters. Throughout his years as director, MacDougal also maintained a close relationship with the New York Botanical Garden and helped to facilitate Nathaniel Britton's and Joseph Rose's taxonomic studies of cacti, which the Carnegie Institution published.

The staff scientists also traveled abroad, to do both field research and laboratory investigation. Livingston visited the Munich laboratory of Karl Goebel, the "prophet of causal morphology," who took an experimental approach to the study of form and adaptation that was closely related to the work being done at the Desert Laboratory.[49] In fact Goebel was one of the few European botanists who took a close interest in the American work. Cannon also explored the deserts of the Algerian Sahara; MacDougal and Sykes made an expedition to Egypt and Libya to gain a comparative perspective on the American deserts.[50]

In presenting his work to the Carnegie Institution and the public in the laboratory's early years, MacDougal's strategy was to highlight the laboratory's work on basic physiological processes, especially those dealing with the water relations of plants. This strategy reinforced the relationship between ecology and physiology, for MacDougal's definition of the laboratory's purpose implied a view of ecology as a science closely related to physiology. He explicitly denied that the purpose of the laboratory was to study desert vegetation per se: the study of life in arid regions as its own branch of science, he said, made no more sense than "mountain astronomy." The scientists were engaged in the study of this region not for its own sake but in order to shed light on fundamental scientific problems. The desert location was simply a convenient one for the analysis of certain kinds of physiological processes.[51]

MacDougal's concern, in advancing the cause of experimental botany, was also to deflect any suggestion that the purpose of the laboratory was mainly economic. Certainly the laboratory's existence was related to the economic development of the region and the need to have rational policies of land use, as MacDougal well knew from his travels with Frederick Coville. The basic research undertaken at the laboratory would eventually find its application as the need for more intensive agriculture increased with population growth in the Southwest.[52] But MacDougal always wanted it to be clear that the laboratory was not an agricultural station. Only one agricultural project was undertaken at the Desert Laboratory, an attempt to cultivate guayule, a wild desert rubber plant, to see whether commercial rubber production would be feasible in the United States.[53] These experiments were not conducted at the Desert Laboratory grounds, however; the breeding work and genetic analysis were done on grounds purchased near Tucson by a private company. In fact, MacDougal recognized that the nature of the agricultural stations was changing, and positions in agriculture demanded qualifications in basic science, especially genetics, chemistry, and physiology, not just practical experience. As basic research advanced, agricultural work would improve apace. But there had to be a niche for "abstract science," and that niche was secured by the Desert Laboratory.

To the extent that analysis of physiological mechanisms meant considering the organism in relation to its environment and in relation to the struggle for existence that affected its distribution and abundance, this research was ecological. Other scientists recognized the innovative features of the laboratory's work. A reviewer in 1910 praised the laboratory's work as the epitome of a "new era" of geographic ecology because it placed the study of plant relations on the same quantitative and experimental basis as the study of plant physiology.[54]

MacDougal himself approached ecological work differently from the more traditional natural historian, who at this time would have been engaged in mapping, not experimenting. Such a traditionalist was Clinton Hart Merriam, a naturalist who worked in the same region of Arizona. His biogeographic approach focused on mapping life zones that showed how species were generally restricted to certain regions.[55] This approach to ecology had its precursors in the mid-nineteenth-century work of Alexander von Humboldt, who influenced a later generation of European naturalists. But MacDougal was far more interested in teasing out the nature of adaptation and coming to grips with the exact nature of environmental influence on plant life. Too often naturalists had assumed that they understood the adaptive significance of given characteristics. MacDougal and his colleagues wanted to call those assumptions into question and submit them to experimental test.

However, the laboratory scientists also engaged in descriptive studies and data-gathering, particularly climatic data that they hoped would explain plant distribution and abundance. Shreve and Livingston devoted several years to mapping vegetation and compiling climatic data, with the goal of relating climate to species adaptation.[56] Ultimately the creation of a long-term record would help to identify which environmental conditions were most important in determining species distributions and abundance. Long-term analysis was needed because even in the earliest stages of research, Shreve noticed signs of trouble. Although MacDougal had earlier dismissed local concerns about destruction of cacti in the region, one of the early results of the laboratory's work was to show that the cactus populations were threatened. Shreve noted that the giant saguaro cactus was not reproducing itself, even in regions of greatest density, although he could not say what was causing the decline.[57] If abundant species were dying out, it was crucial to know why.

Ecological work was not simply about doing experiments in the field, or "field physiology." It involved creating a record of changes in the land, building a historical record from continuous observation that would, over time, become an increasingly valuable source of information for future ecologists. The record included many photographs of vegetation. The camera was one of the most important tools of research, providing a quick way of taking in data about the broad features of the landscape. Ecology could scarcely exist without such a record if it was to go beyond mapping and deal with the causes underlying change. Thus the very act of building a historical record helped to give definition to the subject, even though firm conclusions about causality remained far off.

Science is serious work, but it is also a form of play, and the less guarded and more subjective comments made during play can provide glimpses of the values and attitudes underlying science. The scientific work at Tucson was inextricably linked to the excitement of adventure; play and work were wrapped up together. Ellsworth Huntington, a human geographer whom MacDougal had invited to the Desert Laboratory, recalled the delightful feeling of freedom that traveling through the desert by car evoked, especially at night.[58] (The main nuisance was the speed with which the land was being fenced in for farmland. As fences were built, the roads changed their locations daily to avoid them. A careless driver speeding at night risked a crash—even a beheading—by running into the wire.) The day's travels would wrap up with campfire discussions on the adaptation of plants and animals to the environment and on the similarity of humans and animals.

The blend of science, comradeship, and adventure can be gleaned from an account of one expedition to the Pinacate region of Mexico in 1907 (fig. 4.2). Sykes and MacDougal went with William Hornaday, a prominent conservationist and director of the New York Zoo, and John M. Phillips, an iron manufacturer and Pennsylvania state game commissioner. Phillips had accompanied Hornaday on an earlier trip to the Canadian Rockies. Jefferson Milton, a U.S. immigration inspector, joined the group for part of the trip. Three other men served as guide, wagon master, and packer and cook. Hornaday published a popular account of the expedition, *Camp-Fires on Desert and Lava*, and put his own spin on the meaning of these trips for a generation of Americans who had to adjust both to the rapid western expansion of the United States and to the influx of immigrants that was changing the complexion of the cities on the East Coast.

Hornaday fled west to escape the moiling crowds of immigrant New Yorkers, whose ways threatened his sense of order and of northern Anglo-European supremacy. He advised travelers not to trust Indian guides, who were "enough to depress the spirits of a barometer," and instead go with a small group of "good white men, with red blood in their veins." The appeal of Arizona in winter was the space and sunshine, the "cleanness of the face of Nature" as compared to the "polluted streams, dirty humanity," and wear and tear of big-city life.[59] Describing the train trip across the country to the West, Hornaday reflected on how pleasant it was to see the clusters of small farms. These people, he thought, were "clean and independent," whereas in the East people "pile up like sheep." Reaching the arid Southwest, where the farms gave way to ranches, he remarked on the signs of the "great struggle between Man and Desert" going on in the vast expanse of land from Texas to the Pacific Ocean.[60] He preferred the Mexican language of the carriage driver in Tucson to the "disgusting Bowery English of New

Fig. 4.2. Scientists at play. The expedition to Pinacate Peak was recounted by William Hornaday (*left*) in *Campfires on Desert and Lava.* Jefferson Milton is in the middle; Godfrey Sykes is standing on his head. Courtesy of the Arizona Historical Society / Tucson (AHS 70643).

York in the mouths of swashbuckling drivers, conductors and shop-girls of a hundred kinds." Tucson he found modern and lovely: "True, the absence of ten thousand vacant 'lots' covered with twenty thousand tons of ghastly rubbish makes a resident of New York feel very lonesome; but, then, Tucson is new, and the herds of human cattle from the overcrowded cities of southern Europe have not yet arrived."[61]

With Phillips, MacDougal, and Sykes (whom he called the "Arizona Wonder" and the "Paragon") Hornaday could regain some sense of control and supremacy. The explorers traveled to Pinacate, a volcanic mountain, observing and photographing striking plant specimens and a Papago Indian village that was temporarily vacated. Sykes and MacDougal bestowed names on the landmarks as Sykes mapped them: Hornaday Mountains, MacDougal Pass, MacDougal Crater, MacDougal Volcano, Sykes Crater, Phillips Buttes, and Carnegie Peak, a lower mountain visible from the top of Pinacate Peak. On Pinacate they indulged in "wild revels" like boys let loose from school, and they marched to the peak of the mountain lined up abreast so that each would have the honor of being the first white man to set foot on the summit.[62]

The other purpose of the adventure was to hunt mountain sheep, ideally to bag a ram with large horns, which Hornaday found singularly thrilling. The head was cut off for the trophy and the body kept for meat. But Hornaday made a point of discriminating between the sportsman's approach to hunting, which was restricted to a single trophy kill, and the indiscriminate and wasteful hunting of the uncivilized. His book on the expedition promoted his efforts to conserve big game by imposing a bag limit on hunters and forbidding the sale of game for food. Like Theodore Roosevelt, he considered hunting a thrilling affirmation of manhood; conservation laws were meant to preserve the animals for sportsmen. This philosophy pitted Hornaday against people for whom hunting was a means of subsistence, whether they were Indians or the local laborers who worked in the mines, in the forests, or in railroad construction. His vision of America excluded such people, substituting in their place a society of "jolly good fellows" who mapped, observed, measured, photographed, encountered Indians, reveled in the scenery of mountains, rivers, and craters, told stories around the campfire, and hunted, all the while affirming the ingenuity and moral superiority of the Nordic male.[63]

The Desert Laboratory as a research center was at its height when the Ecological Society of America was organized in December 1915. MacDougal, Cannon, and Shreve were all charter members of the society, which was formed with the intention of improving the connections between plant and animal ecology and which from the start had strong contingents from the New York and Washington, D.C., areas.[64] MacDougal had no single vision of what ecology should be as a science, nor was he worrying about how to define a new discipline of ecology when he took over the laboratory. But by contributing to ecological research along diverse lines through his directorship of the laboratory, he helped to develop the discipline. In 1919 the Desert Laboratory's journal, *Plant World,* was handed over to the young society and continued under the name *Ecology;* it became the leading American journal for ecological research.

By the time ecology acquired disciplinary coherence in the 1920s, the Carnegie's interest in the Desert Laboratory was waning. In the 1920s MacDougal shifted his base to the Carmel laboratory. This choice was partly necessitated by family obligations. His daughter Alice had died shortly after giving birth to a son, her first child. Her husband was Harold Stearns, a writer who was beginning to make his mark as a critic of American culture. He was not settled enough to take over responsibility for the baby. MacDougal and his wife Louise, who had not gone to Tucson but lived in Carmel, adopted the boy and raised him there.

Herman Spoehr had already moved with his family from Tucson to Carmel in 1920. In 1928 Spoehr succeeded MacDougal as chairman of the Carnegie Institution's renamed Division of Plant Biology, and in 1929 Spoehr moved his physiological laboratory from Carmel to Stanford University, where the Carnegie established a new center for its experimental program. He left for one year in 1930 to become director of natural sciences for the Rockefeller Foundation but then returned to the Stanford laboratory and resumed his chairmanship of the Division of Plant Biology. MacDougal remained in Carmel and retired from the Carnegie in 1934. The link between ecology, physiology, and cytogenetics grew closer at the Stanford laboratory, where Spoehr remained until his retirement in 1950. Forrest Shreve took charge of the fieldwork at the Desert Laboratory.

With this split between experimental and field science, and with the geographic isolation of the Tucson laboratory from the more substantial Carnegie laboratory at Stanford, Shreve found it increasingly difficult to convince the Carnegie Institution that the Desert Laboratory was worth supporting as a center for basic research and a locus for cutting-edge science. Shreve's biographer, Janice Bowers, noted that several conditions conspired against the laboratory, including the impact of the depression on Carnegie stocks, waning confidence in the program's vigor, and Shreve's difficulty in conveying the importance of the work.[65] Other problems might have been on the minds of the Carnegie officers. The Burbank project in the end did not yield significant results, the mutation theory of de Vries had been abandoned, and Lamarckian research had by this time been thoroughly discredited as a flawed idea.

But a crucial blow was the separation between experimental work and field studies, which meant that the link between descriptive ecology and the quest for control over life was severed at Tucson. Spoehr, now head of the division and Shreve's boss, was not antagonistic toward ecology or field research. He appreciated the broader view of life and environment that came from working in the desert. He was a conservationist, involved in the Save-the-Redwoods League, which was trying to preserve the giant sequoias of California.[66] At the dedication of the Stanford laboratory in 1929, he gave a sympathetic account of the central problems of the Desert Laboratory and its affiliated stations.[67] However, he also expressed the opinion that biology was only at the beginning stages of an exact approach that would bring it closer to chemistry and physics. Spoehr was trained in chemistry and believed strongly in the analytical methods of experimental science. His dedication speech was an appeal for the application of physical and chemical methods to biological problems.

Spoehr did not stop defending the importance of ecological study even as he pursued the more analytical course of research, but it is not clear how sincerely he held these views, for in defending ecology he largely borrowed his colleagues' words. In 1938, in a Carnegie publication on cooperation in science to celebrate the retirement of Carnegie president John C. Merriam, he argued for the need for ecological study, but portions of his text were taken nearly verbatim from Frederic Edward Clements's writings.[68] His incorporation of Clements's ideas into his essay and his discussion of the value of ecological research as done at the Desert Laboratory may have reflected full support, but it is difficult to judge.

Spoehr did argue that the key to progress was careful coordination and cooperation between the three basic methods of plant biology: the descriptive methods of field research addressing problems in ecological succession and the impact of the environment on plants; the use of experiments in garden and greenhouse to investigate problems in ecology and genetics; and the more exact laboratory methods of chemical and physical measurement that addressed problems of plant function. We can be confident that his heart lay with the third field of research, although he was willing to defend the larger view of science.

This ideal of a close cooperative relationship became increasingly hard to achieve once the experimental and field research was divided between Stanford and Tucson, for Shreve was unable to implement a coordinated research program of the kind that Spoehr envisioned. In 1940 Vannevar Bush, the new president of the Carnegie Institution, finally closed the laboratory, handing its buildings over to the U.S. Forest Service. (In 1956 the University of Arizona bought the laboratory to house its Department of Geochronology. Ecological work continued there under the auspices of the university, and the building remained much as in MacDougal's day, on a reserve of 350 hectares. Concern about development of the land at the expiration of the university's lease prompted a campaign to preserve it and the remarkably diverse vegetation that MacDougal and his colleagues so admired.)

Spoehr's remarks in 1940 encapsulated the dilemma that the Desert Laboratory faced after nearly three decades of work.[69] Setting the scene for the laboratory's closing, he recalled Coville's emphasis on economic considerations back in 1902 and noted that the laboratory's investigations had been aimed at understanding problems involved in rational use of the land. Moreover, he acknowledged, the problems underlying the use of arid and marginal lands had intensified in the 1930s and were now to be considered urgent. But this very urgency had caused the government to increase its expenditure on these problems to such an extent that the Carnegie's commitment to this research was deemed to be no longer required. In the end, with practical problems overwhelming basic science, the laboratory's work became the government's business. Unfortunately, government funding boosted applied fields such as range management but did little to further basic ecological research once the Dust Bowl ended.[70]

In the days when ecology was an embryonic discipline, the close relationship between experimental research and fieldwork, and the fact that people like MacDougal were crossing over easily between experimental study and field study, provided ecological research with meaning and importance. In understanding this complex of related enterprises in science, we can also see how ecology, which is a broad, eclectic subject, became a discipline. It did so because the science addressed larger American goals related to economic development: this was the reason patrons of science were interested in funding the research. Ecology was connected to a larger quest for control over life that was shaping the direction of the life sciences in the early twentieth century.

We cannot claim that all studies deemed ecological dealt directly with problems of control and prediction. What matters is to recognize how these goals were associated with ecological research and why these goals were valued. In addition, the many transformations occurring in the land provided opportunities for ecological studies of how organisms were affected by these changes. Questions concerning what causes affected the distribution and abundance of species, what conditions were favorable to life, and what unfavorable, were very obvious ones to pursue given these social and economic changes. The ecological view gave a vivid picture of the dynamism of nature and suggested ways in which human actions might be altered in the light of this knowledge.

Through their science, Americans were indirectly trying to puzzle through the problem of how they themselves fitted into the landscape. What would be the outcome of that struggle for place in which all organisms, humans included, were engaged? How could the outcome be predicted and controlled? Answering such questions entailed not just gathering facts and performing experiments, but developing theories about nature and the human relationship with nature. How broad and inclusive would ecology be? This question demanded further debate.

CHAPTER FIVE

Visioning Ecology

In 1913 Arthur Tansley, a British botanist with a strong interest in the development of ecology, toured the United States with a small group of European scientists. He came away with a vivid sense of the "earnestness and single-mindedness of American science," especially in the field of ecology.[1] The tour began in New York, where the scientists visited the New York Botanical Garden, Columbia University, and the Brooklyn Botanic Garden. It came to a brilliant close at the Desert Laboratory. MacDougal lavished hospitality on the group, helped by a grant from the Carnegie Institution. The Europeans visited the laboratory's research sites at Carmel, the Salton Sea, and the Santa Catalina mountains. Tansley was impressed by the scientific operations he encountered in America and by the welcome he received from his American hosts, who were eager to show just how far they had come in expanding the range of biological research.

He was equally impressed by how quickly the forests and prairie were being replaced by cornfields and factories. The great beech-maple forests that had once covered the northeastern United States were now dwindling. Tansley visited a remnant of these forests in Michigan, a preserve established by a local manufacturer who had bought the land for its lumber and then decided there was greater value in preserving the forest. As Tansley remarked, appreciation of the spiritual value of nature was "all too rare among the business men of any country."[2] Traveling west from Chicago to Lincoln, Nebraska, Tansley saw the woods gradually degenerating as they were turned into pasture grounds for cattle. In other places, where tree seedlings were not eaten by cattle and the prairie fires had been suppressed, the forest was advancing steadily on the prairie. Across the continent he noted the luxuriant growth of European clover, striking because it seemed to grow far better in North America than in its native habitat.

Change was everywhere. Through the gently rolling agricultural lands of Iowa, he caught not a glimpse of untouched prairie. He knew that the Great Plains were still largely in a natural state, but traveling through Nebraska and Kansas he saw no natural prairie from the train, only the stubble of farmlands. As the train headed from Akron, Colorado, toward Denver, he marveled at the crops growing on irrigated farmlands. Scientists at agricultural stations were studying the native vegetation for clues as to which plants might be grown in drier regions. Visiting the Salton Sea, he observed the rapid growth of the date industry in the region around Mecca, where all the finest strains of North African date palms were being tried out. Tansley realized that it was impossible to halt the pace of economic growth, but he hoped that a few small patches of "untouched nature" might be preserved along the way for future generations to enjoy.

Economic development and transformation of the landscape went hand in hand with the scientific understanding of how life-forms were adapted to environments and how the process of adaptation could be controlled. Discussion of adaptation was not confined to plants and animals. Questions about how people were to adapt to new lands, how the landscape transformed human races, and how humans, in controlling the adaptive process, might control their own evolutionary progress, were posed and pondered in many different ways. What did it mean to move to a new land and fit into it? What transformations of organisms and environment were set into motion by these migrations?

The ecologists who were reconstructing natural history at Tucson were working within the context of the rapid changes that had so impressed Tansley. They too thought about the significance of these changes for the future of American society. Could science help to guide and manage that transformation? Ecology, being scientific natural history, connoted something more than a collection of local histories of the sort that amateur naturalists might make. Scientific natural history suggested that the scientific method might allow Americans to understand not only how the present related to the past, but also how to shape future development.

Ecological questions concerned how natural systems were constructed and how they had evolved and would evolve, but human societies also evolved in relation to their environments. How far could one go in "biologizing" the study of humans and folding human ecology into the broader study of adaptation and evolution that defined general ecology? This question could not have a definitive or permanent answer; it had to be asked periodically and considered anew as the scientific discipline evolved and as the social context changed.

Two closely related visions of ecology presented the relationship between general ecology and human ecology in different ways. Both Ellsworth Huntington, a geographer interested in human ecology, and Frederic Edward Clements, a plant ecologist, tried to define ecology as a discipline and as a point of view. Their paths crossed at the Desert Laboratory at Tucson, where Daniel MacDougal invited them to join his growing community of researchers. The main difference between their visions was in the attention paid to human ecology as a component of the new science. Huntington believed that humans should be placed at center stage, so that the problems of human adaptation to a changing land could be explored as an intrinsic part of the larger study of organic adaptation. As he saw it, problems of medicine and public health and questions of human biology, including eugenics, were all included in the subject matter of ecology. In Clements's view, ecology was mainly a botanical subject, with attention focused on "natural" communities of organisms, communities that might be studied without reference to human activity. In his vision, ecology was not directly concerned with problems of human health and evolution, but the knowledge of how nature was constructed and how it developed was indispensable to human social progress.

Huntington and Clements both advanced provocative ideas about the problems of human adaptation and progress that preoccupied Americans in the early twentieth century. Despite their different conceptions of ecology as a discipline, the two had much in common as they grappled with the broad problem of human adaptation to a fast-changing land. They were united in concluding that ecology was not to be defined simply as a branch of another science, such as physiology or biogeography, but was a philosophical approach or point of view that embraced many subjects. Each man's vision illustrates how concerns about human progress were bound up with the construction of the new discipline of ecology.

Clements's ideas echoed and developed the questions framed by George Perkins Marsh, discussed in the introduction. Clements was strongly motivated to pursue ecology as a way to address the major transformations of the landscape occurring in his lifetime. Marsh, shocked at the realization of how brutally humans were altering the land and causing the extinction of other species, had asked in 1864 whether man was part of nature or separate from it. It was still a provocative question a half century later when Clements was starting his career as an ecologist.[3] Huntington's ideas had their origins in the work of Nathaniel Shaler and the school of geography, with its often racist overtones, that explored the geographic conditions that either fostered human civilization or led to its decay. Huntington saw the problem of human civilization and the conditions of its continued growth as central to ecological study.

Huntington's turned out to be the road not taken, whereas Clements's theory dominated ecological thought for a generation. In both visions, however, the emphasis was on predicting and controlling nature and, through the application of scientific expertise, improving upon human adaptation to the environment. According to Huntington, improvement included the idea of using technology to change the environment so that it better matched human needs; for Clements, improvements centered on applying ecological knowledge to land-management practices. Both men adopted an idealistic, almost utopian vision of future society and saw scientific expertise as the means of achieving that utopian future.

In the early twentieth century, when the ideas of Huntington and Clements were first forming, social Lamarckism permeated American thought. Neo-Lamarckism was at the height of its popularity in the late nineteenth century and greatly influenced leading social theorists of the day. When historian Frederick Jackson Turner framed his seminal "frontier hypothesis" in 1893, linking the formation of American character to the frontier experience, he drew on Lamarckian ideas of evolution to support his views about American society and the challenges it faced after the closing of the frontier in 1890.[4] Turner believed that the experience of pioneers on the frontier created a new kind of man, a strong, energetic, democracy-loving character who shaped the developing nation. The closing of the frontier and the replacement of wilderness by civilization had brought this evolutionary process to an end, and Turner wondered what effect the shift would have on the future of the United States. As the environment changed, the social organism evolved.

Turner's thesis owed a lot to the influence of scientific writers, especially to the new science of evolutionary human geography that was emerging as a discipline in the 1880s. In the United States geography was developing into a rigorous science, building on the European work of men like Alexander von Humboldt and Carl Ritter. William Coleman showed that Turner encountered the new physical geography through his friendship with Charles Van Hise, a geologist and conservationist at the University of Wisconsin, whose teaching focused on the relationship between physical geography and the distribution of resources.[5] Turner also discovered the work of the German anthropologist Friedrich Ratzel around 1895 or 1896, a couple of years after formulating his frontier thesis. Ratzel had studied human migrations and the effects of the environment on the diversity of races. Turner recognized the relevance of Ratzel's work to the American experience.

As Coleman noted in his study of the scientific foundations of Turner's frontier thesis, physical geography suggested ways to analyze precisely how the environment influenced the social organism. From these scientific writings Turner adopted the metaphor of the social organism and the Lamarckian evolutionary thesis that lay behind it and constructed a historical thesis about the remaking of American society on the frontier. Historians such as Turner, as well as later writers such as Walter Prescott Webb, found in evolutionary science an assortment of perspectives and biological justifications that helped them to explain the march of human history.[6]

There was a serious contradiction in Turner's writing, Coleman found, and it came to be repeated in ecological writings. This was the contradiction between the idea that society evolved in response to the environment and the idea that the development of society proceeded along a fixed path, recapitulating the evolutionary history of the entire race. The German Darwinian Ernst Haeckel had elevated the idea of recapitulation into a central doctrine of evolutionary theory, and this doctrine was widely incorporated into biological and social theories by the late nineteenth century. Turner built into his frontier hypothesis the notion that the development of societies was destined to follow a fixed course, analogous to the development of the embryonic organism. He believed he could identify those developmental stages on the frontier, with each successive frontier experiencing the same sequence of changes. As Coleman pointed out, belief in a fixed developmental sequence should have contradicted the notion that the social organism was plastic in its response to environmental conditions. But for Turner the more important point was that society was an adaptive organism: an expanding population meeting new environments would evolve new institutions.

The central idea was that of adaptation to the environment and consequent evolution of the social organism.

Within ecological writing the same questions and the same contradictions arose. Did nature move along a progressive path that could be known and predicted, or were natural processes entirely dependent on local circumstances with no tendency to develop in a given direction? These questions were debated in discussions of the nature of ecological succession, which is the study of how different types of vegetation replaced preceding types over time. The human counterpart of the question was to ask how human societies developed in relation to the land and what caused some societies to advance and others to decline. American ecologists placed the study of succession at center stage: it was the ecological counterpart to contemporary reflections about the development of American society and the uniqueness and value of American institutions.

When the Desert Laboratory was founded in 1903, Ellsworth Huntington was embarking on the path that would make him one of the leading advocates of eugenics and human ecology in the United States between the world wars. Born in 1876, he had grown up in New England as the son of a Congregationalist minister. He studied at Beloit College in Wisconsin, then went as a missionary teacher to Turkey, where he expanded his scientific interests in geology and meteorology. He returned to the United States in 1901 to pursue graduate studies in geology, mineralogy, and paleontology at Harvard University.[7] He taught at Yale University from 1907 to 1917 and then became a research associate in geography at Yale, where he developed the subjects of human geography and human ecology.

Two teachers, Nathaniel Southgate Shaler and William Morris Davis, exposed Huntington to the leading ideas of physical geology, with its emphasis on the relationships between environment, race, and society. Especially in Shaler's writing these ideas were linked to the belief in Anglo-Saxon superiority. For Huntington as well the question of maintaining racial superiority formed the recurrent theme of his life's work. He believed that portions of the North American continent were preeminently suitable for advancing civilization. From his pulpit at Yale he preached the gospel of environmental determinism and the necessity of eugenics to keep human evolution on the path of progress. And he envisioned a new interdisciplinary synthesis combining history, geography, and biology—a new science of human ecology—devoted to the scientific study of these ideas about environment and destiny.

Huntington used Central Asia as the first testing ground for his geographic ideas. In 1903–4 he accompanied the Carnegie-funded expedition led by Raphael Pumpelly to the Trans-Caspian region. Pumpelly was interested in the origin of the Aryan race, whereas Huntington was interested in long-term climatic changes and their effects on human settlement. Out of this and other research came his book *The Pulse of Asia*, published in 1907, in which he developed a geographic theory of history. The people of Central Asia, he argued, were molded by the physiographic environment. Climatic changes in particular had affected not just the migration and distribution of people, but also their "occupations, habits, and even character."[8] In developing these arguments, Huntington was building on theories about the relationship between climate and historical events proposed by other writers.[9] Huntington believed he could go one step further in demonstrating the importance of the impact of climate on civilization.

His book caught the attention of Robert Woodward, president of the Carnegie Institution and formerly chief geographer of the U.S. Geological Survey, and of MacDougal, who was at that time studying the Salton Sea. MacDougal invited Huntington to work at the Tucson laboratory, but three years passed before Huntington was able to take up the offer. In 1910 he accepted the invitation, with the idea of using the southwestern landscape to test his theories about climatic cycles. With Carnegie funding, Huntington spent five spring seasons exploring the southwestern United States, Mexico, and Guatemala.

He examined changes in the thickness of the rings of the giant Sequoia, stumps of which were found abundantly in California, where the ancient trees were being cut for fence posts, shingles, and pencils. Many of the trees were more than one thousand years old at the time they were cut, and three trees were thought to be more than three thousand years old. In those rings Huntington discerned apparent pulsations in climate over a long period of time. The pulses detected from the tree rings appeared to coincide roughly with some of the climatic changes he found in Asia, suggesting a worldwide climatic cycle. Huntington concluded that the cycles must be related in some way to changes in the sun, especially sunspot activity, which might affect the weather on earth. He hypothesized that variations in the eleven-year sunspot cycle might be behind longer-term climatic pulses that occurred over the course of centuries.

When he found evidence of alternating periods of wetter and drier climates, it seemed to him inescapable that such cycles were connected to the rise and fall of civilizations in some way. He was struck by the evidence that the ancient Mayans were a highly civilized, prosperous people. Surely their achievements were the products of a "wideawake, progressive race" very different from the "slow, mild, and unprogressive people" who were the descendants of this population. The ancient Mayans appeared to have a level of accomplishment as high as that seen in any other part of the world; they must have been remarkably inventive and intelligent. "Was it something in the fiber of the original race," he asked, "or was it something in their environment?"[10] Were these ancient people more highly endowed than any modern race, a kind of superrace? Huntington thought not. It must be the case that the climate had been better in the past, producing an environment that was freer of disease, better for agriculture, and in general less enervating and more conducive to productive work. Ancient Mayans were not biologically superior, he concluded; they were simply able to work more efficiently in a climate that brought out the best in human ability.

Huntington developed these themes in his 1915 book *Civilization and Climate*, where he argued that there had been significant climatic pulses since the Ice Age and that the climate of the past had been quite different from that of the present. To explain how climate could affect the progress of civilizations, he zeroed in on the connection between basic features of climate, such as temperature and humidity, and the energy level of human populations, the idea being that people were vigorous only under the right climatic conditions.[11] He collected data on what conditions prevailed when people were thought to work most efficiently or most productively. He studied changes in the output of factory workers and had his class at Yale keep a daily log of their level of activity and feelings. He sent out surveys to collect people's views on which regions represented the highest and lowest levels of civilization, then mapped the results so that the civilized regions could be correlated with climate. Using data on temperature and humidity and his rudimentary knowledge of how climate affected human energy, he constructed a worldwide map of human energy, showing those places where a good climate could be expected to produce high levels of work. The energy map and the civilization map matched well enough to convince him of the correctness of his thesis.

Just as climate affected the physical activity of the body, he believed it also affected one's character, especially those traits related to "industry, honesty, purity, intelligence, and strength of will," because moral qualities were believed to be related to the physical condition of the body. Certain climates, for example, might promote nervous excitation, possibly leading to weakened self-control and hence to vice and "excesses of various kinds." Unsurprisingly, he found that temperate regions such as western and central Europe were blessed with the best overall climate for energetic activity and for the development of civilization. Another region of high energy and civilization was Japan; yet another could be found in portions of the northern United States, with a strip of southern Canada included.[12]

Humans, he argued, were more closely dependent on nature than they had realized. But in acknowledging this environmental determinism, Huntington saw the first step toward freedom from these climatic handicaps. Climate appeared to be the ultimate cause of idleness, dishonesty, immorality, stupidity, and weakness of will. Therefore, he concluded with a flourish, "If we can conquer climate, the whole world will become stronger and nobler."[13] Conquering climate was just one prerequisite for the advance of society; the other was to control the adaptive process, to improve the world for human habitation by bringing it into line with human needs. The experiments on evolution at the Desert Laboratory were an exciting harbinger of the future, even if the botanical transformations that MacDougal described were still limited in scope. But Huntington imagined that "man might soon be able to induce variation in plants almost at will" and that "the wonderful process of evolution, rightly directed, may ultimately double the food supply of the world." All this stemmed from the recognition of the need for research laboratories where scientists had freedom to pursue "pure knowledge apart from any practical result."[14]

In *World-Power and Evolution*, published in 1919, Huntington suggested that environmental extremes produced genetic variation, or mutation, and that this mechanism might explain how climate influenced human society and human evolution.[15] He cited the experiments being done on plants and animals at Tucson, as well as several other kinds of experiments, mostly on lower animals, that showed the effect of temperature extremes in producing mutation. Some of these experiments came from the laboratories of respected geneticists; others, notably the neo-Lamarckian experiments of Paul Kammerer in Austria, were discounted by many scientists as inconclusive. Huntington marshaled data from anthropological studies of head shape and stature in human populations and boldly concluded that the kind of environmental effects seen in insects and amphibians might operate on humans as well. Temperature variations, he believed, could alter bodily form and therefore also mental activity.

His suggestion that racial characteristics were produced by the environment and that people would be altered as a result of migration to different climates might imply that races were malleable to the point where one race could be made like another. But Huntington did not think that millennia of history could be undone. He believed that the racial type, once formed, was largely fixed, and neither environment nor education and training could erase the differences between races. In the 1920s Huntington softened his environmental determinism by giving more emphasis to biological differences between peoples. He increasingly began to stress the importance of genetic endowment, allied himself with the more extreme wing of the eugenics movement (at a time when biologists were starting to be openly critical of these extreme positions), and tried to advance an approach to history that struck a balance between geography, biology, and cultural factors.

Huntington was very active in the formation of the Ecological Society of America, founded in 1915, and served as its second president in 1917. He encouraged geographers to join the society so that their viewpoint would be reflected in the advance of human ecology, which he vigorously promoted through the war years and in the early 1920s.[16] He hoped the society's journal, *Ecology*, which had evolved from the Desert Laboratory's magazine *Plant World*, would have broad scope and include human ecology as a primary subject.

An idea of what he envisioned for ecology can be gleaned from an article he published in the first issue of *Ecology* in 1920, discussing the control of pneumonia and influenza by the weather.[17] Here he raised the question of how the quality of air affected human health and called for statistical analysis of the relationship between air and health. For instance, studies showed that pneumonia and influenza incidence seemed to vary inversely with temperature, with peak times of disease falling when temperatures were the coldest in winter. Humidity might also affect the incidence of disease, he thought. Variability in the weather seemed to bear a close relationship to disease, although these results were far from conclusive. Finally, he thought it possible that air quality might affect the severity of epidemics, such as the influenza epidemic of 1918, but not their onset.

Research of this kind depended on better coordination of data collection, especially from physicians. Huntington noted that during the influenza epidemic record keeping was generally poor about such things as the length of the disease. Even when cities required physicians to fill in health cards, not all provided the data requested. Huntington had to sort through fifteen or twenty thousand of these cards in order to generate a sample of about three thousand cases for analysis. Even with these limitations, he believed that he discerned a relationship between variability in weather and disease, with variability being beneficial to health. The ideal was something like a typical pleasant spring or fall day in the northeast regions, suggesting that indoor climates should be adjusted to mimic conditions found outdoors in May or October.

The final stage of this analysis would be to investigate the cycles of disease-causing organisms and to see whether these cycles were also related to the environment. At this time, the influenza outbreak was thought to have been caused by a bacterium of unknown form, and possibly several types, and Huntington postulated that an environmental condition might have stimulated these bacteria and produced the epidemic. It was, he said, "speculation, but it is controlled speculation." He envisioned a bold project that would test speculation by experiment, to discover principles that would "revolutionize many phases of man's activity" by finding the optimum environment that favored humans: "In this great work there is a part for every ecologist to play, whether he deals with plants, animals, or man; for what we now need most is a determination of the great principles on which the science is founded."[18]

His plan was extremely ambitious, verging on the utopian, but was not without support within the Ecological Society. In an editorial comment published immediately before Huntington's article of 1920, the president of the society, Barrington Moore, had challenged members to be constantly aware of the way individual research problems touched on neighboring fields and disciplines, and therefore to be open to cooperation with people in other fields.[19] Moore's vision, in keeping with Huntington's, was that ecology was not a branch of some other discipline but was a way of thinking superimposed on other sciences. Answers to problems in agriculture, forestry, and range management required ecological insight. Medical problems did as well. He recounted that "during the war one of the army camps was infested with mosquitoes. An ecologist was summoned. He experimented, and found a certain kind of fish, which when introduced into the neighboring ponds destroyed the mosquitoes." All the biological sciences demanded ecological research, Moore said, and "nobody can doubt that this demand will be even greater in the future." He challenged his fellow ecologists to be bold: "Will we be content to remain zoologists, botanists, and foresters, with little understanding of one another's problems, or will we endeavor to become ecologists in the broad sense of the term? The part we will play in science depends upon our reply."[20]

Huntington's vision had equal boldness and breadth, but ultimately these views could not prevail against the tendency toward specialization and separation of research fields. The appeals failed to transform the field of ecology or the Ecological Society of America as Huntington had hoped. The society organized a symposium on the relation of general ecology to human ecology at its annual meeting in December 1923, and a few articles on human ecology were published in *Ecology* in the mid-1920s, but in general the journal became more narrowly focused on biological research with emphasis on botanical subjects.[21] Huntington's expansive vision of the subject as embracing questions of public health did not catch on. Huntington therefore lost interest in the society as a vehicle for the promotion of human ecology. By 1937 he complained to Alfred Emerson, editor of *Ecology*, that the society had totally backed away from an original vision that included all phases of ecology, plant, animal, and human, and had neglected humans in particular. The greatest loss, as Gene Cittadino recounts, was in excluding the study of humans in general ecology; ecologists focused more on the analysis of pristine environments where human impact was light.[22]

Although Huntington was unsuccessful in shaping the Ecological Society to his liking, his work connected elsewhere with growing interest among ecologists and other biologists in the causes of population fluctuations. The appearance of regular cycles of abundance and crash led to more systematic study of population ecology, along with efforts to describe these relationships mathematically and to relate their causes to variations in climate.[23] Biologists had identified cycles of various periods, including four-year, nine-to-ten-year, eighteen-year, and possibly six-year cycles. Economic concerns motivated many of these analyses. Fisheries and the fur trade were subject to extreme variations in population, creating economic hardship in some years. Introduced animals such as the rabbit in Australia and the Canadian muskrat in Europe became troublesome pests as populations exploded. Predicting these cycles and understanding their causes might help to stabilize certain industries and aid in the conservation of resources, as well as cope with the increasing problem of introduced pests.

Under the leadership of its new president, paleontologist John C. Merriam, the Carnegie Institution of Washington sponsored two conferences on cycles in the 1920s. Huntington and several scientists from the Desert Laboratory contributed papers. The hope of these conferences was to illuminate the relationship between physical and biological cycles and those that appeared in human communities. Understanding cyclical processes would lead to better prediction, and hence more intelligent adaptation to extreme conditions.[24] At a conference on biological cycles convened in 1931, Huntington and a group of biological experts explored the evidence for various kinds of cycles, looking for common ground and hoping to encourage more research in this field.[25] Huntington's work on the biological effects of temperature, humidity, and light was considered an important contribution to this growing subject. Realizing that the relationship between population cycles and climate was an expanding area of research, he vowed to press forward in his studies, especially on the mental effects of weather and climate.

Huntington believed that he was shining the hard light of science on his global hypothesis. He appealed to a variety of kinds of evidence—measurements of tree rings, experiments on insects, the productivity of factory workers, the shape of immigrants' heads—believing that such "facts" tilted the scale toward environmental determinism and its attendant eugenic implications. He did not recognize that he had framed his problems in a circular way, defining superiority by criteria that could only be met by those populations he had already decided were superior. Laboriously collecting the data exposed by the wanton destruction of America's most long-lived species, the giant sequoia, Huntington used those measurements to support a historical thesis that predicted a rosy future for the energetic American people who lived where the climate was right or who could control the climate through technology. No trace of irony can be seen in his optimism that future evolution might be directed along the "right" path and that the people who had felled those ancient trees to make pencils and fence posts were the proper ones to determine that path.

Always controversial, Huntington's bold ideas about climate and its effects suggested that one might develop a broad view of how nature operated and that a few general principles might underlie complex phenomena. The issues he raised were thought to be important and interesting, even when they were "unacceptable without further research."[26] Within ecology, his ideas about climatic cycles and environmental determinism found favor with Frederic Clements, who had come to Tucson to write the magnum opus that would set the theoretical basis for the science of ecology.

Frederic Clements was a fervent champion of ecology as a separate and distinctive field of study. He believed with Huntington that the key to understanding the relationship of organisms to environment lay in understanding climate and the short- and long-term cycles that affected temperature and precipitation. He also had ideas similar to Huntington's about how ecology might contribute to the control of nature. This control would be gained by developing the science of ecology along experimental and quantitative lines. A rigorous science, backed by a theoretical framework that explained how nature was organized and how it developed through time, would show humans what they needed to do to fit into these landscapes. He too found a place to study and a welcome reception at the Desert Laboratory.

Clements came out of the Nebraskan school of botany dominated by Charles Bessey, who had been one of the leaders of the new, experimental, European-inspired botany.[27] In exploring the prairie, Bessey and his students also had to grapple with the practical problem of assessing grazing and farm practices to determine how the land should be used. In 1896 Clements took part in a survey of western grasslands and pasture lands done under the auspices of the new Division of Agrostology (the study of grasses) that was established by Congress in 1895 in response to growing concern about the deterioration of grasslands due to overgrazing and poor farming practices. Clements's research was in Colorado, while other field-workers explored extensively throughout the western states and territories. Cornelius L. Shear directed the research and published a report of the division's work in 1901. He drew attention to the dire state of much of the grasslands and the need for better scientific assessments of plant growth in the region.[28] He was calling, in short, for more ecological study.

Clements's earliest work was conducted in a land under siege. The stock-growing industry had expanded rapidly after the transcontinental railroads were built, and the relatively arid western prairie was fast opening up to agriculture. The destruction of prairie, the loss of timber lands, and the resulting erosion demanded more accurate scientific assessments of what species grew well in these regions and what needed to be done to restore the land. The original prairie held a rich flora: on any given square mile perhaps fifty species of grasses and forage plants could be found, yielding one and a half to two tons of hay per acre. By the 1890s both the diversity of species and the overall yield were sharply reduced over most of the land.[29] The carrying capacity of the land, its ability to produce vegetation and hence to support cattle, was dropping dramatically; it had decreased as much as 50 percent in some areas.[30] Yet in regions where there was enough grass left to produce seed, restoration could be fast and effective. Shear remarked that giving the range a brief rest produced results "which astonish one who has never witnessed the experiment."[31] But not all scientific judgments were so optimistic, for Shear was also calling attention to the need for better management of the grasslands. Some recommendations were deemed unacceptable by the government: Clements noted later that a chapter dealing with some of the more serious problems was suppressed, a sign of the "evil trail of politics in the reconstruction of the West."[32]

Nebraska, where Clements did his doctoral studies, had gone through a dramatic cycle of boom and bust in the years before his work in the mid-1890s. Although the region was prone to severe drought, the period from 1875 to 1886 had been very wet. Homesteaders flocked to the land and settled in marginal semiarid regions. The population soared during the 1880s. The idea that "rain follows the plow," or that cultivation of land actually increases rainfall, gained currency. Clements observed that the idea was so popular that university professors managed to find "convincing proofs" of why cultivation might cause improved climate.[33] This was a reference to Samuel Aughey, predecessor to Charles Bessey at the University of Nebraska and one of the state's biggest boosters.[34] But the plow was followed by drought and famine in this instance; a severe drought during 1893–96 brought about an equally dramatic depopulation. Clements reported that eighteen thousand prairie schooners were said to have rolled eastward from Omaha in a single year, part of an exodus of about half a million people from the frontier during the drought. These kinds of changes occurring during Clements's formative years as a student must have left a deep impression. Science showed the way to rescue people from the tragic consequences of trial-and-error approaches to settlement.

Samuel Hays has argued that the early conservation movement was dominated by an ideology of scientific management and efficiency, the hallmarks of the Progressive Era belief in planned development.[35] The new science of ecology was carried forward by people like Clements who espoused a similar faith in science and planning. The ecologist was to be a new kind of leader, armed with knowledge and above the corruptions of business and politics, a manager who would establish the rules for human settlement and land use. Far more than MacDougal had done, Clements emphasized the distinctiveness of ecology as a science with a mission.

MacDougal knew Clements from his early days at the New York Botanical Garden, where Clements and his wife Edith had gone to do research soon after the laboratory opened. While Clements worked on ecological topics, Edith did experimental work on the mutation theory. In 1905 Clements, then on the faculty at the University of Nebraska, gained visibility by publishing a book on research methods in ecology that advocated a more systematic approach to ecology, including the use of experimental, statistical, and graphical methods.[36] That book also included a section on the importance of applying experimental methods to evolutionary problems, a subject dear to MacDougal's heart.[37]

Clements was interested in the mutation theory because of its experimental approach but was not persuaded that mutation was an important cause of evolution, for it seemed too rare. Instead, he was impressed by the ability of plants to be modified by environmental influences: certain species appeared to be quite stable, whereas others seemed very plastic. This subject demanded a systematic experimental analysis. Clements did not describe these views as "Lamarckian"—indeed he made no reference to Lamarck—but he certainly believed that new adaptive forms could arise through environmental influences. Much of his later work would be designed to test how changes in habitat affected species. Clements came to be seen as representing one of the last vestiges of neo-Lamarckian thinking in America, for his views never wavered on this score throughout his career.[38]

MacDougal recognized a kindred spirit and appreciated Clements's discipline-building drive. He saw the advantages of forming an alliance that would free Clements from administrative duties. Clements had moved to the University of Minnesota as chairman of the botany department by 1913, the year MacDougal invited him to join the research group at Tucson. He later was appointed research associate at the Carnegie Institution and remained affiliated with the Institution for the rest of his career, creating an important body of theoretical work that defined and dominated American ecology.[39] Clements had an obsessive nature, combined with a puritanical disdain of smoking and drinking, which must have put a damper on laboratory life. But in MacDougal Clements found a strong supporter, for they shared an intense ambition to build scientific empires and to advance the experimental study of evolution.

Clements was one of the few people who had a clear vision of ecology as a distinct subject combining experimental physiology and the field sciences.[40] He forged this link between experiment and field observation by developing the idea that plant associations were "complex organisms," which interacted with their environments and had developmental histories like individual organisms.[41] The organismic concept followed from the premise that the central task of ecology was to study ecological succession, or the replacement of one plant association by another over time. Clements argued that the organization and self-equilibrating properties of plant associations made it legitimate to consider them to be a special type of organism. The "complex organism" was not therefore a figure of speech but a term denoting a new class of organism, although in structure and function it differed greatly from an individual organism. But like the individual organism, it had a definite developmental sequence, Clements imagined: it evolved over time in a direction that could be predicted. The idea of a fixed developmental sequence followed directly from his organismic concept.

Clements's monumental study *Plant Succession: An Analysis of the Development of Vegetation* was written in his early years with the Desert Laboratory. This treatise was his magnum opus of ecological theory. The hefty manuscript earned MacDougal's highest endorsement and was published by the Carnegie Institution in 1916.[42] With MacDougal's backing Clements then became a full-time Carnegie researcher, working partly at Tucson, partly at an alpine laboratory in the Rocky Mountains, and later at Santa Barbara, California. During these years he became the leading theoretician and spokesman of American ecology, publishing on a wide range of ecological problems involving community functions, adaptation, and the evolution of plant communities. His work was exceptional in its boldness; he stood out for his willingness to argue that there could be a general theory of the evolution of plant formations. His goal was a theory that transcended the vagaries of local circumstances, an overarching theory of the evolution of plant communities. His approach was justified, he argued, because his fieldwork covered a vast geographic range in the western United States and therefore provided a broader sampling than most ecologists had made to date.

The importance of analyzing units of vegetation in dynamic terms was already well recognized in European plant geography. In the United States Henry Chandler Cowles, geologist and ecologist at the University of Chicago, took the lead in showing that the study of succession was a fruitful way to understand ecological processes. However, he did not have a fixed view of what the process entailed. Having been trained in geology, he related ecological changes to the continual transformations of the physical landscape, rather than to climatic conditions. He argued that there was a constant dynamic interaction between plant formations and underlying geologic formations. Studying the sand dunes in the Chicago region, he saw the flora of a landscape as an ever-changing panorama. The ecologist had to discover the laws governing these changes, but the changes themselves did not have a fixed course comparable to the development of an embryo.[43]

Cowles's research on the Indiana dunes enabled him to work out a complete successional series, that is, the steps by which one plant association replaced another, or succeeded another, over time. Cowles assumed that vegetational changes in space paralleled changes in time. Therefore as he walked inland from Lake Michigan, the ecologist also walked backward in time. By putting together the spatial sequences of plant formations, Cowles reconstructed the temporal development of plant associations.[44]

Looking at the ever-shifting landscape of the dunes, Cowles was not inclined to view the process of succession too rigidly, although he did see succession as a generally progressive process, with vegetation passing through a period of "youth" and "maturity" and finally reaching "old age."[45] He believed that succession tended toward a stable final form, the climax, but did not believe that complete stability was ever achieved over an entire area. Moreover, the successional stages leading to the climax community—the final stage of succession—were never in a straight line but could regress in the normal course of events before regaining their path forward.

His concept of the community included not only the idea of a continuous, never-ending process of change but also the idea that organisms were connected in a vast, complicated symbiosis. Gene Cittadino suggests that Cowles approached ecological analysis in a manner parallel to Frederick Jackson Turner's discussion of social evolution on the frontier. Both were concerned with the way landforms challenged organisms and forced them to adapt quickly to survive.[46] But in Cowles's case the frontier did not close, for natural forces created local variations that stimulated new cycles of succession all the time.

Clements studied the more stable grasslands and conifer forests of the western prairie and lacked the strong geologic perspective that shaped Cowles's ideas. His view of nature was consequently very different. In taking the complex organism as his field of research, he was trying to understand how plants came together and stayed together. The processes of plant succession depended on continual interactions between the habitat and the life-forms of the community. Habitat and populations acted upon one another in a reciprocal way until finally a stable state, the climax, was reached. The form of the climax was determined by the region's climate. If the climate remained stable and no humans intervened, the climax might persist indefinitely, maybe for millions of years.

Clements envisioned successional processes as largely driven by competition between plants with similar needs, and much of his experimental work was directed toward the analysis of different ways in which plants competed.[47] He took advantage of the natural experiments provided by well-worn roadways like the old California trail that had carried thousands of pioneers and their animals west.[48] As the roads were abandoned and repopulated by plants, they created a series of stages in succession, allowing Clements to study plant growth when competition was first eliminated by the wearing of the road, then restored as plants invaded the roadways. Clements later used the fenced right-of-way enclosing the railroad as a type of control site, showing what the "natural" state of the land was. The enclosure created an unintended prairie reserve, a strip of vegetation hundreds of miles long cutting across the country, illustrating the way to reconstruct the grassland in cultivated and grazed regions.[49]

The term *competition* applied mainly to relationships between similar organisms. Species of trees were said to compete sharply when together because of the similarities in their form and requirements, whereas the relation of trees to shrubs was said to be one of dominance and subordination, rather than competition. Clements thought that as succession progressed, competition was gradually reduced, although it would never disappear entirely. That is, the outcome of competition was to establish dominance hierarchies among the plants of the community, as it moved toward the final climax stage. In keeping with the concept of the complex organism, succession was always progressive: in any region, a plant association should progress always to the same climax. In contrast to Cowles's understanding, the process did not run in the opposite direction any more than an embryo could develop backward—unless, that is, some disturbing force like human activity intervened to prevent nature from taking its course. It was this idea that there was a fixed sequence to the developmental process, a direction that could be predicted and that was determined by climate, that distinguished Clements most sharply from his contemporaries.

As Joel Hagen suggests, the concept of the complex organism helped Clements to apply experimental methods to ecology and to define the subject matter of ecology.[50] The physiologist used experimental methods to study the individual organism; the ecologist, a kind of outdoor physiologist, used similar experimental methods to study the "complex organism." Universal principles, or generalizations that applied to all kinds of vegetational forms, could be articulated based on an understanding of the shared properties of all complex organisms. Without this organismic concept there was a danger that ecology could not be a distinct science but would be nothing more than a collection of local natural histories. Clements abhorred that prospect.

Hagen points out that organismic analogies were common at the time and that Clements's appropriation of the metaphor could be seen as unexceptional. Clements himself, when challenged by his critics, was quick to point out that others had anticipated his ecological viewpoint or professed similar organismic views.[51] He recommended that readers consult the writings of Herbert Spencer, who had articulated the concept of the social organism and used it as the basis for a far-reaching evolutionary philosophy. Alexander von Humboldt and Eugen Warming, intellectual leaders of ecological thought, had also thought of the plant community as showing both cooperation and division of labor, Clements emphasized. Analogies between animal and plant communities and human societies were common in late-nineteenth-century ecological writings.[52] However, Clements himself brought up the idea in 1905 as a concept that "may seem strange at first," so we have to assume that he believed he was introducing a novel point of view and expected opposition to this concept.[53]

The dogmatism that lay behind Clements's insistence on using the complex-organism concept and his unwillingness to abandon it in the face of criticism needs explanation. As Arthur Tansley argued, calling the plant formation a complex organism seemed exaggerated and rigid.[54] At most it might be compared to a "quasi-organism," but even that level of comparison could not be pushed too far. Clements was not satisfied with this compromise, though, and insisted that the plant association must be understood as a type of organism. He was trying to draw attention to the "interplay of development and structure in the life-history of the community."[55]

Although there is no single explanation for Clements's high degree of rigidity, we might understand his frame of mind by recalling that the land he studied as a student was under siege and that in such lands human society could become lawless and morally bankrupt. Donald Worster has explained Clements's concept of the ecological community by relating it to his experience on a "virgin prairie" that would soon be destroyed, as though Clements's whole approach derived from his encounters with a stable natural world.[56] I would suggest that the opposite is more likely, that he was deeply impressed by the degradation of the land and what this impoverishment implied for human society.

In Clements's writing from the 1930s, at the time of the Dust Bowl, there are indications that his formative experiences as a young man were not his contacts with "virgin prairie" as much as with lands under attack. I suggest that he connected the destruction of the land to moral degeneracy and the breakdown of society. Even late in his career, he recalled the "ruined countryside" torn by the "struggles between cattle-rustlers and vigilantes" on which he had made his first observations.[57] The disturbance of the land, because it led to moral decay and threatened the social order, was profoundly disturbing to him.

Prior to this modern onslaught, the prairie seemed remarkably stable, enduring fire, flood, drought, and animal incursions for untold centuries, responding to climatic fluctuations but always able to repair itself. Even on land broken by the plow and overgrazed, nature's recuperative powers were still astonishing to the human witness.[58] When human actions went too far, intelligent agricultural practices guided by scientific knowledge could rescue the land and get American society moving along a healthy progressive path, in harmony with the natural direction. Ecology was a means of fighting the corrupt influences of money and politics that had created abuses of the land and taken such a toll in human life and happiness.

Clements was a moralist, but his vision, although mindful of the destructive excesses of rampant capitalism, was not to return to a pristine wilderness as much as to foster what would later be called "sustainable development." This goal was in tune with Progressive Era ideas of efficient science-based management, and it meshed well with Franklin Roosevelt's New Deal policies during the Dust Bowl. Rescuing the land meant knowing the difference between a natural climax community, that is, one that was controlled by the climate, and a false climax that was actually the result of human intervention on the land. Commonsense observation simply did not allow one to distinguish between natural and artificial environments.

Certain regions of the short-grass plains, for instance, were thought to represent climax formations, but Clements disagreed. He argued that the short grass, although extending over vast areas, was a result of overgrazing, which had altered competitive relations between species and driven out taller grasses. The dominant short grasses in certain regions were not natural at all but were evidence of human disturbance and overgrazing: this was a "disturbance climax" or "disclimax," not a climatic climax. Knowing the difference mattered, for the natural climax was thought to support a more productive type of vegetation. Ultimately this knowledge would help to create better grazing lands. Scientific management, in short, would directly benefit society by keeping the land as productive as it could be.

Forming these kinds of judgments required that a distinction be made between the natural developmental process of the "complex organism" and the unnatural or disturbed development that revealed that humans were not properly adjusting to the land. Only by learning the difference between the natural and the disturbed environment could the ecologist diagnose whether human impact was having a deleterious effect. If succession appeared to be going backward, toward an "earlier" stage and away from the climax, that meant the area was being misused and remedial action was needed. If the community was moving properly toward the climatic climax, then the grazing plan was satisfactory.[59]

The "complex organism" provided a way of seeing nature, a way of interpreting history, and a way of answering George Perkins Marsh's question: Were humans part of nature or separate from it? Like Marsh, Clements looked at the evidence of human effects on the landscape, at the loss of species and the destruction of the land, and felt that humans stood outside nature, often as disturbers and destroyers. Yet in another sense Clements realized that humans were not outside nature, for they were subject to the same controls of climate and landforms that determined what plants could survive together in a region. Problems arose when people were ignorant of those controls.

Just as Marsh had looked to human ingenuity to solve problems, so Clements also believed that science could guide development. The urgency he felt to create that science was, I suggest, the main reason for his insistence on the complex-organism concept. If humans could figure out how nature behaved normally, they could correct for the pathological effects of human ignorance. To achieve this understanding and control, an association could not be considered a random assemblage of species. The plant community had to have structure, and it had to have powers of development and regeneration. If nature was not meant to develop along a predictable course, then how was it possible to identify the human impact on the environment? If natural succession did not automatically result in restoration of damaged landscapes, how could the scientist figure out a way to mitigate the destructive forces of humans and their animals? How could the ecologist function as an expert guide? Ecology, like other experimental sciences, had value to the extent that it could predict and control nature.

In 1935 Clements took up the theme of the social role of ecology in the middle of the worst ecological disaster in the history of the United States, the creation of the Dust Bowl in the Great Plains.[60] That spring had been an especially bad season for dust storms; in Kansas from mid-March through April there were storms on half the days, and sometimes the air was filled with dust for several days at a time.[61] Drought and wind were part of the normal scene, but the destruction of vegetation was, as far as anyone knew, unprecedented.

Clements had for years emphasized the relevance of ecology to problems of land use and resource management, especially after the First World War when agricultural problems in arid regions were becoming more pressing. The Carnegie Institution was interested in practical ventures in cooperation with other research groups. In 1918 the Desert Laboratory entered into joint research in grazing and range management with the Forest Service, the Biological Survey, and the University of Arizona. Clements touted this work as demonstrating ecology's relevance to agriculture, although he also saw these cooperative projects as a way to build up a body of knowledge about ecological succession and hence contribute to basic research. In his annual reports to the Carnegie trustees, he rarely missed a chance to highlight the practical importance of ecology and the need for more research.

By 1935 his arguments for the importance of ecology were taking on an almost religious tone: ecology embraced everything, "all problems in which life and its environment are concerned," and far from being a specialized field like physiology or morphology, ecology was "a point of view and a plan of attack."[62] Like Huntington and Moore, Clements had shifted the vision of ecology from its status as a subspecialty of biology to an approach or perspective that was superimposed on a variety of problems. The ability of ecology to serve as a general guide to the gamut of problems involving life and the environment depended a great deal on the organismic perspective. If the plant community was developing like an organism, then history had a known direction, a direction that could be predicted with accuracy. Evidence that the community was not developing along its proper path was evidence that disturbing forces were at work, and humans above all other causes represented disturbing forces. Left undisturbed, every bare area would begin a slow but inevitable movement to its climax, and the final form was as constant and predictable as that of the individual developing embryo.

This level of predictability also meant that the ecologist was potentially in a position to control the process. Ecological knowledge meant power over nature. It was not just a question of working in harmony with nature; that was a bit too passive. Once understood, the natural process could be retarded, accelerated, telescoped, held in one stage indefinitely, or deflected along another course, perhaps even destroyed in order to allow the process to start again. It could be manipulated and modified by inserting new species. It could be protected from all but climatic change. "In short," Clements argued, "as an instrument for the control of the entire range of human uses of vegetation and the land, succession is wholly unrivalled."[63] The value of the complex-organism concept lay in its power to confer on the human observer, the scientist, the ecologist, total control of a landscape to the extent permitted by the climate. Faith in the possibility of control stemmed from belief in the objectivity of science, the objectivity of quantitative, experimental methods, and the applicability of the organismic concept, which was for Clements no metaphor but a statement of objective fact. Solving problems of the kind presented by the Dust Bowl, which were foreshadowed in the degraded lands that Clements studied as a student, required the ability to predict exactly what would happen when humans entered a landscape and disturbed its natural progress.

Trial-and-error development had to be replaced by scientific method. The ecologist was like the physician, applying scientific knowledge to diagnose the ills of the complex organism and bring it back to a state of health. Studying how nature reclothed bare areas had to be the starting point of landscaping to prevent erosion, but with knowledge of how nature took its course the ecologist could speed up or telescope the process to get fast results. Above all, ecology taught—contrary to common wisdom—that settlement and cultivation did not improve the climate. The perceptions of settlers that winters were less cold and rainfall more abundant and snow less deep had no relation to the effects of human settlement. All was under the control of climate, and for Clements climatic cycles were not influenced by human activity but were under the control of even more remote causes such as sunspot cycles, as Huntington had also thought.

Without the belief in the reality of the complex organism, this vision of civilized progress guided by scientific expertise collapsed into uncertainty, leaving some room perhaps for scientific knowledge, but also a lot of room for trial and error, because scientific knowledge could not lead to prediction of the future. Clements's ecology was designed to give humans better control over the landscape, limited only by the ultimate causes that determined the climate. "In a changing world," he wrote, "one immutable fact remains. Man will never approach mastery of his environment and hence of his destiny until he understands the universal ebb and flow of processes and uses them to his own advantage." No wonder he could not give up the organismic analogy, for giving it up was to give up the idea that ecology was one of the most important forms of science, one of the most important points of view, of modern times. It was to abandon the hope that ecology and ecologists would save the world. The dilemma faced by Americans, he believed, was their failure "to meet change with change and to apply it to the problems of readjustment with scientific foresight."[64] Humans needed to adapt: the concept of the complex organism provided the basis for acquiring the knowledge that would enable them to adapt.

Clements's ideas influenced the directions of much of American ecology well past his death in 1945. His prolific writings helped to define the discipline of ecology in America, gave it purpose, and showed that it was a science designed for a rapidly changing land. As Tansley remarked in 1926, addressing an international group of botanists in Ithaca, New York, the importance of American work was in its focus on succession—a focus that was entirely appropriate given the enormous changes that were occurring in the land. Even if it was conceded that the climax community was not eternal and that nature was probably always changing, still it was important to realize that some types of communities were relatively stable compared with others, and this stability could be the basis for ecological classifications and judgments. Most tellingly, he noted that the successional point of view was proving fruitful in applications of ecology to problems of land use, and this was especially the case in "new" countries, where the changes were occurring quickly and dramatically.[65]

The study of succession was not something that ecologists automatically took as their primary task. What seemed obvious to Americans was not obvious to others. Henry Gleason, a younger contemporary of Tansley's, remembered that while many Americans, influenced by Cowles and Clements, quickly took up the study of successional series and had no trouble whatsoever recognizing such series in the field, European plant ecologists were either reluctant to admit the existence of succession or even persistently denied it.[66] He further observed that initially the study of succession was oriented toward future change rather than past history. When Gleason was starting his career in the early twentieth century, ecologists mostly tried to project where succession was leading, rather than to look backward.

This impulse to look forward rather than backward can also be seen as a reaction to rapid change in the land. Coping with change began by identifying what was stable and "natural" and determining how the future ought to proceed. Gleason believed that he was the first ecologist to look backward when in 1908 he tried to reconstruct vegetational history from contemporary evidence in studies of forest and sand prairie in northern Illinois. Gleason's unusual sensitivity to history eventually led him to the most far-reaching challenge of Clements's theories, and in making this challenge he was nearly alone (see chapter 6).

The American landscape had been changing with astonishing speed after the advent of the railway. Responding to those changes, Americans created a new science that might help them to adapt and to control the ecological interactions occurring between organisms and their environment. The concepts and the vision of ecological science, and the sense of its purpose in relation to human progress, were scientific responses to these startlingly rapid transformations of the landscape. The overlapping research fields defined by ecology, human ecology, and experimental evolution were products of economic development and reactions to human ecological change even on the most basic level. The sea that was accidentally created when an irrigation ditch overflowed, the felled trees that revealed evidence of climate cycles, the fenced railways and abandoned pioneer roadways that gave a cross-section of nature and provided experimental plots for the analysis of competition—all of these human disruptions furnished opportunities for investigation and provoked questions about how nature worked and what the future held.

The challenge for ecologists was to convince the American public that this science deserved a place in American society, that by showing how to meet change with change, ecology would guide development and solve problems along all fronts. As Huntington found, selling a bold vision of ecology was not easy despite its promise. American ecologists, largely content to think of themselves as zoologists and botanists, turned away from the broadest vision of ecology, as projected by Huntington and Moore, which would have brought medicine and public health under the umbrella of ecology. But even the narrower, biologically grounded vision that Clements propounded was strikingly bold in its ambition and could not bring about the larger revolution in values that Clements passionately desired. Paul Sears, a grassland ecologist who wrote about the Dust Bowl in 1935, lamented that local governments failed to take advantage of ecological expertise and too many universities failed to recognize the subject.[67]

Ecology was slowly maturing as a field of science, but ecologists were not finding employment as expert problem-solvers. The field was still struggling to find its place in American society despite the impressive developments to which Tansley drew attention. In order to gain popular recognition, ecologists had to return to the central problem suggested by Huntington, which was how to relate ecology to human goals and needs. How should one think about nature, about humans in nature, and about the history of this relationship?

CHAPTER SIX

Science, History, and Progress

By identifying the central problem in ecology as the study of succession, American ecologists had raised the larger question of how to view natural history: did it have a direction, and if so could that direction be predicted? Was the climax formation a relatively stable final form toward which nature progressed? Frederic Clements had argued persuasively that these questions were to be answered in the affirmative. Natural history had a direction, and Americans were interested in knowing both what the future held and how to control that future development through scientific knowledge.

There were, however, some dissenting views by midcentury. The dissent began with a critique of Clements's theory of succession by botanist Henry Allan Gleason, whose ideas were picked up and elaborated by James Malin, a maverick historian and critic of environmental determinism. Malin's critique was one voice among many reflecting on the past and future of society in the atomic age. I introduce the themes of this broader debate by considering the ideas of human geographer Carl Ortwin Sauer, then by reviewing a conference that Sauer organized in 1955 on the subject of man's role in changing the face of the earth. In this forum we can see the close relationship between debates about ecological concepts and concerns about human society, its past and its future. I end the chapter with an ecological study conducted in the American Southwest, which drew on historical records to try to determine the relative impact of human activity and climate on species diversity and abundance. Uniting these works was, on the one hand, a strong belief in the importance of taking a long-term historical perspective on problems of general and human ecology and, on the other, a rejection of environmental determinism.

The discussions reviewed here dealt with the problem of whether there existed a "state of nature," that is, a natural world that was separate from humans and was relatively stable, and that modern humans inevitably disturbed as their populations grew. This was a historical problem. People began to ask how far back in time human impact could be considered significant. Could we ever know whether something was truly "natural" in the sense of being without human imprint? Was the "natural" state, if one existed, really characterized by stability or balance? As ecologists probed the organization of the natural world in midcentury, they had to confront these large problems. They found that others within the social sciences and humanities were considering similar ones.

Discussions about the nature of the ecological community, succession, and the climax therefore led to broader questions about how to view humans in relation to nature. Ensuing discussions in the postwar decades followed two paths, representing different perceptions of science. One was forward-looking and abstract; it removed ecology from its natural-history traditions and gave it a more modern scientific cast. It began by looking for a better way of characterizing the natural world. In place of the organismic concept, a new idea, the "ecosystem," was substituted. The ecosystem was then defined in an increasingly precise way in the hope of making ecology into a more predictive science. The development and application of the ecosystem concept in the postwar decades is the subject of chapters 7 and 8.

The other line of thought, this chapter's topic, was more in line with the idea of ecology as "scientific natural history" and sought to bridge the growing cultural divide between science and the humanities. I focus here on a group of critics who emphasized the importance of remaining sensitive to historical contingency and local variation. They warned that in dealing with historical causation, processes of such complexity were involved that it would be difficult, if not actually impossible, for the scientist to assess the relative weight of these causes without a great deal of careful study. See chapter 9 for an examination of how these two lines of discussion converged toward the end of the twentieth century.

Frederic Clements, as the chief ecological theorist of the early twentieth century, had a great impact on American ecology, but his ideas were always controversial.[1] European ecologists, as Arthur Tansley noted, could be dismissive, speaking of Clements's "fantasies," "fairy tales," and "laughable absurdities."[2] Clements's adherence to Lamarckian evolution based on the inheritance of acquired characters, though common in the early twentieth century, grew increasingly out of step with scientific opinion as the knowledge of genetics advanced. Even Clements's colleagues at the Desert Laboratory were unconvinced by his generalizations. To Forrest Shreve and Burton Livingston his writings appeared dogmatic, jargon-laden, and incorrect in details, especially when Clements dealt with the desert vegetation that they knew.[3]

Other American ecologists, although they were stimulated by Clements's ideas and his vision of ecology, preferred to take a broader view of succession as simply the study of vegetational change, without the organismic concept and the rigid classification of developmental sequences that was the corollary to that concept. Many ecologists had the data to challenge aspects of Clements's theories on empirical grounds after accumulating research for a few years. An obvious response to the more doctrinaire aspects of Clements's theory was simply to deny that nature was as well programmed and as predictable as Clements claimed.

The difference between Clements and his critics was in part a difference in perspective. It was one thing to conceive of vegetation in the broadest sense and think in terms of climatic climaxes lasting for millions of years, and quite another thing to be closely focused on the details of the dynamic processes going on within communities. If your interest was the variability of local conditions within a given climatic region, then what for Clements was merely "noise" became the object of close attention. Clements was always trying to transcend the local in favor of a general theory that could apply to all communities. As with any theory of historical development, detailed case studies will often reveal that the theory does not hold except perhaps in the most general sense. Such generalities may not allow for very accurate prediction in the end, even though the theory was motivated by the desire to predict the course of history.

Henry Allan Gleason was Clements's most radical American critic in denying the possibility of making universal predictions in ecology. Gleason adopted a point of view diametrically opposed to Clements's organicism. He argued that the distribution of species was a matter of chance and probability. The plant association, far from being an organism, was merely a coincidence. By drawing this conclusion, Gleason undercut the value judgments that were the basis of Clements's ecological prescriptions.

Gleason was raised in Illinois and graduated from the University of Illinois in 1901, just as the ecological ideas of Henry Cowles, the physiological studies of Wilhelm Pfeffer, and the evolutionary theories of Hugo de Vries were starting to make an impact in the United States. Gleason was interested in all these new trends. At the University of Illinois animal ecology, still a young science, was slowly taking shape. Charles Coulson Adams, who would become one of the foremost animal ecologists in the country, was Gleason's instructor. Gleason delved into the new literature and decided to pursue ecological research. Hoping for a permanent faculty position at Illinois, he went to the New York Botanical Garden in 1905 for a doctoral degree. He met Britton, whom he remembered as "a little scrawny man with unkempt beard, who greeted me with his usual limp handshake but with a twinkle in his eyes" and who set him to work on a taxonomic revision of the Veronieae (ironweed) of North America.[4] Daniel MacDougal was then running the laboratory, and Lucien Underwood was head of the Botany Department at Columbia.

Within a year Gleason had his degree and headed back to the University of Illinois, intending to specialize in ecology. He was greatly impressed by Clements's new book, *Research Methods in Ecology* (1905), which he studied closely. Four years later he moved to the Michigan Biological Station, where he taught plant geography, taxonomy, and ecology. The prairies and forests of Illinois and Michigan provided good testing grounds for Cowles's and Clements's ideas about succession.[5]

A year of field research in the Philippines, Java, and Ceylon (Sri Lanka) in 1913–14 gave him some expertise in the ecology of tropical grasslands, rain forests, and tropical mountains. After hearing him lecture on his travels at the Torrey Botanical Club, Britton offered him a permanent staff position at the New York Botanical Garden, starting in 1919. Gleason remained at the Garden for thirty years as curator and assistant director, with a brief stint as acting director between 1936 and 1938. Britton got him interested in South American plants, including work on the specimens collected by Henry Rusby, the very collection that had prompted the founding of the Garden decades earlier. Collections of South American plants had continued to pour into the Garden, and it fell to Gleason to work them up.

As a sideline, Gleason also worked on ecological problems, his first interest. He was not granted permission to divert his attention from taxonomy to ecology on the Garden's nickel.[6] In his first decade at the Garden, most of his ecological work had to be done during vacations and at his own expense, apart from a quick ecological survey done over four months in 1926 in Puerto Rico, as part of the Garden's ongoing scientific survey of Puerto Rico and the Virgin Islands.[7] The constraints on Gleason in the 1920s confirm the argument made earlier in this book, that Britton was not deliberately trying to develop ecology as a discipline but that the synergistic relationship between taxonomy, exploration, and experimental science in the early years of the Garden boosted ecological work and gave the nascent discipline credibility. By the late 1920s Gleason had largely given up ecological studies. After Britton's death in 1934, Gleason had administrative duties at the Garden and may not have had time for ecological research over and above his primary work in taxonomy.

In his first few years with the Garden, however, Gleason was still interested in ecological problems. He had been impressed by Clements's earlier work and had at one time accepted Clements's ideas about succession. The more experience he acquired, the more his faith eroded. He published a brief critique of Clements's ideas in 1917, based on his research in Illinois and Michigan.[8] That article had little impact, but he returned to the subject in 1926, after spending several years in detailed quantitative studies of plant associations. These studies led him to propose a radically new interpretation of plant associations. In an article that critiqued ecological assumptions about plant associations, and without mentioning Clements directly, Gleason advanced what he called the "individualistic concept" of the plant association.[9]

While working in southern Illinois, Gleason had uncovered what he thought were clear examples of retrogressive succession. Later, comparing plant associations in different parts of the world, he noticed that very different environments might support similar plant associations, whereas environments similar in physiography and climate might have very different plant associations. He accepted the idea that plants formed associations and that succession occurred, but he was impressed by how much diversity could be found among comparable associations. Clements had refused to heed the significance of local variations for the sake of drawing out the general features of plant formations, but Gleason focused on the evidence of local diversity or local variations in plant populations. Local variation was the key observation, not noise to be ignored.

The plant association, far from being an organism, was scarcely even a vegetational unit, he concluded: its makeup was a matter of coincidence, governed by the laws of probability. To understand how plant communities were created, attention needed to be paid to how individual plants migrated into an environment and settled into it. The distribution of plants depended on their individual characteristics and environmental requirements. But how particular plants ended up in particular locations was largely the result of chance events, the product of a "fluctuating and fortuitous immigration of plants and an equally fluctuating and variable environment."[10] Because so much depended on chance events, efforts to classify plant communities too rigidly were inherently doomed. The controversies that raged in ecology over whose system was best could never really be settled. Gleason flatly rejected the idea that nature's progress could be known with the kind of confidence that Clements displayed. The faith that the future could be predicted was misplaced.

Gleason's thesis generated debate at the International Plant Congress held at Ithaca, New York, in 1926. Botanist George Nichols presented a lengthy review and critique of Gleason's article. Although he found much to agree with in Gleason's work, Nichols clearly felt uncomfortable with his emphasis on chance. The extreme notion that the plant community was an organism could be readily rejected, but giving up the classification of plant formations was tantamount to abandoning the hope of doing science, that is, of producing "order out of chaos." Gleason had gone too far in making associations contingent on chance events, on historical accidents, and few could follow him down that road. As Gleason recalled, "I know of no other paper in ecology which produced so much commotion among ecologists." Only one European botanist, Alwar Palmgren of Finland, accepted Gleason's ideas fully.[11]

Gleason wrote one follow-up article, an application of the individualistic concept to succession, published in 1927.[12] From then on, his scientific research was purely taxonomic, although he did reiterate his individualistic thesis in 1939.[13] The article of 1927 tackled Clements's theories more directly. Here Gleason identified the main problem confronting ecologists who sought a social role for their science: the social value of ecology depended on how well ecologists could predict vegetational change. In challenging the ability of ecologists either to predict the future or to interpret past history, Gleason was undercutting ecology's usefulness. Gleason had no soothing words to offer: given that multiple causes of ecological change were at work, operating on different scales, most predictions were flawed. As opposed to the certainty of Clements's theory, he denied that it was possible to know in which direction succession would proceed in all cases. When two plant associations were found next to each other, it was not possible to say positively which one was succeeding the other.

To illustrate how uncertain the direction of succession could be, Gleason drew attention to the importance of fire, including fires set by humans, in the prairie-forest landscapes he had studied. Fires favored prairie over forest, whereas in areas without fire, climate permitting, the forest would advance over the prairie. Gleason saw that both successional processes were occurring at the same time within the same region. On one side of a stream, he noted that the wind direction favored the spread of fires, which meant that prairie was taking over the forest. On the other side fires were less frequent and the forest was advancing over the prairie. Gleason did not distinguish one as more natural than the other: the fact that humans might set fires did not mean that one direction was "natural" and the other "disturbed." His point was that succession could go in either direction.

Gleason meant to cast doubt on the construction of elaborate successional series, which ecologists used to interpret the past and predict the future. He challenged the common belief that the phases of vegetation repeated themselves in different places and times. As a result, he completely undercut the confident value judgments that were at the core of Clements's theory: the idea that when succession appeared to go backward, something was wrong and needed fixing. The importance of judging nature—of deciding whether nature was taking its proper course and advocating appropriate policies to ensure that it did—had been woven tightly into the fabric of Clements's work. For Gleason, who pointed out how difficult it was to predict the course of history, such value judgments could not be made easily.

Gleason himself did not develop the implications of his critique for the usefulness of ecology or the ability of ecologists to act as nature's physician. But the issues he raised were relevant to current discussions about how ecological ideas could or should be translated into land-use policies. Gleason's approach to ecology was attractive to those who felt that the concept of the climax was problematic and that ecological theories were unproven hypotheses. If ecological theories of succession and the climax were called into question, then perhaps the idea that humans were such destructive forces had been overstated. James Malin, a historian at the University of Kansas, drew out the political implications of critiques like Gleason's. His reasons for challenging the concept of the climax community stemmed from his disagreement with the policy recommendations that emanated from those ecological theories.

Malin, a maverick thinker in the historical profession, pioneered an ecological approach to the writing of history in various studies of the grasslands beginning in the 1930s. He believed it was important to establish the physical and biological context of the region's history. To this end he drew extensively on ecology as well as on farming knowledge and various branches of agricultural science, never hesitating to condemn scientific ideas he thought were wrong. He objected strenuously to Clements's ecological assumptions and to his "doctrinaire" attitude. In *The Grassland of North America: Prolegomena to Its History*, a study privately printed in 1947, and in several articles published in the 1940s and 1950s, he critiqued Clements's ideas about the nature of the prairie climax, pointing out where the ecological literature diverged from Clements's work. He praised Gleason's "down-to-earth realism" and noted that his sharply contrasting view of nature (which was reprinted in 1939) had never been properly answered.[14]

Malin was not just interested in getting the facts right and challenging the inconsistencies he spotted in Clements's writing. He was also combating the policy implications of Clements's ideas, implications with which he strongly disagreed. Clements believed his ecological framework would help Americans to reorganize their society and perhaps even to suppress what he viewed as their destructive individualism. Farms, ranches, and communities also could be seen as complex organisms whose parts had to be coordinated to bring about the optimum results. Human communities had to cooperate to achieve an ecologically sound and properly coordinated policy of land use: "Grazing, irrigation, agriculture and urban communities are all bound together in such an intricate and vital bond that the interests of any one must be harmonized with the interests of all." The rancher who overgrazed or set fire to the chaparral unwittingly provoked flooding and erosion and jeopardized the water supply needed for urban areas and irrigation projects. These interdependencies needed to be understood: people needed to work together.[15]

Clements had suspected that because of "myopic individualism," such cooperation would not occur automatically. It had to be "evolved under the stimulus of outside forces," in other words under the impetus of government agencies that could determine and direct how land would be developed. Society was also a complex organism evolving in harmony with its environment: it was imperative therefore that the environment "be so fashioned as to call forth progress and not retrogression." Ultimately Clements envisioned coordination on all levels reaching up to the nation as a whole, producing a state where the ecological ideal of "wholeness," or of "organs working in unison within a great organism, prevails over partial and partisan viewpoints."[16]

Malin found this political vision abhorrent. He noted the historical links between organicism and totalitarianism and hinted that biologists were somewhat naive in allowing their ideas to be translated into social doctrines without being aware of where such ideas might lead. He warned against the danger of applying science too quickly to social problems, without adequate understanding of the facts, and of imposing any kind of determinism on the world. Clements's organicism implied that nature unfolded in a deterministic way and supported social doctrines that, as far as Malin was concerned, clearly threatened human freedom. The programs and bureaucratic structures of the New Deal, which Clements had welcomed as signaling a political climate favorable to the ecological principles he espoused, were for Malin the harbingers of totalitarianism.

Much of this discussion revolved around the problem of how to assess the impact of recent colonists in comparison to Indians or more ancient peoples who inhabited the continent long ago. Clements believed that Native Americans had an impact similar to that of Europeans but to a much lesser degree, and he had relatively little to say about it. He emphasized the difference between the state of nature and the disturbing tendencies of modern humans who were upsetting the balance between climate and climax. Clements's most influential student, John Ernest Weaver, the leading ecologist of the grasslands in the 1950s, similarly played down the role of Indians as ecological agents. Weaver dispensed with the organismic metaphor but retained Clements's point of view and his conclusions concerning the stages of succession, the climax, and the distinction between a natural state and the modern disturbed state.[17]

The problem, as Malin noted, lay in the assumption that a "state of nature" even existed after humans came on the scene. One could speak of a state of nature only prior to human occupancy of the land, he thought. Recent archaeological finds in the American Southwest suggested that human occupancy went back several thousand years earlier than had been believed, perhaps extending eight thousand years into the past. Furthermore, the idea that nature was normally balanced was, he thought, exaggerated. He saw no justification for the concept of a balance between climate and climax as Clements conceived it, arguing that intermittent disturbances, including even dust storms, were part of the normal course of events. He argued that it was incumbent on scientists to take account of the antiquity of humans and the possibility that there had been continual disturbances, due to "impersonal forces" as well as to human activities, for many thousands of years. The common conceit that settlers were breaking "virgin prairie" was, he thought, mistaken. At the very least, more research was needed with a longer-term historical perspective. "Beware," he warned, "of the egocentric present-mindedness of the dominant thought of the mid-twentieth century!" The modern viewpoint was arrogant in attempting to escape from history by "ignoring or ridiculing the past." Malin's critique included sharp criticism of how research was being directed and funded in the postwar years. He disparaged "technological research of a short-term character to achieve functional ends" and argued that creative fundamental thought was needed more than "quickie research."[18]

Malin's viewpoint was not in the least nostalgic. In disagreeing with the idea that human relations with the earth were always destructive, he was also attacking the militant conservationist view that urgent decisions needed to be made to forestall disaster. The doomsday scenario was overstated, he believed, and the solution was certainly not to return the land to a former "state of equilibrium" that might never have existed. Nor did he view Indians as having any superior wisdom that enabled them to live in harmony with nature. Rather, the modern challenge was to recognize and accept the new responsibilities that the changed circumstances of the atomic age had brought. In the larger context of the development of an infrastructure for defense in the cold war, the grasslands took on new significance in the North American continent. Being in the geographic center of the continent, the grassland "contained the nerve centers of the military communication systems that defend or strike in its behalf." "Instead of a return to the simplicity of a grazing country," he argued, "the challenges of atomic power indicate a further incorporation into the complex network of areal and cultural interdependence."[19] Malin looked toward that new era, not backward in nostalgia to a lost past.

The comment about the need to develop new ideas for the modern age was made at a symposium titled "Man's Role in Changing the Face of the Earth" that was held in 1955 at Princeton, New Jersey. Dedicated to George Perkins Marsh, the symposium brought together some of the leading thinkers in the natural and social sciences in order to explore all the ways in which humans affected their own evolution and altered their physical and biological environment. The symposium was largely the work of Carl Ortwin Sauer, a human geographer at the University of California, Berkeley. Sauer had long pondered the kinds of questions that Marsh had raised a century earlier and had encouraged an interdisciplinary approach to human ecology that drew heavily on anthropology and history. Like Malin, he believed that a long historical perspective was essential to understanding and evaluating human prospects. He developed a set of hypothetical scenarios of human ecological impact that depicted humans as significant agents of change even in the distant past.

Human ecology at this time was a poorly defined field without disciplinary coherence. Ellsworth Huntington, as we saw in chapter 5, envisioned ecology as embracing subjects in medicine and public health, but his ideas did not transform the Ecological Society of America as he had hoped. Ecology had developed largely as a biological subject, in which plants and animals were studied but humans were ignored. Nonetheless, leading ecologists recognized the importance of "humanizing" ecology by taking humans—especially their actions—into account as much as any other kind of organism. Stephen Alfred Forbes, an animal ecologist at the University of Illinois, made this point in his presidential address to the Ecological Society of America in 1921. He was trying to counter the perception that applied science was not part of ecology, a position that struck him as illogical. Charles Adams made a similar argument in 1935 and admitted that the borderland between general ecology and human ecology had been neglected.[20]

Despite this lack of systematic connection between ecology and human ecology and the absence of disciplinary coherence within human ecology, the problems at the core of human ecology, problems relating to how human history had been shaped by interaction with the environment, were increasingly recognized in midcentury as needing a fresh approach. Carl Sauer was one of the leading revisionists in this period. Like Malin, he challenged the crude environmental determinism popular at the time as well as the assumption that one could readily separate modern human impacts from those of earlier peoples. He applied history, geography, ecology, and anthropology to reconstructions of the human past in North America, pushing the story of human disturbance back thousands of years to the end of the last glacial period. His radical hypothesis was that human culture in America, for all its accomplishments, was marked by a history of dramatic impact even at its earliest stages. In this respect he differed from Malin, for he never shied away from admitting how devastating the human impact on nature had been, both in the present and in the distant past. He also differed in having a strong feeling for what was being lost in the rapid onward march of the modern age.

Sauer was born in Missouri in 1889 but was schooled in Germany before attending Central Wesleyan College in Missouri, a bilingual liberal arts college founded by German refugees of the failed revolutions of 1848. At the time Sauer graduated in 1908, it was still a strongly Germanic institution. Sauer's father, a composer and musician who emigrated to America from Germany in 1865, was head of the college's Department of Music. Sauer developed an interest in geography and obtained his Ph.D. from the University of Chicago, studying under renowned geographer Rollin D. Salisbury, a much loved and inspiring teacher, and learning plant ecology from Henry Cowles, who had also been a student of Salisbury. Sauer's dissertation was a study of the Ozark highlands of his native Missouri; he also did fieldwork in Illinois and Kentucky. Sauer joined the faculty at the University of Michigan in 1915, moving up to the rank of professor, and left to join the Department of Geography of the University of California at Berkeley in 1923. At Berkeley he developed a friendship and collaborative relationship with anthropologist Alfred Kroeber and at the same time redirected his fieldwork to the American Southwest and Mexico. He retired in 1957 but continued to teach courses and remained active until his death in 1975.[21]

Sauer's interest in historical geography was shaped by his reading of the German geographic literature, which was then not well known in the United States. He was influenced especially by the work of Eduard Hahn and Friedrich Ratzel.[22] Hahn provided an important corrective to the idea, propounded by Frederick Jackson Turner among others, that cultural development followed a general, predictable course. Hahn was opposed to deductive approaches to cultural evolution, arguing that the specific historical steps of each cultural change had to be investigated.

Ratzel was a remarkably prolific naturalist and journalist who turned to geography and ethnology and wrote extensively on human geography. As a journalist he wrote a series of sketches of America based on observations made during travels in the United States in 1873.[23] He later wrote an exhaustive handbook of the United States that included an analysis of the natural conditions of cultural development. Ratzel's general theory of human development, *Anthropogeographie*, published in two volumes in 1882 and 1891, was popularized in the early twentieth century by geographer Ellen Churchill Semple, who depicted him as an environmental determinist.[24] Sauer thought she had misinterpreted Ratzel by overlooking the ideas in the second volume, which dealt with how cultural traits were diffused from one place to another.

The question was why some cultures innovated and others continued in their old ways. Ellsworth Huntington had theorized that environment and climate determined civilization, and it appeared that Ratzel's ideas were of a piece with the environmentalism that was popular in the early century but was increasingly discredited by midcentury. Sauer perceived that Ratzel was more subtle than that. The key was to examine how traits were transmitted from one culture to another; it was not the case that similar environments produced similar cultures. Geographers like Ratzel who had been maligned as environmentalists actually were quite critical both of the idea that cultural evolution followed a fixed series of stages and of the notion that the environment determined culture.

Sauer realized that ecology also harbored some of the same deterministic ideas that he wanted to challenge in the study of human culture. He had at first accepted the idea that climax communities were stable formations controlled by climate. His fieldwork showed him, however, that vegetation cover depended more on the nature of the landscape, for example the roughness of the terrain, than on the limits imposed by climate. Stated simply, grasslands occupied plains, whereas broken terrain was often wooded. Part of the reason for the prevalence of grasslands on flat plains, he realized, was that fires swept easily over such areas. Like Gleason before him, Sauer began to pay more attention to how the shape of the land affected its vegetation and how fire or the suppression of fire could affect plant communities.[25]

He concluded that plant ecologists had erred in not taking account of the role of animals, and especially of humans, in shaping plant associations. As a result their idea of nature was too static. Sauer emphasized that the environment was continuously developing, but not along any preordained path. Close attention to details of time and place would show that change was constant and that neither nature nor mankind followed any general law of progress. In place of the idea of stable, indefinitely reproducing climax formations, he drew attention to the constant deformations of the landscape that had occurred over millennia. The fact that fire was such an important cause of deformation meant that human disturbance had a long history, measured at least in tens of thousands of years, if not in hundreds of thousands.

Questions about the role of fire in maintaining ecological communities were becoming more important in applied ecology in the midcentury. The Forest Service had adopted a policy of suppressing fire, but protection against fire set in motion successional processes that were reducing grassland. If the grazing lands had been modified by fire in the past, and if the modern challenge was to work out a modus vivendi for humans in a changing environment, then the use of fire might still be important. Sauer's point was to take "the longest possible view" of ecological activity by humans.[26]

His method was historical more than it was scientific, as he acknowledged. He preferred the term *natural history* to *ecology* and the term *cultural history* to sociology or social science, in both cases emphasizing history over the scientific approach that ignored history. Combining his interests in geography and cultural development with his strong historical bent, Sauer discerned that ecological questions concerning the balance of nature and its destruction by modern humans could benefit from a longer historical perspective. Specifically, analysis of how humans interacted with the environment required recognition of how they developed new tools and how they might have used these tools. Humans not only changed their environment; they were also agents affecting plant and animal evolution. Given the sustained selective pressure that humans had exerted on nature, it did not make sense to speak of a "natural balance" without humans. Sauer concluded that undisturbed or natural vegetation might not exist in many places. He developed these ideas in a series of books and articles published from the 1930s through the 1960s.

Sauer's approach to human cultural evolution depended a lot on imaginative reconstruction of how early humans might have acted, based on present-day knowledge of human behavior, biology, and social patterns. He put himself back into the time of Ice Age humans and tried to imagine how they would have behaved and the possible consequences of that behavior. Since there was no record of human behavior that long ago, the stories he created were hypotheses, intended to stimulate thought and further study. One notable feature of his writing was that he placed women at the forefront of cultural evolution because of their role as tenders of hearth and family. The long dependency of children on their mothers would have encouraged social grouping into clusters of households. Sauer did not visualize early humans as wandering and homeless, but as relatively sedentary, storing provisions where possible and learning to use tools in various ways to secure food and otherwise improve on their conditions of life. As populations spread outward, he saw them as venturesome and creative, trying out their skills in new environments. Man (including woman) would have increasingly "imposed himself on his animal competitors and impressed his mark on the lands he inhabited."[27]

His argument depended on important new evidence regarding the antiquity of humans in America. This had been a highly controversial, emotionally charged subject of debate for several decades. The weight of authority had favored the recent advent of people in America, perhaps as recent as three thousand years ago. These beliefs were dealt a major blow by studying archaeological materials found in the 1920s and 1930s in New Mexico, Colorado, and Nebraska. Using radiocarbon dating methods developed in the 1940s and 1950s, along with other dating techniques, scientists estimated these finds to be eight thousand to ten thousand years old. Moreover, there was continuing controversy over whether these represented the earliest human cultures in North America, or whether human occupation could be pushed back even further.[28] Sauer himself believed that humans might have lived in the New World for much longer than ten thousand years, maybe as long as forty to sixty thousand years. If that was true, there could have been an extremely long history of human disturbance of landscape and vegetation.

Convinced of the long existence of humans in the Americas, Sauer advanced a provocative hypothesis in 1944 concerning its impact.[29] Scientists knew that large mammals such as the mastodon, the giant beaver, the horse, the giant bison, the tapir, and the giant elk had once ranged in the interior of North America, such that by the mid-Pleistocene North America had the most diverse large-game fauna in its history. Yet all these large mammals had become extinct, apparently not long after the arrival of humans on the scene, and some of their remains were found in association with human sites. Explanations of the extinctions included disease, climate change, human impact, and even the idea that species, like individuals, had a normal "life span" and died out at the end of their allotted time.

Sauer agreed that humans were probably involved in these extinctions.[30] The question was how a small group of weak men could have destroyed so much with the few stone tools that had been uncovered. Sauer argued that indeed they could not have done so, but they possessed in addition a weapon of mass destruction: fire. By driving game with fire, especially on the interior plains with their inflammable ground cover, early humans could destroy much larger numbers of animals than they needed for survival. Apart from directly killing animals, such fires would set in motion many ecological changes that could bring about mass extinctions. With fire as a weapon, humans could have had a devastating impact on animal populations, Sauer believed. His portrait of Ice Age humans paralleled George Perkins Marsh's description of the American conquest of nature in the nineteenth century, the main difference being the time span involved in bringing about these extinctions. And perhaps in his mind were other comparisons to the destruction of the current war.

Sauer looked to the deep past to understand the present and assess the future. To do this he needed to put people back into the story of the history of the environment, so that it became almost impossible to try to conceive of a "state of nature" without human beings. But his view of ancient humans was not wholly negative; he also appreciated the accomplishments of people, their ability to adjust to new environments, to innovate, and to create. In fact Sauer had great respect for the achievements of past cultures, for the diverse ways in which people had learned to live on the land harmoniously. He viewed the economic and technological changes that were threatening those traditional ways of life with dismay. He regretted the passing of the family farm in favor of large-scale mechanized agriculture. He felt "amazed and bewildered" by the fast-paced development of technology after the First World War, a pace that continued to accelerate after the Second World War. He valued the diversity of human cultures and bemoaned the loss of native wisdom as countries like the United States introduced mechanization to other nations, with all too little respect for what appeared to be their backward ways. He did not share the modern faith in continued progress through technological advance.[31] To him modernization was an alienating force.

To probe the nature of these rapid changes and their cultural impact, Sauer sought the expertise of scientists and humanists, bringing them together to reflect and learn from each other. The conference he organized in 1955, "Man's Role in Changing the Face of the Earth," was an exploration of these concerns about modernization and its impact. It paid homage to the work of George Perkins Marsh and became known as the "Marsh festival." The conference was broad in its disciplinary coverage, a virtual Who's Who of creative thinkers in the natural sciences, social sciences, and humanities.

Most participants were from the United States, though a few were from Britain, Europe, India, and Africa. They included biologists, human ecologists, geographers, scholars of urban development, economists, sociologists, anthropologists, engineers, and humanists. Presentations included overviews of human history from ancient times; analysis of how humans had shaped their environment, including changes in climate, soil, and impact on plants and animals; analysis of the problem of human wastes, including radioactive wastes; consideration of urban and industrial growth and its impacts; and reflections on the looming problems of overpopulation and exploitation of resources.[32]

These conversations about the past and the future of humankind embraced a host of ecological and evolutionary issues. They ranged across many biological topics, including food production and the domestication of plants and animals, control of population as a biological as well as social problem, the ecology of disease, and the ecological impact of humans. Several leading ecologists and evolutionary biologists participated in the conference: Marston Bates from the University of Michigan, Frank G. Egler from the American Museum of Natural History in New York, F. Fraser Darling from the University of Edinburgh, Paul B. Sears from Yale University, John T. Curtis from the University of Wisconsin, and Edgar Anderson, a geneticist and director of the Missouri Botanical Garden.

Several of the papers tackled the problems of human ecology by giving more emphasis to the long-term impact of human populations on nature, on the one hand, and encouraging ecological analysis of human-dominated environments, on the other. John Curtis, a plant ecologist, described how completely humans had transformed the grasslands and forests of the Midwest and in passing remarked on the impact this activity had on the study of ecology.[33] One cause of human disturbance—the railroad—was also the chief means by which the prairie was being preserved, as Clements had also realized. The fenced right-of-way of the railroad constituted the only strip of original grassland available for ecologists to study. Under Curtis's leadership the Wisconsin school developed in the 1950s the "continuum" theory of the ecological community, which drew on Gleason's individualistic concept.[34] It was one of the main challenges to Clements's theory. The main idea was that there were gradual or continuous changes of vegetation over a given region and that vegetational units did not have discrete boundaries.

Nonetheless, Curtis did believe in the existence of climax communities existing in dynamic steady states, although he saw them becoming rarer as human disturbance spread. He distinguished between what could be called nonexploited communities and those that were heavily influenced by humans. He believed that ecologists could assess human impact only by studying nonexploited communities and in particular by analyzing how such communities maintained organization. In many respects his concept of ecology was not so much a direct challenge to Clements as a shift in focus, with more emphasis on studies of the flow of energy and materials, the hallmarks of the ecosystem approach that is discussed in chapter 7.[35] His conference address asserted that climax communities like the tropical rain forest or the climax deciduous forest were "incomprehensibly improbable" phenomena, since there were many forces at work reducing these high levels of stability.[36] Humans were among the most powerful destabilizers. It was precisely the human impact that had reduced the number of major communities and had blurred the boundaries between them by altering the distribution of species. Humans had decreased the complexity and level of organization of nature and had introduced randomness into the system.

Other speakers took up related themes concerning humans as ecological agents. Omer C. Stewart, an anthropologist at the University of Colorado, developed the thesis that humans had long controlled their landscape through the use of fire; he argued that burning by primitive peoples was important in determining the condition of the North American grasslands. Lewis Mumford, professor of city planning at the University of Pennsylvania, campaigned for more research on the natural history of urbanization. The field he imagined included not only studies of modern cities but also historical analysis going back to Neolithic times and even earlier to cave dwellers. It would seek alternatives to the unsound practices that were degrading both urban and rural environments and try to regain an ecological balance between city and country. Edgar Anderson showed how profoundly humans had affected evolutionary changes in plants and animals since the Pleistocene. Vegetation complexes that appeared to be native and "natural" were in fact relatively recent results of human disturbance. He chided biologists for wanting to study plant and animal communities on mountaintops and in jungles rather than on doorsteps and in gardens, where human influence could be directly observed.[37]

These scientific discussions also had a philosophical side, for the way nature was perceived and how one should consider ideas such as the "balance of nature" were constantly highlighted and applied to the larger issues debated at the conference. In discussion, ecologists pointed out that climax communities were generally understood to be in equilibrium, a steady state, or in balance. But older ideas of succession proceeding to a stable climax were increasingly challenged: there was no conclusive evidence to support the idea, for example, that the North American vegetation was in a stable climax when white men entered. As opposed to the idea of stable climax, the newer ideas stressed that change was universal. This did not suggest that the concept of the "balance of nature" was dead, however. Far from it: the new idea was to keep in mind the dynamic processes that underlay such balances.

Nature existed not in a static equilibrium but in a constantly shifting one, or, following another way of thinking, a succession of short-term equilibriums. The more humans were placed in the picture, the more likely it was that the equilibrium would shift or that a disequilibrium would result. Change and equilibrium were the two ends of the seesaw of debate. Change was constant, nature was in flux, yet not all change was desirable. There were also processes that maintained equilibrium, which humans needed to understand so as to regain balanced states that had been lost.[38]

How did one deal with ideas of change and equilibrium at the same time? In part the problem was the point of view taken. One could hold time constant, examine the way the parts of nature fitted together, and be impressed by the appearance of ecological balance. But the element of time brought constant change into the picture, change that destabilized the world and left one uncertain how it was really meant to be. As Marston Bates expressed it in summary: "We are impressed everywhere in nature, when we look in a given cross-section of time, with the balances, the buffer mechanisms, the cycles, that maintain equilibriums. But, when we look longitudinally in time, the changes, whether apparently random or directed, impress us the most; and the system of nature appears to be in disequilibrium rather than in equilibrium."[39] Perhaps, he concluded, the equilibriums were illusory; perhaps we saw equilibrium because we wanted to see it. The idea of equilibrium possibly fitted "more neatly into the way the human mind works." But humans also wanted to survive; they wanted to preserve certain conditions that they felt were natural and desirable, and discussions of survival kept returning to the question of maintaining equilibrium. Bates realized that the words used to discuss these large issues—words like *equilibrium* and *change*, *culture* and *environment*—were inadequate, possibly false and misleading. But limited by these inadequate tools, humans had to figure out creative ways to think about their relationship to the environment, to get beyond the dichotomy of individual and environment and, ultimately, the false dichotomy of the sciences versus the humanities.

Repeatedly throughout the conference speakers said that people, especially in the United States, had an exaggerated faith in progress, in the idea that growth could continue unabated because somehow human ingenuity would come to the rescue in the form of new applications of science and technology. At the forefront of discussion was the acute awareness of how seriously humans were interfering with nature, whether for good or ill. The wholesale destruction of insects and the long-term use of antibiotics implied levels of disturbance whose effects were only dimly understood. Environmental engineering, along with higher standards of public health, would have a major impact on patterns of disease in populations. Increased use of atomic energy prompted questions about risks of contamination of air, soil, and water. The possibility of harnessing atomic energy also opened the way to the spread of technology and, as a consequence, raised the specter of an even higher rate of population growth. Could humans adjust to these rapid changes? What kind of world did they want? Would people be able to withstand the stresses of the environments they were creating? As Malin remarked, the goal was not to recreate a lost environment of the past, but to meet the responsibilities and the new demands of the atomic age. The issues were ethical and philosophical as much as scientific, technical, and economic.[40]

Humans were already becoming deeply alienated from the natural environment, as illustrated by the sterile uniformity of fast-spreading suburbia; the unattractive, polluted, and impoverished towns that clustered around the industries they served; and the mindless consumerism of postwar society. More and more, people were living in an artificial world, paved-over, climate-controlled, with animals bred to grotesque degrees so that they might better serve humans. Human relationships with the environment lacked stability. What might be involved in regaining stability? Perhaps a higher level of social organization, perhaps a new ethic, perhaps some revolutionary technological discovery or transformation that could not be predicted. The participants went back and forth on these issues, trying to steer a course between the extreme pessimism of the conservationist and the overconfidence of the technocrat. The two extremes were captured in a bit of doggerel by economist Kenneth Boulding in which the conservationist lamented:

> The world is finite, resources are scarce,
> Things are bad and will be worse . . .
> The evolutionary plan
> Went astray by evolving Man.

The technologist replied:

> Man's potential is quite terrific,
> You can't go back to the Neolithic . . .
> Man's a nuisance, Man's a crackpot,
> But only Man can hit the jackpot.[41]

The conference revealed that even within science there was growing disaffection with the belief that knowledge was power. Marston Bates observed that although the world needed science and its technological applications for survival, the self-important attitudes commonly heard among scientists needed to be balanced by a humanistic outlook. In modern times the sciences and the humanities formed a false dichotomy, one that Charles Percy Snow was soon to describe, in a now-classic work, as "two cultures." These two cultures had to understand their connection to each other; they had to be able to speak across these boundaries.[42]

Lewis Mumford followed up on this theme and closed the conference on a humanistic note. He also was critical of scientific tendencies to view the main problem as involving human control over nature. As he pointed out, the deeper issue was how humans exercised control over themselves, how they made decisions, how they established priorities for the applications of technology, and whether they gave sufficient thought to the consequences before making those applications. Alluding to Aldous Huxley's *Brave New World,* he pointed out that if the goal of our culture were to produce a "uniform type of man, reproducing at a uniform rate, in a uniform environment, kept at a constant temperature, pressure, and humidity, living a uniform life, without internal change or choice, from incubator to incinerator," then many of the problems concerning our relation to the environment would disappear. However, no one would prefer such a life. Mumford deplored the mechanization of modern life and its tendency toward uniformity, which meant a loss of cultural diversity, individuality, and environmental richness.[43]

Preserving the richness of the environment and admiring and preserving natural diversity were inseparable from the goal of preserving human individuality and bringing out the best of human life. Not to value the richness of nature was to devalue the human spirit. The challenge was to be creative in reinventing the self for the new world, to evolve a culture that avoided dehumanizing standardization and encouraged individualism. Carl Sauer also felt that the "brave new world" of a "faceless, mindless, countless multitude managed from the cradle to the grave by a brilliant elite of madmen obsessed with accelerating technological progress" was becoming all too real in the United States. In common with other critics of modernism like Huxley, he believed there was an urgent need to question the technological-economic system in the making.[44]

How persuasive such a critique of modern culture would be depended a great deal on how well one could document the human impact on the environment. The questions about human impact were, however, still obscure. By midcentury there were two dominant points of view. One emphasized the determining role of climate in shaping land and life-forms, with the added idea that climatic changes were cyclical, perhaps controlled by distant events such as sunspot cycles. Clements and Huntington exemplified this viewpoint within ecology. On the other side was the cultural explanation that gave a greater role to human agency in altering the landscape. Complicating both perspectives was the idea, expressed in its most extreme form in Gleason's work, that random events and processes were at work, operating in an irregular way, creating many variations on the local level depending on circumstances.

Competing hypotheses had to be tested. One way to test a hypothesis was by experiment, perhaps by destroying or otherwise disturbing the organisms in a given area and seeing how the area became repopulated. But such an experiment would not necessarily answer the question of whether current environmental states were the result of past human activity or of climatic change. That question was more historical in nature, and answering it required the existence of a historical record. One such record, albeit an incomplete one, could be found for the American Southwest and northern Mexico, the region studied by the scientists at the Desert Botanical Laboratory, which was the nation's first laboratory devoted to basic research in ecology. A long-term investigation begun there made it possible, decades later, to formulate some answers to the questions of ecology and human ecology discussed here.

The study in question was by James R. Hastings of the University of Arizona and Raymond M. Turner of the U.S. Geological Survey.[45] Published in 1965, *The Changing Mile* dealt with the ecological changes that had occurred in the desert, the desert grassland, and the oak woodland of the Southwest: these landscapes made up the first mile above sea level. The authors were especially interested in the changes that had occurred since 1880. Their work was motivated by the perception that the region was becoming drier. The abundance of species was changing in various ways: the giant saguaro cactus was declining, while mesquite thickets were invading many areas and taking over the grassland. Many of these changes were already evident in the early twentieth century and were noted by Forrest Shreve, who concluded in 1910 that the saguaro was failing to maintain itself. With changes in the vegetation cover came problems of erosion. Hastings and Turner considered various causes, including the role of fire, climate, and effects of human habitation such as cattle-grazing. The question was to what extent humans were responsible for these dramatic changes.

To assess the condition of the land in the 1880s, Hastings and Turner turned to much older historical records and to recent anthropological research in the area, including work by Sauer. In keeping with many of the arguments in the Marsh symposium papers, they rejected the idea that modern Americans had disrupted a "natural" order. They viewed the land as a "fluid environment shifting with the centuries under the impact of a succession of cultures, each differing somewhat from the preceding in its relation to the life around it."[46] Well before 1880, human disturbance in the area was significant. Sifting through various archaeological records and written accounts of travelers and settlers in the region, they considered the relative impact of Indian hunting, gathering, and agriculture and Spanish and Mexican mining and cattle-grazing. For the more recent history, from the late nineteenth century to the 1960s, they relied on the extensive photographic record. Some of the photographs had been taken by Daniel MacDougal during his explorations from the Desert Laboratory. By visiting the same sites and rephotographing them, Hastings and Turner created a series of "then and now" views that allowed them to pinpoint exactly what changes had taken place. Often they could identify from the early photographs individual plants that were still there six decades later. The Desert Laboratory scientists' extensive study of the region, especially their photographs, provided the exact historical record that was needed to document the changes in the land.

On the debated issue of fire and its impact, Hastings and Turner concluded that fires, although they might well control the spread of shrubs, did not occur often in the desert grassland and were not responsible for the maintenance of the grassland over a wide area. Of the several hypotheses they started with, by the end of the study they had eliminated all but two important causes of change: climate and cattle. The impact of grazing animals was well known to the first scientists at the Desert Laboratory, who had to erect fences to restore the original vegetation cover. That climate change was important was evident in such sites as MacDougal Crater and MacDougal Pass, areas undisturbed or only slightly influenced by humans but showing changes in plant cover that clearly pointed to drier conditions. But cattle-grazing was undeniably a strong factor in other areas. Hastings and Turner concluded that a combination of climatic change and overgrazing together had acted to produce the changes of the past century.

By putting these two causes together, they were able to draw a further conclusion that was important for any attempt to remedy the degradation of the land. Whereas a climatic cause acting alone might suggest a cyclical process and the possibility that the land could return to its earlier state when the climate became wetter and cooler, Hastings and Turner thought the added impact of overgrazing meant that they were not witnessing a cyclical process. Human impact had set in process a permanent evolution of the landscape, one not yet completed and whose endpoint was unclear. Nature's cycle, if such existed, had been broken by human action.

This historical analysis eschewed generalizations that went beyond the evidence. The authors stated that their observations applied only to the areas studied. Generalizations derived from other grasslands were not necessarily applicable to the desert grassland. Each area was different. They realized how difficult it was to determine the type of cause behind observed changes in species distribution: "To what extent is this ebb and flow of plant life part of the normal rhythm of the desert region? To what extent is it a unique event?"[47] They knew that their conclusions had to be considered tentative, consistent only with the evidence they had to date and open to correction with new evidence. The subject needed more study. (An updated study, *The Changing Mile Revisited*, was published in 2003).[48]

This was a work of ecology conceived as scientific natural history. Hastings and Turner's work included experiments, which were needed to study the adaptive limits of species, but the experimental results were cast in the light of history. Their study represented an approach and a genre that shared the vision of people like Gleason, Malin, and Sauer, even while it disagreed with specific conclusions they had drawn. It was made possible by the exceptional historical record that was started when people like MacDougal and others carried their enthusiasm for science into the southwestern desert. The book was filled with pictures of plants and landscapes: no models, no diagrams, no equations. It represented an approach that was soon to be challenged by new scientific enthusiasms: fascinations with mathematics, modeling, and generalizations of the most sweeping kind, coupled with a disdain for natural history.

CHAPTER SEVEN

A Subversive Science?

Ecosystem ecology was a thoroughly modern response to the challenges of the postwar period, an effort to convert the "soft" science of ecology into a "hard" science and show that the subject could command intellectual respect. Whereas Carl Sauer had preferred to describe his approach as "natural history" rather than "ecology," ecologists now were trying to overcome the stigma of being considered glorified bird watchers and bug collectors. The last thing they wanted was to be described as natural historians. Ecosystem ecology sought to move beyond general conceptions of ecological processes by adding exact measurements, experiments, and tests of hypotheses. It welcomed applications of new techniques in applied mathematics, looked to physics to provide basic principles for ecology, and pushed ecology into the emerging computer age with enthusiasm. It was cautiously optimistic about the possibilities of environmental engineering. And it too sought a way to incorporate human activity into the analysis of ecological processes, both on the local level and in the biosphere.

One of the objectives of the ecosystem viewpoint was to gain a better assessment of how human actions were affecting the planet. A related goal was to make a place for ecology in the nuclear age. Ecosystem ecology was nurtured in the bosom of the atomic age, funded largely by government agencies that evolved from the Manhattan Project. Ecologists had to find a purpose and a place in a world of rapid growth, where the idea that the main purpose of science was to control nature still dominated. The ecosystem concept provided a way of responding to this challenge as ecologists adapted to the postwar environment and reassessed their social role.

By the mid-1960s ecology developed the appearance of a subversive subject whose role was to criticize the system of which it was a part. Ecologists had to deal with a contradiction. On the one hand they wanted to contribute their expertise to society, to have recognized and valued social roles, which meant for the most part guiding development so that it was more efficient in the use of resources and less harmful to health. On the other hand ecologists provided knowledge that undermined the ethos of growth. The role and usefulness of ecology became a subject of perennial discussion as ecology expanded during the late twentieth century.

At the symposium "Man's Role in Changing the Face of the Earth" in 1955, Marston Bates observed that war had been given no formal place on the agenda. Yet war was one of the most dramatic ways in which humans had changed the landscape. Bates noted that the subject of war had hung over their minds throughout the meeting, "as it hangs over the minds of all men in the Western world these days."[1]

Americans were learning to live with the Bomb. In 1949 Ralph E. Lapp, a nuclear physicist who later became a prominent writer on nuclear science and its social impact, aimed to give Americans a clear view of what they might expect in a nuclear war, under the assumption that any major future war would involve nuclear weapons. He believed that the country could survive a nuclear war if it prepared for such war, but preparation meant nothing short of a major reorganization of American society. Lapp devoted a portion of his book to the subject of "atomic bomb geography," or the problem of redistributing residential and industrial centers so that they would make less attractive targets. Cities like New York and Chicago were, he thought, "cities of the past": with their concentrated populations and skyscrapers they were highly vulnerable targets. He proposed that a program be devised to decentralize cities over the course of a decade or two. Among his ideas was the "doughnut city," built as a series of concentric rings containing residences and businesses, with an airport located in the otherwise empty center.[2] Although one reviewer thought these proposals were highly unrealistic, not to mention fantastically expensive, Lapp's book was generally praised as a sober, scientific assessment of how to adjust to the atomic age.[3]

While scientists worked on the development of more powerful nuclear weapons, which rendered Lapp's ideas of adjustment to nuclear war obsolete, they also explored new peaceful applications of atomic energy. The Atomic Energy Act of 1946 authorized the newly formed Atomic Energy Commission (AEC) to promote scientific research on radioactive materials for medical, biological, health, and military purposes.[4] In 1947 the AEC appointed a permanent advisory committee for biology and medicine, consisting of some of the top biologists and medical scientists in the country. Government laboratories that had been involved in the Manhattan Project became research centers for both military and peaceful uses of atomic energy. Initial research focused on the health consequences of exposure to radioactivity, including possible genetic effects. Handling of radioactive wastes and the health hazards of working with radioactive materials were major concerns.

The benefits of radioactivity as a tool for science were touted with enthusiasm. Radioactive tracers transformed the study of physiological processes. Agricultural scientists envisioned using radioisotopes for treatment of plant diseases, studies of nutrient uptake in plants, and even such things as stimulating growth through radioactivity or producing gene mutations to improve breeds. Very little of this research was directed toward ecological analysis initially, although scientists did study the level of radioactivity in sands, plants, and animals in the vicinity of nuclear tests in the United States and in the South Pacific. Studies noted that radioactivity could accumulate through the food chain but did not consider the radioactivity from isolated test sites to be hazardous over the long term.[5]

Discussions about the effects of nuclear testing often emphasized that some level of radioactivity was natural, a normal part of the environment. Nonetheless, the level of fallout from bomb-testing led scientists to conclude that when it came to radioactivity, the "natural world" no longer existed. Frederick P. Cowan, head of the health physics division at Brookhaven National Laboratory, noted in 1952 that continuing atomic tests meant that the days of undisturbed natural background radiation were gone. But he thought the increased level of radioactivity was "hardly of ecological significance except in certain small areas."[6] In the days when people commonly used x-ray shoe-fitting machines and wore watches with radium-painted dials, the spread of radioactive substances through the biosphere did not cause much concern.

This perception changed with the development of thermonuclear weapons.[7] The United States tested the new hydrogen bomb at Enewetak Atoll in the Marshall Islands in late 1952. The test vaporized an island and sent 80 million tons of solid material into the air and around the world. A new series of bomb tests began at Bikini Atoll in March 1954. The first test was exceptionally and unexpectedly powerful: it created a crater 250 feet deep and 6,500 feet in diameter. Fallout spread over a large area as a result of winds at high altitudes, exposing hundreds of people in the Marshall Islands to significant levels of radiation. Fallout landing on a Japanese fishing boat eighty-two nautical miles to the east caused the crew to become sick and provoked protests from the Japanese.[8] Tests continued until 1958, when the United States, the United Kingdom, and the Soviet Union agreed to a temporary suspension. The Partial Test Ban Treaty, which put an end to atmospheric tests, was enacted in 1963.

In the United States, weather reports alerted people to the location of drifting fallout after tests. On March 11, 1955, newspapers reported a radioactive sea of clouds 1,000 miles long and 200 miles wide hanging "harmlessly" over the eastern United States, the result of the "Big Shot" atomic explosion in Nevada four days earlier. The explosion was so large that it was seen from Mexico to the Canadian border, and more tests were scheduled for later that month.[9] As the cloud drifted eastward, citizens in Baltimore, in a state of near panic, phoned the Weather Bureau to report a downpour of reddish, pea-sized particles during a rainstorm. The rain dropped a greasy, gritty film that had to be scrubbed off. People feared they were being bombarded with radioactive dirt from the nuclear test. The Weather Bureau reassured the people of Baltimore by explaining that the source of the dirt was not in fact the bomb test, but a dust storm in the Texas panhandle, which was experiencing severe drought. The bureau forecast a "beautiful spring-like day" after the storm.[10]

Such was the measure of how the world had changed, when a major dust storm like those of the "dirty thirties" could appear insignificant beside the new hazards of the nuclear age. As prominent scientists such as Linus Pauling in the United States and Frederick Soddy in England called for a halt to further tests, the Atomic Energy Commission sought to calm an increasingly skittish public. Willard F. Libby, a member of the AEC, assured Americans that apart from the areas close to a bomb blast, radioactive fallout posed no health hazard. Studies of fallout conducted under the name "Project Sunshine" showed, he argued, that worldwide health hazards of bomb testing were "insignificant."[11] At the 1955 Princeton conference on man's role in changing the face of the earth, John C. Bugher, director of the AEC's Division of Biology and Medicine, drew on the same studies and played down the biological significance of radiation, except in the event of general atomic war or a serious reactor accident.[12] Instead he stressed the scientific knowledge that could be gained by using radioactive tracers to study the oceans and atmosphere, knowledge that could help advance nuclear power production.

The potential for the development and application of nuclear power, especially if the process of fusion could be controlled, fostered hope that the world's hunger for energy could be satisfied. Although the nuclear age brought fear of annihilation, it also held out the vision of societies transformed by almost limitless energy and liberated from dependence on the fossil fuels that choked industrial regions with smog. In 1954 a revision of the Atomic Energy Act allowed the AEC to cooperate with industry in the development of nuclear power reactors. In 1955 the United Nations, prompted by President Dwight D. Eisenhower, sponsored a conference to explore the development of peaceful uses of atomic energy and to look for ways to foster international scientific cooperation for the betterment of mankind.[13]

As plant ecologist John N. Wolfe observed, the postwar years were a time of "monstrous growth" in science and technology, which created a dizzying sense of progress in the control of nature. The chemical control of disease-transmitting organisms promised to solve world health problems. Industrialization of food production promised to emancipate people from the limits imposed by the amount of land available for agriculture. Air-conditioning would give cities a climate independent of nature, and atomic power would free societies from dependence on fossil fuels. What is more, Wolfe imagined that deciphering the genetic code would enable people to create plants and animals and even races of people to their own liking.[14]

Ecosystem ecology emerged in this postwar context of deep anxieties and high hopes. The new concept of the ecosystem, the way the term was defined and applied in ecology, and the perception of ecology as a science with a new mission all reflected the changing priorities of American science in the atomic age. Historians and scientists have discussed many features of this postwar development in detail. I will summarize this literature briefly, focusing on two issues: how the ecosystem was depicted as an object of analysis and how ecosystem ecology provided a forum for continuing discussion of human impact on the natural world.

To set the stage, we'll return to the prewar years and the critiques of the complex-organism concept, and then bring ecology forward into the atomic age.

Arthur Tansley coined the word *ecosystem* in 1935. He had long questioned the aptness of the complex-organism concept. By the 1930s he felt it was being turned into rigid orthodoxy, so much so that balanced analysis of alternative views was not even attempted. Tansley was responding specifically to the use of organismic concepts by the South African ecologist John Phillips, who was enamored of the holistic philosophy of General Jan Smuts, formerly a Boer War commander and now South Africa's elder statesman. Tansley chose the occasion of a festschrift published in 1935 in honor of Henry Cowles to deliver a full critique of the organismic analogy. He accepted the idea that a climax community was a "complex whole with more or less definite structure," thus taking a very different approach from that of Henry Gleason, whose individualistic viewpoint he did not accept.[15] The issue was not to challenge the idea of wholeness, as Gleason did, but to express the concept of wholeness without falling into the circumlocutions of organicism.

Tansley always considered the "complex organism" to be a poorly chosen term. Clements's insistence that the term was not a metaphor but was a legitimate way to characterize the community was difficult to grasp and created more problems than it solved. Tansley proposed a new term, *ecosystem*, which he thought better described the world that ecologists were studying. Ecologists not only studied the organisms of a community but also the physical factors that formed the environment in which they lived, and from which they could not be separated. The term *ecosystem* was meant to express the totality of that system, embracing organisms and physical environment, and to acknowledge that exchanges were occurring between the physical and the biological components. Ecosystems developed toward a state of greater integration and stability, which was the climax. However, climaxes varied in their stability, and some would disintegrate over time. The main point was that plants and animals were components of a system that included climate and soil and that it was important to recognize the existence of these systems. Tansley did not agree with Gleason's idea that the world was made up of random associations that shifted according to the laws of probability.

He addressed the thorny issue that was at the heart of Clements's organicism: "Is man part of 'nature' or not? Can his existence be harmonised with the conception of the 'complex organism'?"[16] The ecological viewpoint expressed by

Clements and his followers assumed that humans were not components of the "complex organism" and that it was possible to identify a "natural" state even after humans entered the scene. The separation between humans and nature arose from Clements's focus on the plant community, but by the 1920s Clements agreed that one might consider the community as a "biotic" unit consisting of plants and animals. He realized that redefining the community to include animals raised "pertinent questions as to the place and rôle of man in it."[17] Moreover, he thought that human communities could also be considered complex organisms. Tansley preferred to get rid of the complex-organism concept rather than redefine it. It created a distinction between low-impact tribal societies that were seen as part of the natural community and modern humans who, being more destructive, were not part of the ecological community. Tansley argued that one could not easily make such a distinction.

In Tansley's view there might be different kinds of ecosystems, and some might be created by humans, but nonetheless the processes that resulted in human-dominated systems were the same as the processes that produced "natural" systems. Humans might well have significant impact on certain landscapes and on the ecological relations between other organisms. The "complex organism" was simply inadequate for the analysis of these processes, in large part because it was too narrowly defined. It gave an incomplete view of the world. The ecosystem concept, encompassing everything, made it easier to see how humans were part of the system, not standing outside it. Tansley had a holistic perspective not unlike that of Clements, but he wanted to envision the whole in a way that included humans as intrinsic parts of the system.

The new term was picked up sporadically but did not come into general use until the 1950s. When *ecosystem* resurfaced at that time, the concept included certain ideas that were not mentioned in Tansley's article. One was an emphasis on the cycling of nutrients, such as carbon or nitrogen, through the biotic and abiotic components of the system in a roughly circular pathway. The other was a focus on the one-way energy flow through the system, where energy flow referred not only to what organisms ate but also to how they expended energy in various vital activities. Analysis of the processes that defined ecosystems concentrated on these two ideas, the cycling of chemicals within the system and the flow of energy through the system.

Studies of nutrient cycling entered ecology from several sources in the early twentieth century through the work of geochemists, mineralogists, soil scientists, physiologists, and chemists.[18] A key step in the development of the ecosystem concept after Tansley's article was articulation of the biogeochemical approach to ecological systems, an approach that in the United States owed a great deal to the work and teaching of George Evelyn Hutchinson at Yale University. Hutchinson, an Englishman who came to the United States in 1928, had an enormous impact on American ecology through his own work and through the training of students who became leaders in the discipline. He possessed remarkable breadth, with interests that included freshwater ecology, population ecology, and biogeochemistry. He also had a knack for framing hypotheses. An article titled "Circular Causal Systems in Ecology," published in 1948, illustrated how the study of biogeochemical processes could be used to raise problems of fundamental significance.[19] If we take this piece to represent what Hutchinson was teaching to his students in the 1940s, then we can get a clearer picture of his intellectual impact.

The article was part of a symposium held in 1946 on teleological mechanisms, or the mechanisms that underlie apparently purposeful systems. The topic revealed growing interest in what became the new science of cybernetics, the study of systems in which there were feedback mechanisms that regulated or stabilized the system. Norbert Wiener, who also participated in the conference, coined the word *cybernetics* in 1947 to describe a general class of problems dealing with control and communication.[20] The concept of such a field arose from consideration of engineering problems that Wiener had encountered during the war, such as guiding a missile so that it would hit a moving target, and the development of computing machines. In discussing these matters with colleagues, he realized that the mathematical techniques developed to deal with engineering and computational problems could be extended to biological problems. For example, a rapid-computing machine could be used as a model of the nervous system. Cybernetics involved extending and exploring analogies between living systems and machines, with the result that the living system was analyzed as though it was machinelike. These living machines appeared to act purposefully as a result of the various feedback mechanisms that governed their operation.

A building in which temperature is controlled by a thermostat relaying a signal to a furnace is an example of a cybernetic system. An organism with multiple physiological mechanisms for maintaining homeostasis is a more complicated example. In ecology the transfer of substances between organisms and environment, forming a cycle, also could be thought of as a feedback mechanism that stabilized the system. An example was the carbon cycle. Green plants took up carbon from the atmosphere by photosynthesis; the plants gave back some of it by respiration, and the rest was returned through the respiration of animals eating the plants and by bacterial decomposers. Qualitatively the cycle was well known, but a quantitative assessment was rarely tried and little understood. Hutchinson focused on this question: How could this cycle be understood in exact, quantitative terms, and what would this knowledge tell us about how the biosphere functions and how human activity influences this function?

As Hutchinson worked through calculations of how much carbon was moving through the system, he related this research to the larger problem of how the ocean acted to regulate carbon dioxide by mopping it up and what effect human activity had on this regulatory mechanism. In the twentieth century scientists noted that the level of carbon dioxide was increasing in the atmosphere in Europe and eastern North America. Did this mean the normal regulatory mechanisms were not functioning properly, and if so, what was the cause? One thesis advanced in 1940 was that the ocean's regulatory mechanism operated too slowly, so that as humans kept adding more carbon dioxide from industry, the carbon dioxide finally reached a level where it accumulated in the atmosphere.

Hutchinson suggested that the increase in carbon dioxide might be caused by changes in the biological mechanisms of the cycle rather than by increased industrial output per se. He argued that the real problem might be a decreased efficiency in how the plant cover on the land absorbed carbon by photosynthesis, a decrease perhaps caused by converting forested areas into farmland. As he remarked, it was possible that deforestation, along with erosion and loss of nutrients to the sea, had changed the composition of the atmosphere in certain areas. His article went on to consider other examples of feedback cycles in ecology, including some in population dynamics, but these details do not concern us. The importance of the article for our purposes lies in the way it illustrates how Hutchinson related specific measurements to broad, global patterns.

Hutchinson showed how, by paying attention to quantitative analysis—getting measurements of such things as the efficiency of photosynthesis, the production of carbon dioxide, and the absorption of carbon dioxide—scientists could understand how human activity was affecting the biosphere. The history of life on earth had resulted in the creation of regulatory mechanisms that maintained the stability of the composition of the atmosphere. If human activity was altering the atmosphere, it was crucial to understand exactly how that was happening. Ecological analysis provided a way to link local human actions to large-scale global processes. It also provided a way of measuring human impact in comparison with other kinds of naturally occurring processes, such as volcanic activity. As Hutchinson said, this kind of quantitative assessment was rarely undertaken. Different hypotheses needed to be compared and evaluated. The article did not solve the problem definitively, but it showed that understanding ecological relations helped to identify plausible alternative hypotheses. The point was to bring ideas forward, test them, and in testing them to gain better understanding of processes and relationships.

Although Hutchinson is not considered to have been an ecosystem ecologist, his students elaborated on his ideas and in the process brought the ecosystem concept to the forefront of ecological thought. Two men in particular were important in developing this concept, Raymond Lindeman and Howard Thomas (Tom) Odum. Lindeman worked with Hutchinson as a postdoctoral fellow after completing a Ph.D. on the food cycles in a small lake in Minnesota. During his postdoctoral year Lindeman refined the dissertation and developed many of the ideas that became central to ecosystem analysis. The main novelty of his approach was the interpretation of ecological processes in terms of energy transfers across different levels of the food chain (called trophic levels).

Lindeman argued that the ecosystem concept had fundamental importance in ecology because it allowed one to depict the system as a complex of biotic and abiotic components. Although he had studied one particular lake, Lindeman was trying to characterize the basic features of all systems and advance a general theory of successional processes understood from an energetic standpoint. Lindeman died while his article, published in 1942, was in press, but Hutchinson drew attention to Lindeman's work in his own writing, thereby ensuring that ecologists would take note of it.[21] As Lindeman's work was cited and discussed, his approach came to be recognized as an important theoretical advance in ecology.

Hutchinson encouraged his students to explore new ideas, look for patterns, and not shy away from theory. He believed that when one had the glimmer of an idea that might prove to be new or interesting, one should drop everything and pursue that idea. In Lindeman's case, the support he provided went beyond making suggestions about how to analyze ecological processes in terms of energy transfers. He supported Lindeman's willingness to theorize, to generalize when he did not have enough data to back up his claims, and to publish his ideas without spending a lifetime collecting the data.[22] Hutchinson was willing to launch a theory to see what reaction it got. Theoretical generalizations might turn out to be wrong, but that was not the point. The point was that the research stimulated by new hypotheses would produce more knowledge than would have been produced in the absence of those hypotheses.

Evelyn Hutchinson presented ecology as an intellectually exciting discipline that was generating insightful hypotheses and tackling basic questions about how humans were affecting the biosphere. Eugene and Tom Odum, brothers who both became prominent ecologists, showed how these ideas could be used to reorient ecology as a science of systems. Separately and together, they redefined ecology in the 1950s, setting the concept of the ecosystem at the center of ecological study. This viewpoint was being articulated just as interest in ecology was picking up in the mid-1950s. The new problems raised in the atomic age created incentives for funding and expanding ecological research. Ecosystem ecology was a product of this conjunction of ideas and opportunities.

Tom Odum was Hutchinson's student; his doctoral dissertation, published in 1951, was a study of the strontium cycle.[23] Strontium later became the focus of debate about the hazards of radioactive fallout. Eugene Odum had done his doctoral work in physiological ecology at the University of Illinois in the 1930s and had joined the faculty of the University of Georgia in 1940. He learned about Hutchinson's teaching through his brother's lecture notes but also corresponded directly with Hutchinson. In 1953 Eugene published a textbook called *Fundamentals of Ecology*, which was periodically revised and expanded and became one of the most successful ecology textbooks in the world.[24] Tom Odum contributed to several chapters, especially to those discussing energy relations and biogeochemical processes. The text was unusual in that it opened with a definition of the ecosystem and used this concept to characterize the subject matter of ecology. In later editions the focus on the ecosystem became even more pronounced.

Eugene Odum's concept of the ecosystem embraced both a concrete and an abstract definition.[25] The concrete, commonsense approach would be easily understood by anyone accustomed to thinking of nature as made up of communities or natural groupings of organisms. An ecosystem was a natural unit including living and nonliving parts that interacted to produce a stable system (fig. 7.1). Identifying such natural units was easy, although they might vary in size: a lake would be an example, and the entire biosphere was another example. Odum's own course introduced students to the ecosystem idea by having them study a pond and an agricultural field. Tansley's concept was in line with this definition.

But Odum added something else: the exchange of materials between the living and nonliving sectors of the ecosystem followed roughly circular paths

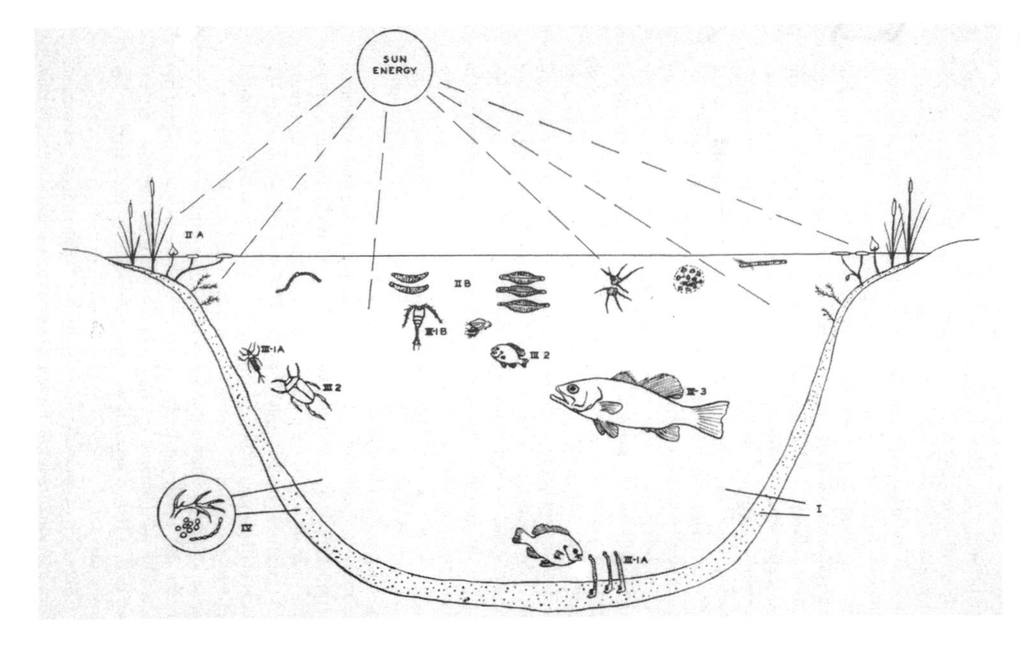

Fig. 7.1. The energy chain. In Eugene Odum's depiction of a pond ecosystem, plants absorb energy from the sun and transfer it to plankton, invertebrates, and vertebrates. From *Fundamentals of Ecology*, 3rd edition, by ODUM © 1971. Reprinted with permission of Brooks/Cole, a division of Thomson Learning: www.thomsonrights.com. Fax 800-730-2215.

(fig. 7.2). Again, the sizes of the systems might vary, but all systems were defined by "circulations" of material: "Thus, the entire biosphere may be one vast ecosystem with numerous more or less circular systems within it."[26] This definition, which reveals the direct influence of Hutchinson's biogeochemical approach, illustrates what ecologists found attractive in the ecosystem concept: it implied that regulatory mechanisms operated to maintain the ecological community. The term *ecosystem*, as a later writer put it, signified a "highly ordered functional pattern of flow in nature."[27] The circular paths characteristic of communities were necessary to maintain life on the earth.

As elaborated by Eugene Odum, the ecosystem concept provided a way of depicting a natural unit, whether lake or grassland or forest or coral reef, in abstract physicochemical terms, as a series of chemical cycles whose operations were driven by the energy captured from the sun and flowing through the system. Because the ecosystem was defined in this way, it was possible to do ecosystem ecology without detailed knowledge of all the species that made up the system. In

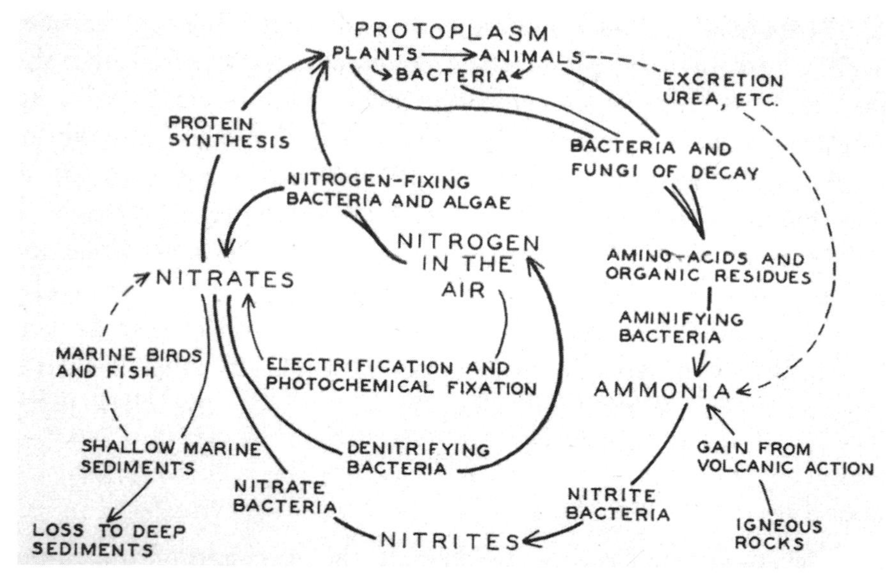

Fig. 7.2. The nitrogen cycle. This relatively perfect, self-regulating cycle illustrates the biogeochemical approach to the study of ecosystems. From *Fundamentals of Ecology*, 3rd edition, by ODUM © 1971. Reprinted with permission of Brooks/Cole, a division of Thomson Learning: www.thomsonrights.com. Fax 800-730-2215.

general, one needed to know what kinds of organisms dominated in the system, but not the specific identity of every component. The ecosystem concept was a way of stepping back from the system, looking at it through a "macroscope," as Tom Odum put it, and bringing out its general features.[28]

The concept of the ecosystem could be used to organize and focus ecological research, but at the same time it suggested ways to link ecology to the physical sciences on one side and the social sciences on the other side. Not everyone understood how this might be an advantage in developing ecology. Francis Evans, an ecologist at the University of Michigan, where ecosystem studies were under way in the 1950s, noted that some ecologists rejected the ecosystem concept because it seemed unrealistic.[29] Ecosystems did not have easily identifiable boundaries; they were open systems that connected to myriad other systems. Nature was not in reality divided up into ecosystems, it appeared. But Evans believed that the concept helped the ecologist to focus on the problem of regulation, on the mechanisms that existed to preserve systems and maintain their identity.

The ecosystem concept was best appreciated as a way of viewing nature abstractly. Ironically, the abstract definition that focused on feedback loops and self-regulation had to override the commonsense definition before one could appreciate the usefulness of the concept. Achieving an understanding of how systems were regulated was even more important in the human-dominated environment, where maladjustments and disturbance were all too evident. Evans noted that the disharmonies of modern urban life pointed to a lack of effective regulation. The ecosystem approach could be extended from the natural to the human community, providing a viewpoint that would synthesize the biological and social sciences.[30]

Efforts to develop and promote the concept of the ecosystem at this time coincided with new funding opportunities that boosted ecological research in the 1950s. With this new support ecosystem ecology began to develop rapidly. The main sources of funding were the Office of Naval Research and the Atomic Energy Commission, which both poured money into ecological studies after 1954, when concern about radioactive contamination was increasing along with the growth of the atomic industry.[31] In 1955 John N. Wolfe, a botanist at Ohio State University, became the first ecologist to serve on the advisory committee of the Division of Biology and Medicine at the AEC. He went back to Ohio in 1957–58, then returned to the Atomic Energy Commission as chief of the new Environmental Sciences Division, which expanded the scope of ecological research funded by the commission.

The national laboratories at Oak Ridge in Tennessee and Brookhaven in New York became growth centers for ecosystem ecology and radiation ecology. At the Hanford Laboratory on the Columbia River (now the Pacific Northwest Laboratory), radiation ecologists studied the effluent released from plutonium-producing plants.[32] The AEC also stepped up its funding of ecological research at bomb test sites. All of this activity helped to advance the idea that ecologists should connect their work more closely to physics and chemistry; that is, they should study ecosystem functions by looking at how material and energy moved through the system.

The success of ecology at the national laboratories depended on shrewd adaptation to an environment dominated by physicists. At Oak Ridge, for example, Stanley Auerbach successfully exploited the perception that ecology had a bearing on the new field of "health physics" that had grown up in response to the health hazards of radioactivity. Inasmuch as health physics included the risks posed by contaminated environments, ecological research fell within its scope. Auerbach gradually built the ecological program and extended the scope of ecological research, making Oak Ridge one of the largest centers of ecosystem and radiation ecology in the country by the end of the 1960s.

Stephen Bocking has pointed out that although there was a certain degree of individual freedom to choose research topics at Oak Ridge, the broader objectives of the AEC determined what choices were available.[33] One of the highest goals in the 1950s was to develop atomic technology, which entailed understanding and minimizing the risks of that technology. The ecology done at Oak Ridge had to complement this goal, at least in part. It also had to mesh with the other scientific projects at Oak Ridge, whose director, Alvin Weinberg, firmly believed in the importance of nuclear power as the solution to modern concerns about dwindling sources of energy in the face of growing population.[34] But nuclear energy meant radioactive pollution, and this was where Weinberg fitted ecologists into the scheme of things. They would help to cope with noxious effluents and mitigate the effects of contamination in the biosphere. Bocking shows how skillfully Auerbach developed an ecological perspective that capitalized on the opportunities available at Oak Ridge, while working within the constraints set by the mission-oriented government laboratory. A science that began as a marginal part of the Oak Ridge enterprise grew into a major area of research over a period of twenty years.

Auerbach was influenced by Eugene Odum's ideas about the ecosystem, which he found could be adapted well to a quantitative, physicochemical research focus that would sit well with the physicists at Oak Ridge.[35] Odum, who was on the faculty of the University of Georgia, had been associated with the AEC since 1951 and later served on the advisory committee for the Health Physics Division at Oak Ridge. He had received one of the first AEC contracts in ecology for research on the land surrounding the Savannah River Atomic Energy Plant in South Carolina. When the nuclear facility was constructed in 1952, hundreds of farm fields, about three hundred square miles, had to be abandoned. This provided an opportunity to study succession as the forests reinhabited the fields. Here and at a salt-marsh reserve at Sapelo Island, off the Georgia coast, Odum and his students used radionuclide tracers to study food-chain relationships.[36] This was a general ecological problem, but the research also had a more immediate practical side. It could be used to predict the fate of radioactive contaminants in the environment. With continued AEC funding, Odum undertook a variety of long-term ecosystem studies. Under his leadership the University of Georgia became a major center of ecological research.

Odum's holistic approach to the ecosystem was especially evident in a study undertaken in 1954 when he and his brother, funded by the AEC, studied a coral reef on Enewetak Atoll, one of the bomb-testing sites in the Marshall Islands. Their study was designed to test ways of measuring total community structure and function. Such measurements were needed in order to judge the total effects of radiation or other pollutants on ecological systems and to determine what the tolerance levels were for entire systems. Just as it was possible to determine what dosages of radiation were acceptable for individuals, the Odums believed it should be possible to find tolerance levels for ecological systems. As Eugene Odum pointed out, the ecosystem approach, which aimed at broad measurement of the metabolism of whole systems and of disturbances to those systems, seemed well suited to environmental impact studies of the kind that would be urgently needed for large nuclear power plants.[37]

The ecological system, or ecosystem, was treated as a self-regulated entity, like an individual organism, which had a metabolism that could be measured. It was possible to think of the reef as an organism because it was located in a relatively constant environment and was stable, or in a climax state. It was an isolated system with definite boundaries. The coral reef study investigated the biological productivity of the reef by looking at changes in the oxygen content at different stations during the day and night, which gave a measure of "total reef metabolism." The Odums concluded that productivity of the system was enhanced by the relationship between corals and algae. This idea—that systems benefited from symbiotic linkages that improved nutrient cycling and the efficiency of energy flows—became a recurring theme of their work. As organisms of a system became linked in their life functions, the system became more efficient and less wasteful.[38]

Recalling that Eugene Odum's background was in physiological ecology, one can see how he transferred ideas relating to individual physiology to the physiology of the ecosystem as a whole. This transfer helped to make the idea of the ecosystem as a self-regulating entity more concrete. His strategy was similar to that adopted by Frederic Clements when he described the plant community as a complex organism, but the Odums went further, by trying to measure the metabolic functions of the system and by thinking of the ecosystem in broader terms, as Tansley had recommended.

In the second edition of the textbook, published in 1959 with Tom Odum listed as coauthor, Eugene reiterated his commitment to the ecosystem or "whole-before-the-parts" approach.[39] In part this was a pedagogical device, a way of explaining what ecology was about without getting lost in a lot of detail.[40] But the idea that the ecological community had a metabolism that, like an individual organism's metabolism, could be measured and studied, that the community existed by virtue of the cycling of materials and flow of energy, made sense as a way to gauge the impact of human activity on the biosphere. Anything that disturbed those cycles or altered the efficiency of the flow of energy would change the metabolism of the entire system. Humans could assess their impact according to whether they increased or lessened the efficiency of these processes. Drawing lessons from nature depended on seeing the natural system as a functional whole, not just an assemblage of species.

The new edition of Odum's textbook added a review of the burgeoning literature on radiation ecology, which included the tracking of radioactive substances released into the environment. Understanding how the larger biogeochemical cycles worked and how food chains concentrated radioactivity showed that humans could not ignore what went on in the parts of the biosphere far removed from human communities. As Odum explained, "We could give 'nature' an apparently innocuous amount of radioactivity and have her give it back to us in a lethal package!"[41] Disposal of nuclear wastes, he believed, loomed large among future environmental problems.

Tom Odum meanwhile was making a study of Silver Springs, Florida, also a well-defined stable system where temperature and water volume were fairly constant.[42] This research was funded by the Office of Naval Research. Treating the system as a black box, Odum analyzed the chemistry of the water entering and leaving the system, thereby deriving a measurement of the metabolism of the system. But whereas Eugene thought of the ecosystem in organic terms, as though it were an organism in a state of homeostasis, Tom deviated from this organic analogy and increasingly thought of the ecosystem as a machine governed by feedback mechanisms.[43] His approach echoed the cybernetic way of thinking in its fascination with the similarity between organisms and machines. He took the ecosystem concept in a more mechanistic direction, first developing a hydraulic model of ecosystem function and then in 1960 introducing symbols from electrical engineering to depict the main functions within the system using energy circuit diagrams (figs. 7.3, 7.4, 7.5).[44] Odum advocated the use of symbolic language to depict ecosystems because they were too complicated to describe verbally.[45] His energy circuit diagrams served to simplify the system, bringing out its main features and allowing one to compare systems. Virtually any system could be depicted in this way, as long as one could translate processes

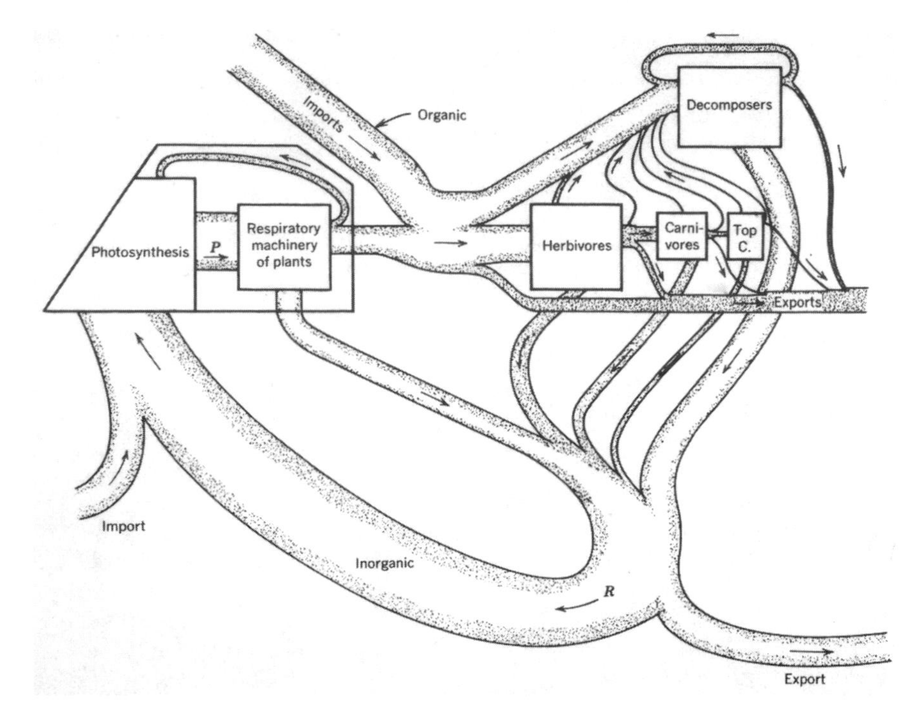

Fig. 7.3. A hydraulic model. Organic and inorganic material as well as energy flow through the ecosystem. From *Environment, Power, and Society*, by Howard T. Odum, p. 2. Copyright © 1971, John Wiley & Sons. Reprinted by permission of John Wiley & Sons, Inc.

into energetic terms. These models were surrogates of the ecosystem, based on the analogy between the flow of electrons in a circuit and the flow of material in a system.[46]

Common to both organismic and machine metaphors was the idea that some ecological systems, but not all of them, were in a steady state, a state of dynamic equilibrium where inflows of material and energy balanced outflows. The steady state was equated with the "climax" state of ecological succession. The state of equilibrium might be temporary, especially in systems prone to external disturbances, for example by violent storms.[47] Other ecosystems were actually maintained by fluctuations in the environment. The Georgia salt marshes and estuarine channels were stabilized by tidal flows, which enhanced productivity. In other areas whole biotas were adapted to periodic fires. The idea of ecosystem equilibrium did not exclude periodic disturbance if it could be shown that the constituent organisms were adapted to the disturbances.

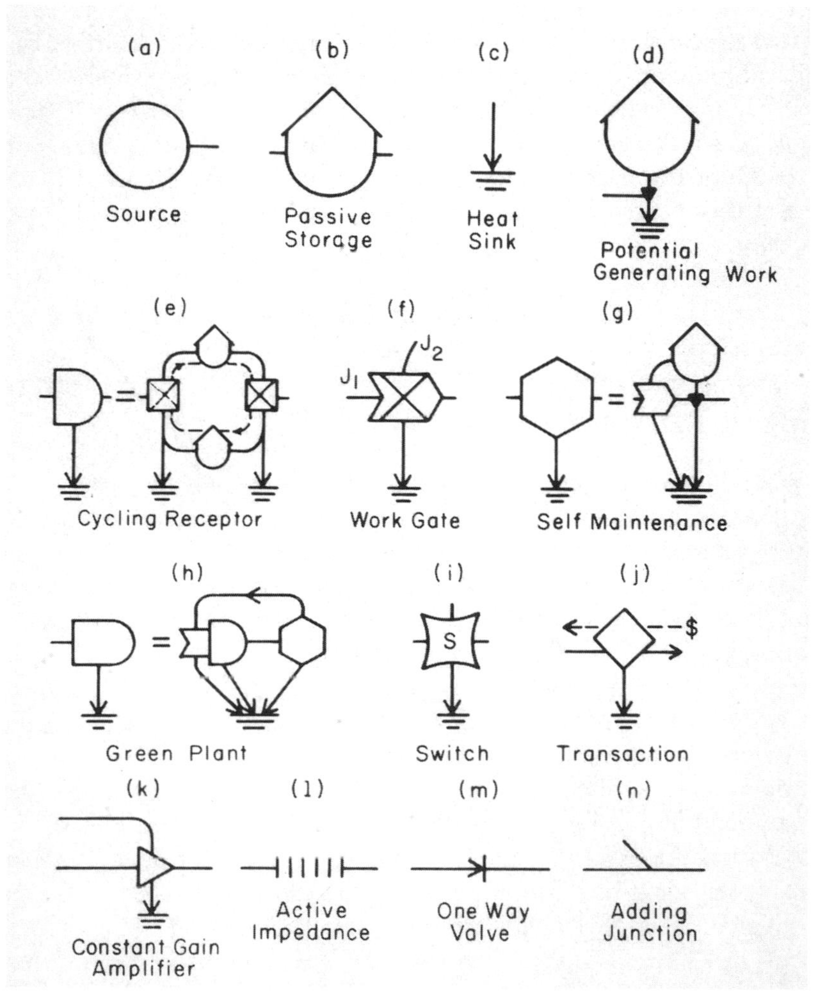

Fig. 7.4. Ecosystem circuits. H. T. Odum used electrical circuit symbols to construct diagrams of ecosystems. *Environment, Power and Society*, by Howard T. Odum, p. 38. Copyright © 1971, John Wiley & Sons. Reprinted by permission of John Wiley & Sons, Inc.

This focus on conditions of equilibrium and disruption of equilibrium invited consideration of how humans might affect natural equilibriums and also whether human-dominated systems bore any resemblance to natural systems. An ecosystem could be a lake, a forest, or a grassland; but it could also be the entire biosphere, or a city, a nuclear power plant and surrounding region, or even a single

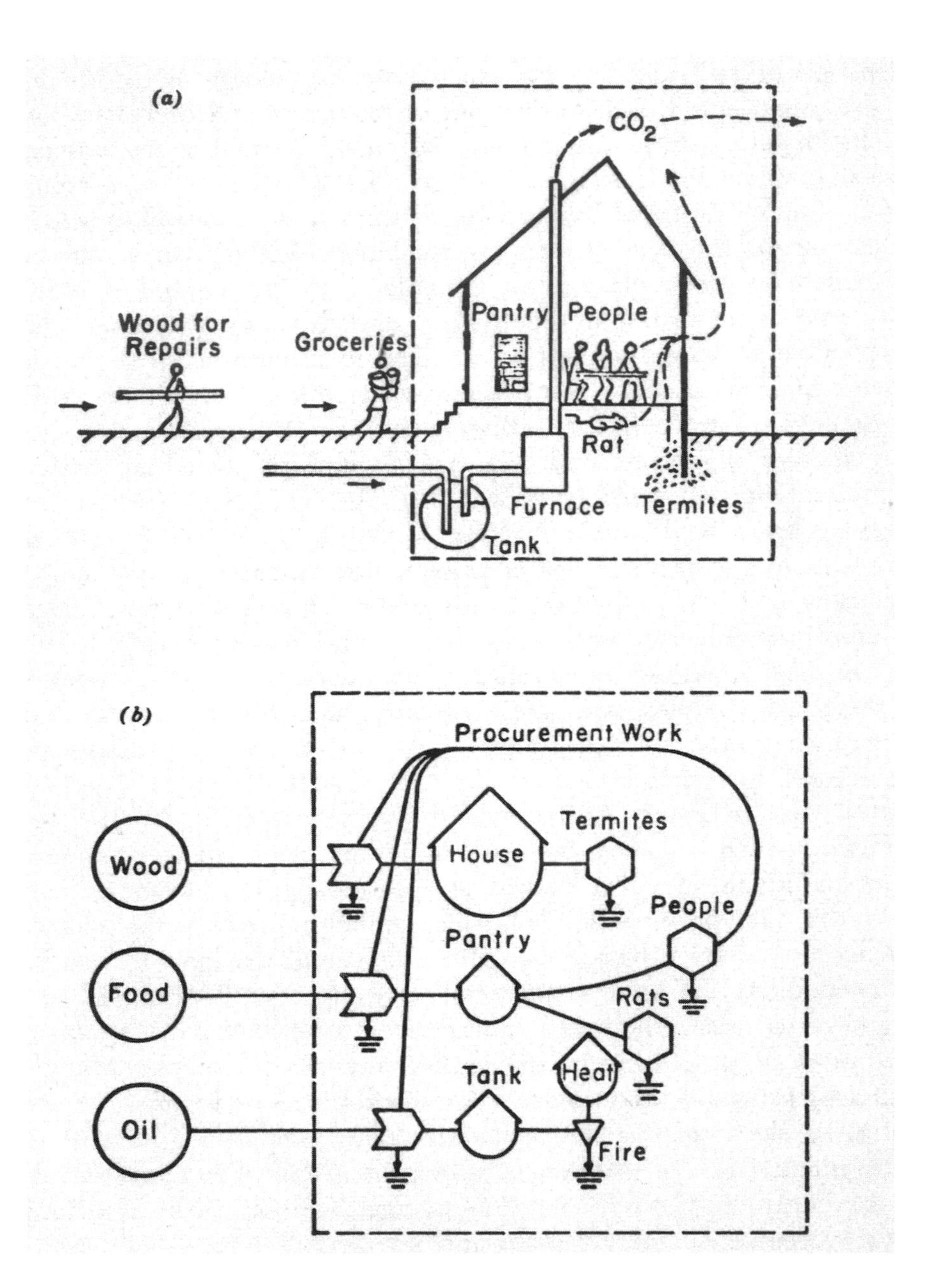

Fig. 7.5. A household ecosystem. Energy and material transfers in the household ecosystem are represented as an electrical circuit. From *Environment, Power, and Society,* by Howard T. Odum, p. 40. Copyright © 1971, John Wiley & Sons. Reprinted by permission of John Wiley & Sons, Inc.

household. How did these kinds of systems work, and how could one compare natural with human-impacted systems? Beneath the problem of understanding how systems worked was a larger implicit question: Did the ecosystem perspective, the ecological perspective, yield a different kind of knowledge from that obtainable by other sciences? What was special about the ecosystem point of view?

Two ideas suggest themselves when one thinks about the world as a series of ecosystems nested one within another, progressing from the smallest spaces up to the biosphere as a whole. One is that "spaceship earth" is a small place: something added in one spot, a small disturbance, affects what goes on elsewhere. Effects can even be magnified as substances are concentrated along food chains. Scientists were starting to recognize and track the extent of these connections and to realize that human impacts were not negligible. The second idea is that since humans affect the operation of ecosystems, understanding ecosystems must involve understanding humans, including how humans relate to nature and how societies function. The study of how societies function must include how science is perceived and used by societies. Therefore ideas, which influence behavior, are also part of ecosystems.

One person, Rachel Carson, grasped these two ideas fully, recognized their implications, and acted on that knowledge in a way that spoke directly to the public. Her book *Silent Spring*, published in 1962, set off a storm of controversy by pointing out the ecological and health hazards posed by pesticides.[48] She used science to attack the optimism of the postwar era of monstrous growth, with its faith in control over nature. Carson researched the scientific literature exhaustively; she synthesized technical information buried beyond public view and made it comprehensible to the lay public, aided by a few scientists who critiqued her drafts, fed her information, and made sure her work was up-to-date.

But Carson's book did not just discuss how chemicals moved along food chains. It showed that the dangers of pesticide fallout were akin to those of radioactive fallout, a peril to which people were already sensitized. Carson exposed the deceptive practices used by industry, which prevented ordinary people from learning what was being done to their environment. The deceptions began with the very word *pesticide*, which lulled people into thinking that only undesirable creatures were killed, whereas in fact these poisons were biocides, killers of life. She drew out the stark consequences of human disruption of the balance of nature. While the hired guns of the chemical industry bombarded her for failing to grasp that modern science was all about controlling nature, she echoed the sentiment expressed in the "Marsh festival" of 1955 that the more serious problem we faced was to control ourselves.

Until this time, ecologists had largely avoided getting involved in public controversies that challenged the gospel of growth as directly as Carson did. In the 1950s ecologists had found new roles, but they were largely supporting roles within the system. They recognized the problem of radioactive wastes from nuclear plants, but the Atomic Energy Commission admitted that this was a problem that justified more funding for ecological research. Ecologists had not led the debate about the dangers of radioactive fallout, leaving that job to more outspoken scientists such as Linus Pauling. Studies of pesticides funded by chemical companies were open to the criticism that they were subject to conflicts of interest. There were similar questions about whether government scientists could openly challenge government policies. Carson's book was exceptional in challenging the system. As Paul Shepard and Daniel McKinley commented, "Her brother biologists, almost to a man, did excellent imitations of people frightened by big money and authority and deserted her before the Establishment which controls the funds that keep scientists fat."[49] But there were exceptions, and in the aftermath of *Silent Spring*, ecologists began to write openly about their social responsibility and the need to connect, through organizations like the Ecological Society of America, to the public.

One of Carson's strongest supporters, Frank Egler, a specialist on the ecological impact of herbicides, considered her book to be the most important single study in ecosystematics (the study of whole ecosystems) that had been written to date.[50] Garrett Hardin, a biologist who was a controversial advocate of the need to relinquish the "freedom to breed" to ensure the conservation of resources, thought her book one of the two most important publications in twentieth-century biology, the other being James Watson and Francis Crick's article on the structure of DNA published in 1953.[51] As Paul Sears wrote in 1964, ecology by its nature was a subversive subject, providing the grounds for a "continuing critique of man's operations within the ecosystem."[52]

Thirty years earlier, in the middle of the Dust Bowl, Sears had written an indictment of the practices that were responsible for creating deserts. He lamented the relatively low status of ecology at that time, when few universities offered courses in the subject and there seemed little demand for this kind of expertise, despite such serious problems as the Dust Bowl. Now in 1964 Sears found himself trying to counter the perception that ecology was of "limited interest and utility."[53] Lamont Cole, writing in the same journal in 1964, also commented on the ivory-tower tendencies of ecologists, who were only recently becoming aware of the need to apply their expertise to solving important public problems and bringing the public up-to-date on what they had learned.[54]

As Sears argued, ecology did have the big picture, and it offered lessons that challenged the "current glib emphasis on economic 'growth' as the solution of all ills." The reason ecology could have this role was that it adopted a holistic perspective: it studied the entire ecosystem and the impact of the world's dominant organism, humans, on that system. The ecosystem concept, with its focus on equilibrium and the disruption of equilibrium, became linked with the emerging environmental movement and its critique of modern industry, the growth mentality, and faith in technology as a panacea. Sears pointed out that although there always was a need for more research, the real problem was that existing knowledge was simply not being used.[55]

It took a writer, a humanist like Rachel Carson, who understood science and had the courage to act independently in the face of virulent opposition from industry, to demonstrate how to convey this knowledge to the public and the government. Her solution involved bridging the two cultures of science and humanities, as Marston Bates had advocated at the "Marsh festival," using imaginative scenarios as fables to convey the meaning of these ecological studies. As scientists like Frank Egler well understood, the development of ecology, its ability to find a niche within American society, depended on the work of people who could explain its importance and its relevance to the general public. But ecologists themselves had to question the relevance and role of their science. They needed to make the case that ecology had to be taken seriously "as an instrument for the long-run welfare of mankind."[56]

In Tom Odum's case, the justification of ecology took a highly idiosyncratic turn into ecological ethics in his book *Environment, Power, and Society*, published in 1971 for a general audience.[57] His symbolic language became the vehicle for developing the larger point that from ecosystem studies came ethical principles. His approach became increasingly broad, embracing literally everything under the sun. To include human economic activities in the system, an energy value (in calories) could be assigned to money, depending on how it was used to buy goods and services. Food was received in exchange for money; fertilizers were purchased to increase crop yields (fig. 7.6). Money circulated in the opposite direction from energy flow. An ecosystem model could therefore incorporate economic exchanges and illustrate how humans increased or altered the flow of energy in the system by spending money. If energy sources were cut off, the dol-

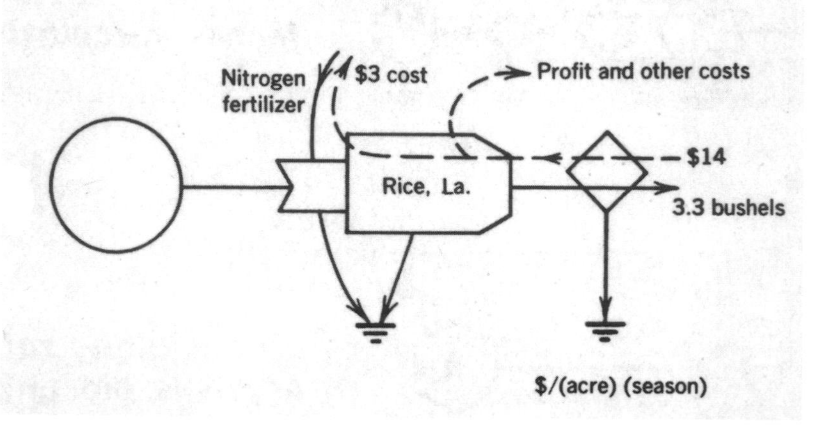

Fig. 7.6. Ecological economics. H. T. Odum's circuit diagram shows the relationship of energy flow to money in an economic transaction. Here nitrogen fertilizer is purchased to stimulate rice production. Money and energy circulate in opposite directions. From *Environment, Power, and Society*, by Howard T. Odum, p. 175. Copyright © 1971, John Wiley & Sons. Reprinted by permission of John Wiley & Sons, Inc.

lar would lose value because nothing would be produced. Any kind of human activity, including religious, political, and economic activity, could potentially be incorporated into these symbolic representations of ecosystems.

Odum also believed that the process of natural selection favored systems that maximized "power," or the rate of flow of useful energy.[58] The use of symbolic language was not only a way to picture the basic processes of a system, but also a way to state objectively what criteria might distinguish healthy and adaptive systems from less adaptive ones, that is, to impose value judgments. Having such criteria was important when humans were brought into the picture, for humans had the ability to adjust their behavior and to modify the system, adding more loops, directing the flow of energy along new paths, and so forth. Ecological energetics pointed toward ecological engineering. In systems combining humans and nature, the goal would be to avoid unnecessary waste and to develop patterns "that emphasize partnerships, symbioses, and diverse functions."[59] Ultimately these models were meant to provide guidelines on how to engineer a better-functioning world. The culmination of Odum's idiosyncratic approach was to root ethics in the laws of energy and the principles of ecology.

Peter Taylor has characterized Tom Odum's approach to ecosystem analysis as an exemplar of "technocratic optimism," and he traces its roots to Odum's interest in the technocracy movement of the 1930s, which similarly sought a way to rationalize human behavior by measuring material and work in units of energy.[60] Systems ecology provided a way to express the "essence" of ecosystem dynamics and showed what kinds of interventions would improve their operation, as though the systems were machines whose efficiency could be improved with a bit of tinkering. As Taylor notes, the overarching goal was to design systems of man and nature, but this was not simply an engineering project: systems science for Odum functioned as a new religion.

Both Eugene and Tom Odum believed that securing a niche for ecology required articulation of a few "big ideas."[61] Chief among them was the notion that new system properties emerged over the course of ecological development and that these new properties accounted for the changes that occurred in the system.[62] As Eugene Odum admitted, the idea that there was a "holistic strategy for ecosystem development" was controversial. Accepting that there was a "strategy" of succession meant accepting that it was an orderly process, reasonably directional and predictable, ending in a stabilized ecosystem. As Frederic Clements had also believed, it was necessary to see nature as having organization and evolving in an orderly way in order to assess human impact on nature. Ecosystem analysis might include humans as part of the environment, but the ecosystem ecologist perceived those humans as operating mostly in opposition to nature's strategy. Humans created stresses that robbed systems of their protective mechanisms and allowed "irruptive, cancerous growths of certain species to occur."[63]

Although they believed that humans were at times the metaphorical equivalent of pathogenic organisms, the Odums were judicious in approaching the subversive implications of ecology. Their aim was not to subvert the system nor to redesign ecosystems completely, but to modify human excesses, to show people (as Clements had done) that conservation was in their interest and that ecology could lead to more effective assessment of risks and more efficient management of resources. People had to understand the extent to which ecosystems provided services that sustained the environment on which we depend for our existence: ecosystems recycled nutrients, purified water, maintained the atmosphere, and in general protected our environment. Eugene Odum argued that the application of ecology to environmental problems demanded a systems-level point of view to provide a way of evaluating how human-induced stresses were imposing energy drains on systems.[64] Finally, the systems viewpoint would enable better integration of ecology and economics, because it would suggest ways of valuing the natural environment and showing that prudent long-term strategies of resource use had tangible benefits for humans.

Ecology was subversive in its critique of the growth mentality; at the same time it was part of the mainstream social desire to make the environment healthier and more pleasant for ordinary people without stopping economic growth or making resources too expensive. The dominant note of American society was its enthusiasm for technological fixes, and this love of technology in turn created a challenge for ecology. Alvin Weinberg, director of the Oak Ridge National Laboratory, put his finger on the dilemma in a 1972 address to the Ecological Society of America.[65] He explained that ecology, if it was to expect continued support, had little choice but to ally itself with the technologist's dream of creating a better environment.

Describing himself as a technologist, an optimist, and a conservative, he rejected calls for social revolution to solve environmental problems. A believer in the technological fix, Weinberg admitted that technology responded very slowly, but he did think it was responding to environmental requirements. His own laboratory at Oak Ridge was developing methods for treating sewage and industrial wastes, demonstrating that scientific skill could clean up polluting technologies. He argued that properly operating nuclear power plants were environmentally less burdensome than coal-burning power plants, although reactor accidents or problems of waste disposal did pose real dangers, at least potentially. On the whole he welcomed new technologies.

But making a commitment to technology required vigilant oversight, he recognized, and continual efforts to improve that technology. As he reviewed the rigorous environmental impact statements that were now required for nuclear power plants, he reflected on the effect of these kinds of multidisciplinary reviews, which combined the expertise of nuclear engineers, meteorologists, hydrologists, demographers, health physicists, ecologists, and economists. His conclusion was that ecology as an academic discipline could not afford to remain separate from other disciplines. If modern technologies were to be environmentally acceptable, then "there must be intense cooperation between ecologist and engineer from the very beginning."[66]

In short, he believed that the critical challenge for ecologists was to close the gap between basic and applied research, to enter the "hurly-burly world of engineering" and provide advice before, rather than after, costly engineering mistakes were made. He raised doubts that a movement toward general theory was really what ecology needed, and argued instead that greater focus on specific ecological questions concerning technological impacts might be more useful and that theories should be formulated in ways that could be applied to specific cases. Speaking as a "technologist who happens to have about 50 ecologists working for him," Weinberg argued for "continuing and close dialogue between the academic ecologist and his applied colleague" who had to deal with the demands of the engineer. Not that the relationship would be trouble-free: "Ecology, the jealous wife, will keep technology honest; technology, the ambition-driven husband, will insist on relevance from ecology." And, he added, lest his remarks appear chauvinistic, "both partners need to be liberated from the constraints which their past history has imposed on them. What kind of hybrid vigor will issue from this union is a question to whose answer both partners must contribute equally in the future."[67]

The problem of communication and justification was very basic to ecology, for it could never be taken for granted that society would accept the need for ecological science or listen to the message when it became even slightly subversive. Studying ecological relationships plunged one into an exceedingly complex tangle of causes, and ecologists went in search of general theories that might reduce this complexity to a manageable level. Weinberg warned that the search should not be conducted without close attention to the needs of society and to the fact that this particular society had chosen to embrace technology as the means of material progress. If society desperately needed the expertise of ecologists, ecologists needed to make sure that the lines of communication on all fronts were open.

CHAPTER EIGHT

Defining the Ecosystem

The concept of the ecosystem focused attention on how feedback mechanisms regulated natural systems, which in turn suggested that ecosystems were normally in a state of dynamic equilibrium with the environment unless disturbed by an outside force. In the 1960s and 1970s the problem of what it meant to speak of a system in equilibrium appeared often in the ecological literature. Ecologists thought of natural systems as evolving over time an ability to withstand or recover from a disturbance. A state of equilibrium, that is, a state of relative stability in a system that was maintaining itself, was a desirable goal. It signaled that the system had evolved a degree of resilience to disturbance that was analogous to the homeostatic equilibrium of an individual organism.

Stability was generally equated with other positive characteristics, such as increase of primary production (capture of energy), increase of biomass, greater species diversity, and increase of pattern diversity.[1] But to argue that certain natural states were preferable to others on the basis of their stability or diversity implied value judgments about what states of nature ought to exist or be preserved. Indeed one might say that ecology as a science was founded on the desire to establish a scientific basis for making judgments of this kind.

Generally speaking, the accepted "standard of excellence" for an environment was to have a relatively low level of human impact. The problem was to determine at what point human impact crossed a threshold that might trigger large and undesirable changes in the ecosystem, which ecologists would interpret as a collapse of the system. The frequently expressed concern about ecological stability, or the "balance of nature" in common parlance, reflected the fear that in the twentieth century that threshold had been crossed. Uncontrolled population growth, combined with the modern gospel of economic growth, was seen as a threat to the world. Ecologists feared that humans were in an unstable and unsustainable relationship with their environment and needed to regain something that had been lost: a "state of equilibrium with nature."[2] These ideas were direct responses to the rapid growth of the postwar period. Emphasis on the problem of stability or equilibrium reflected growing alarm that we lived on a planet with finite resources and that technology would not enable us to evade the consequences of that fact. But demonstrating exactly how human activity was affecting whole systems was, and still is, a highly challenging problem.

Ecological science could not be the sole basis of judgments about what states of nature ought to exist, or what it meant to achieve a "state of equilibrium with nature." Such decisions had to reflect human interests and human values. Thus the negative view of humans as disturbers of nature could be balanced by a positive view of humans as problem-solvers and creators of improved landscapes, where "improvement" had to be understood in relation to human needs. This was the point that George Perkins Marsh had made in 1864. If humans, in their judgment, were causing the degradation of an environment that they depended on, then they had to take action and fix the problem. Farmers might want fields and pastures, but Americans also needed forests, and therefore it was up to them to find the proper balance between the different land uses. In their ignorance farmers hunted birds, not realizing that birds were controlling the insects that they regarded as pests. Again, lacking knowledge of ecological relations, people were harming themselves. Human activity was not inherently harmful; it was more a question of establishing a balance between different functions and landforms and figuring out what had value and why. Nathaniel Shaler recognized that part of the value was aesthetic: we needed certain kinds of aesthetic experiences with nature for the sake of our moral development.

From the start there was a blend of utilitarian and aesthetic considerations embedded in discussions of human impact on the balance of nature. It was understood that humans themselves had to maintain some balances, depending on how they valued resources and in relation to the kinds of landscapes they wanted. The modern debate about ecological stability continued this century-old discussion. Some have interpreted the debate as an example of the improper intrusion into science of an outmoded popular concept of the balance of nature and have argued that the balance of nature should be replaced by a different interpretation of nature that emphasizes nonequilibrium processes.[3] However, ideas of ecological equilibrium are not simply popular notions, but scientific ideas firmly embedded in ecological thought. It is probably unrealistic to try to play down, let alone discard, concepts of balance and equilibrium in ecology, because they have continuing validity and an important role to play in ecological discussions. Even using the term *nonequilibrium* implies that there is an equilibrium state for comparison. Ecological constructs of nature should not be reduced to choices between two opposing views, in any event. Allowing that nature is not in equilibrium in one place does not preclude thinking about stability or equilibrium in other contexts.

I will not digress into this debate, which involves how scientists apply metaphors in their descriptions of nature and whether balance of nature is a useful or misleading metaphor. Instead I will direct our attention to what I see as the underlying problem, namely how we see ourselves in relation to nature and how the purpose of science is viewed in relation to modern needs and demands for expert opinion. In the early twentieth century, ecological science, in tandem with other areas of research, including taxonomy, evolutionary ecology, and physiology, grew in response to the expanding needs of American society as it occupied new lands, sought to control new resources, and competed with European institutions for scientific authority. A subtext in the scientific and popular literature dealt with the human problem of how to adapt to new environments, that is, how Americans thought about and responded to the changes they were causing as they took control over the land. The scientific viewpoint, which emphasized the goal of prediction and control of nature, conformed in large measure to the broader social goal of colonizing new lands and increasing the wealth of the country. Ecology was subversive only to the point of occasionally running up against bad policies that stemmed from governmental corruption. As Clements had pointed out, scientific findings might be suppressed when not in line with government policies. Otherwise ecology's message was the importance of using scientific expertise to guide management of the land and ensure the continued progressive development of society.

The idea that ecology was a subversive science allied to broader critiques of modernity came to the fore in the 1960s as the environmental movement got under way. Ecology now had two messages to offer, one in keeping with American ideas about progress and the other subversive. The clash between these viewpoints made it necessary to think more deeply about the extent to which ecology can or should be human-centered. This question in turn relates to the problem of how we view ourselves in relation to the environment, whether we are outsiders and disturbers or intrinsic parts of a system, whose health depends on our cooperation.

This chapter examines these questions through a history of the development and application of the ecosystem concept from the 1960s to the 1980s. I explore the philosophical or popular treatment of these problems through the writings of René Dubos, who came to environmentalism through a research career in medicine. He articulated the philosophical argument for a positive, humanistic perspective on ecological problems. Dubos continued the conversations about science and modernity that had occurred during the "Marsh festival" in 1955.

Interest in the problem of stability also reflected the impact of applied mathematics and engineering on ecology during these postwar decades. Development of new methods helped to expand the application of ecology to practical problems, but at the same time ecologists were searching for general theory, for a way to unify ecology intellectually. At the end of the twentieth century, ecologists still struggled with some of the problems that Frederic Clements had recognized. Did ecology offer a perspective, point of view, or theoretical framework that made it more than a collection of local natural histories? If it did, how should that theoretical framework be developed and what bearing might theory have on practical problems?

Modeling ecosystem dynamics was a fast-growing enterprise in the 1960s and 1970s, aided by the development of the electronic computer. The expansion of ecosystem ecology into new fields of applied mathematics in the mid-1960s was especially evident at Oak Ridge National Laboratory, where ecologists promoted the new "systems ecology" as a way to synthesize biological and engineering approaches.[4] At Oak Ridge, modelers George Van Dyne, Bernard C. Patten, and Jerry S. Olson developed a yearlong graduate-level course in systems ecology for students at the University of Tennessee.[5] The course introduced students to the latest mathematical techniques and methods of computer modeling and surveyed fields in applied mathematics, such as cybernetics, that had already shown promise for ecological problems. The crossover between cybernetics and ecology was apparent in the conference on teleological mechanisms in which Evelyn Hutchinson participated in 1946 (see chapter 7). Theoreticians such as Patten developed these connections more formally starting in the 1950s, inaugurating a phase of theory construction that turned ecology increasingly toward the mathematical sciences.[6]

One of the leading theorists of the era, and an acute commentator on modern trends in ecology, was Ramón Margalef, a Spanish ecologist who believed that cybernetics offered ecology the chance to develop theoretical rigor. He found that his ideas were welcomed in the United States largely because the cybernetic viewpoint already fitted in well with American thinking about the ecosystem.[7] Margalef delivered a series of lectures on cybernetics and ecology at the University of Chicago in 1966 and acknowledged his indebtedness to the ideas of Evelyn Hutchinson and Eugene and Tom Odum.[8]

The application of cybernetics to ecology was not confined to the idea of feedback mechanisms, which had been the focus of Hutchinson's earlier discussion. Cybernetics offered an analogical argument that helped ecologists to articulate some basic ideas about how nature worked. The argument depended on a three-way analogy between the concept of information (from information theory), the concept of entropy (from thermodynamics), and the concept of the ecosystem. Ecosystems could be seen as providing channels for energy; the pathways might be shorter or longer depending on the structure of the system. The length and number of these energy pathways could be taken as a measure of the degree of order of the system, which in turn suggested an analogy with the concept of entropy in thermodynamics.

Entropy measures the disorder of a system in energetic terms and is so defined that as systems become more disordered, their entropy is said to increase. A closed system that is not receiving energy from the outside will "run down"; entropy increases in such systems. In communications theory, information, like entropy, was technically defined as a measure of the degree of disorder or randomness of a system. The concept of information, in this technical context, was roughly analogous to the concept of entropy. Completing the three-way analogy between the ecosystem, entropy, and information, one could also imagine the ecosystem as a channel for information, in that the current state of a system limited and determined what the future state would be. The genetic information contained in organisms, for example, was channeled into future generations and determined what those populations would be.[9]

The analogies relating the ecosystem concept to the concepts of entropy and information were a little strained and were not applied with any great consistency by ecologists. Nonetheless, Margalef defended their use. He argued that by developing these analogies it was possible to apply ideas from thermodynamics and information theory to the analysis of how the ecosystem operated. The purpose of using these analogies was to see whether the reasoning developed in communications theory might be useful in deriving ecological principles. For example, one conclusion from cybernetics was that "a system formed by more elements with greater diversity is less subject to fluctuations." In the analogous case in ecology, the deduction would be that the stability of ecosystems depended on their diversity, although it remained unsettled exactly how "stability" and "diversity" were to be understood. Another deduction was that "the structures that endure through time are those most able to influence the future with the least expense of energy."[10] This idea suggested that ecosystems developed by following certain laws of energetics.

One goal of this analogical method was to develop the predictive power of ecology. A general theory of ecology should "pass the test of being explanatory and predictive," Margalef said. But as he also pointed out, such predictions only applied on the macroscopic level. They allowed for very general depictions of systems, but nothing at all could be predicted in detail, because ecosystems were products of myriad historical processes and were "so complex that any actual state has a negligible a priori probability."[11] Ecological prediction had to be understood as confined to general trends, as with evolutionary biology: one could not expect to predict how the future would unfold in a given location.

Mathematical sciences such as cybernetics gave ecologists an imaginative way to conceptualize the relationship between ecological stability and the diversity or complexity of the system. "Perhaps," Margalef wrote, "in the depths of his psyche, the ecologist may be grateful to cybernetics for throwing a tenuous mantle of respectability over the discredited method of reasoning by analogy."[12] The weak point of such arguments was their dependence on analogies that, though suggestive, were both general and ambiguous, leaving room for ecologists to apply them in different ways. A problem that Margalef did not mention was that analogical reasoning carried a danger of circular logic. One looks for an analogy that supports what one already believes to be true. The use of analogy thus becomes a rhetorical or heuristic method, a way of justifying one's views to another person, and not a way of testing one's own belief. The cybernetic analogy thus appealed most to people who were already thinking of the ecosystem in physical or engineering terms, as though it were a machine. And it resonated with people who were willing to adopt the "macroscopic" view of the ecosystem and were not wrapped up in the detailed analysis of nature, that is, in those myriad processes that defied prediction.

Margalef knew that some people would deny the validity of his ideas because their own work had not predisposed them to think in broad terms. He linked these predispositions to the landscapes that ecologists studied, commenting that "ecologists reflect the properties of the ecosystems in which they have grown and matured." "All schools of ecology," he continued, "are strongly influenced by a genius loci that goes back to the local landscape." People working in arid regions were more likely to think that weather fluctuations controlled populations and that communities were poorly organized. In the Mediterranean and Alpine countries, where humans had influenced the environment for thousands of years, ecologists had developed a distinctive school of plant sociology to study the vegetation there. Scandinavia, with its poor flora, had "produced ecologists who count every shoot and sprout."[13] Those who were fixated on the smallest details of community composition were not likely to value the very general approach he was expounding.

But he clearly recognized that the United States held a receptive audience. In North America and Russia, he noted, the vast regions had suggested a dynamic approach to ecology and the concept of succession leading to the climax community, a theory especially associated with American ecology. As he acknowledged, Henry Cowles at the University of Chicago was a leader in developing the theory of succession. Margalef argued that the concept of succession, referring to an orderly and directed change, was one of the most important generalizations of ecological research, comparable to the theory of evolution in general biology.[14] The problem was that the concept was expressed in imprecise language: what did it mean to speak of "progressive" trends? Reframing ideas of "progressive development" in terms of energetic principles could allow for a more precise statement of why succession proceeded in the way it did. In fact the initial enthusiasm over cybernetics in the postwar years, not just in ecology but in fields such as neuroscience and psychology, stemmed from the way it enabled scientists to reframe their discussions of teleological or goal-directed processes so that vitalistic language could be completely banished. Clements's "complex organism" gave way to the cybernetic system.

In ecology, cybernetics offered a way of removing from the old notion of progressive successional development, an idea that had originally reflected Americans' views of their national development, the mysticism implied in the theory that nature was programmed to unfold in a certain way. Fundamentally, Margalef was offering Americans an improved version of something they already believed. But adopting the more exact language of mathematics did not mean that the assumptions underlying the theory had actually been tested or validated. This baggage of underlying premises was now somewhat hidden beneath the mathematical formalism of ecology. Within a couple of decades, ecologists who had grown disenchanted with the use of equilibrium models would unpack that baggage and in some cases recommend chucking it overboard.

In addition to cybernetics, ecologists began to delve into other fields of applied mathematics, such as operations research and systems analysis. These fields focused on problems, generally of a military nature, in which decisions had to be made. Operations research, an applied science whose roots went back to the early twentieth century, was developed during the Second World War as a method of providing a quantitative basis for making decisions in a military context.[15] Operations researchers were successful in solving military problems, and after the war the methods were extended to various other domains. Operations research was distinct from other scientific research in that it involved urgent problems that had to be solved quickly and often without adequate data.

Operations research was closely related to systems analysis or systems engineering. The techniques of these disciplines were used to analyze complex problems that arose in large systems, with the goal of helping to make decisions. Like operations research, systems analysis is a method of simplifying a problem by seeking generality. As described by one systems ecologist, Bernard Patten, the method "accepts as an operating principle that no complex system can be fully known in all of its interactive details, and accordingly seeks to elucidate global properties that characterize 'core' dynamics."[16] The analytical tools at the core of operations research and systems analysis were designed to develop the trade-offs between objectives in complex problems, with the aim of assisting decision makers with their deliberations. One of the inventors of mathematical programming, George Dantzig, explained that the analysis started with a mathematical model of a system and an objective that could be quantified, then evolved a computational method for choosing among alternative actions.[17]

Military problems remained an important stimulus for the advance of these techniques after the war. The RAND Corporation in Santa Monica, California, became a center for the development of applied mathematics. RAND, a non-profit corporation formed in 1948, evolved from Project RAND, which dealt with the design of satellites after the Second World War.[18] Through RAND the U.S. government involved hundreds of scientists, mathematicians, and engineers in military-related research and planning after the war. Systems analysis, for example, was applied to problems in strategic bombing, and from there it catalyzed the development of other methods and technologies, such as game theory, dynamic programming, and improved computer capabilities. From the early stages these innovations, aided by the rapid development of the electronic computer, also found applications in various problems in government, industry, economics, and commerce—and in ecology.

When applying these mathematical techniques to ecological problems, modelers framed questions in a way more common to operations research than to traditional ecology, beginning with statements of objectives. Objectives could be quite specific, such as minimizing the effects of radionuclide contamination in an ecosystem, or very general, such as stabilizing a system.[19] Determining the solution would depend on how the objectives were stated, what resources were available, and what constraints operated on the system. Ecological problems could be framed in a way comparable to engineering or economic problems, such as those encountered in fields of urban planning that dealt with the provision of health services, fire protection, law enforcement, housing, water resources, and so forth.

These kinds of problems required computers. In the early 1960s Kenneth Watt wrote about the role computers would soon have in ecology as ecologists turned to detailed simulations of ecological interactions.[20] Watt, then working on insect pest-control problems in Canada, pointed out that ecology could benefit from using the mathematical methods designed to cope with complex systems in areas such as economics, transportation, the military, and engineering. In a discussion of systems analysis in ecology in 1966, Watt argued that computers would free ecologists from the drudgery of data collection and analysis and give them more time for high-level thought.[21] He learned about mathematical programming with the help of Richard Bellman at the RAND Corporation. Bellman had developed the discipline of dynamic programming in the early 1950s.[22]

Ecologists were interested in both digital and analog computers, depending on the nature of the problem. Analog computers were electronic models of the systems being studied, an example being Tom Odum's electrical circuit analogues (see chapter 7). Analog simulation was particularly useful in solving sets of simultaneous differential equations. It was applied to problems in population ecology where, for example, predator-prey interactions could be modeled using electronic components. By the 1970s ecologists were getting interested in hybrid computers (developed in the aerospace industry during the 1960s), which combined features of analog and digital computers.[23]

Computer modeling as conceived by Tom Odum and Bernard Patten was a way of dealing with complexity by reducing a system to its essential features. As Odum explained, these were macroscopic models, that is, overview models. One of the first steps was to simplify the problem, draw out its main features, and then diagram it in a way that conformed to the operation of a machine (as illustrated in figs. 8.1 and 8.2 from Patten and Van Dyne). As the ecosystem grew increasingly machinelike, the "web of life" was transformed into an abstract network or electrical circuit. Frank Golley, an ecosystem ecologist at the University of Georgia, observed in his history of the ecosystem concept that the very popularity of the term *ecosystem* (as opposed to alternatives in use in Europe, such as *biocoenosis*), owed something to the fact that the word conveyed the idea of an ecological machine.[24]

It is ironic that the development of ecology, a science aiming at the understanding of nature, should seem to advance by depicting the "natural" as though it were artificial, forcing nature, with all its diversity, uncertainty, and historical contingency, into the mold of a manmade object. But as molecular biology advanced full throttle toward the goal of controlling life, the need to link ecology to engineering must have seemed not just beneficial but almost necessary, with no lingering nostalgia for what was being lost. Already by the mid-1960s, one advocate of systems biology dismissed the "archaic" categories characteristic of traditional biology. The living world had become "a manifestation of directed energy transformations, information replication, storage, transfer, and retrieval mechanisms."[25] Modelers looked forward to an age when the computer would be an indispensable part of ecological work.

The computer could be used to perform simulation experiments, allowing the ecologist to ask what would occur if the system were altered in some way.[26] Modeling these "experiments" enabled ecologists to judge whether their assumptions about how the system worked were correct, refine the assumptions, and gradually work toward more realistic models. In most cases, however, the ability to model was severely limited by lack of data as well as by the bounds of the computer technology itself. Because of these limitations, many ecologists, even those who welcomed theoretical approaches to ecology, were skeptical about simulation models. Nonetheless, these early efforts were pointing toward ecology's future as a highly technical, mathematical subject.

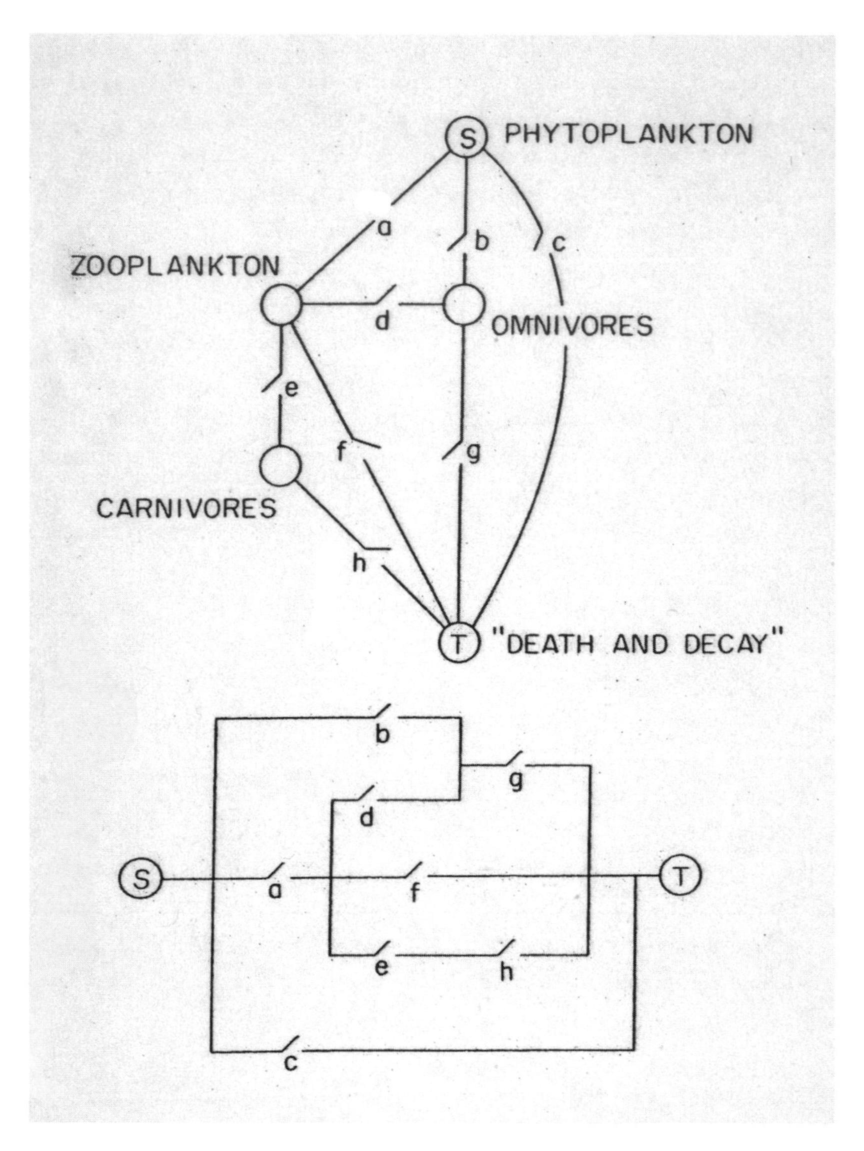

Fig. 8.1. Systems ecology diagrams. Bernard Patten used network diagrams (*top*) and circuit diagrams (*bottom*) to depict the transfers of material and energy in an ecosystem, from the source (S) to the terminal (T). From Bernard Patten, "Systems Ecology: A Course Sequence in Mathematical Ecology," *BioScience* 16 (1966): 595. Copyright © 1966 by American Institute of Biological Sciences. Reproduced with permission of American Institute of Biological Sciences via Copyright Clearance Center, Inc.

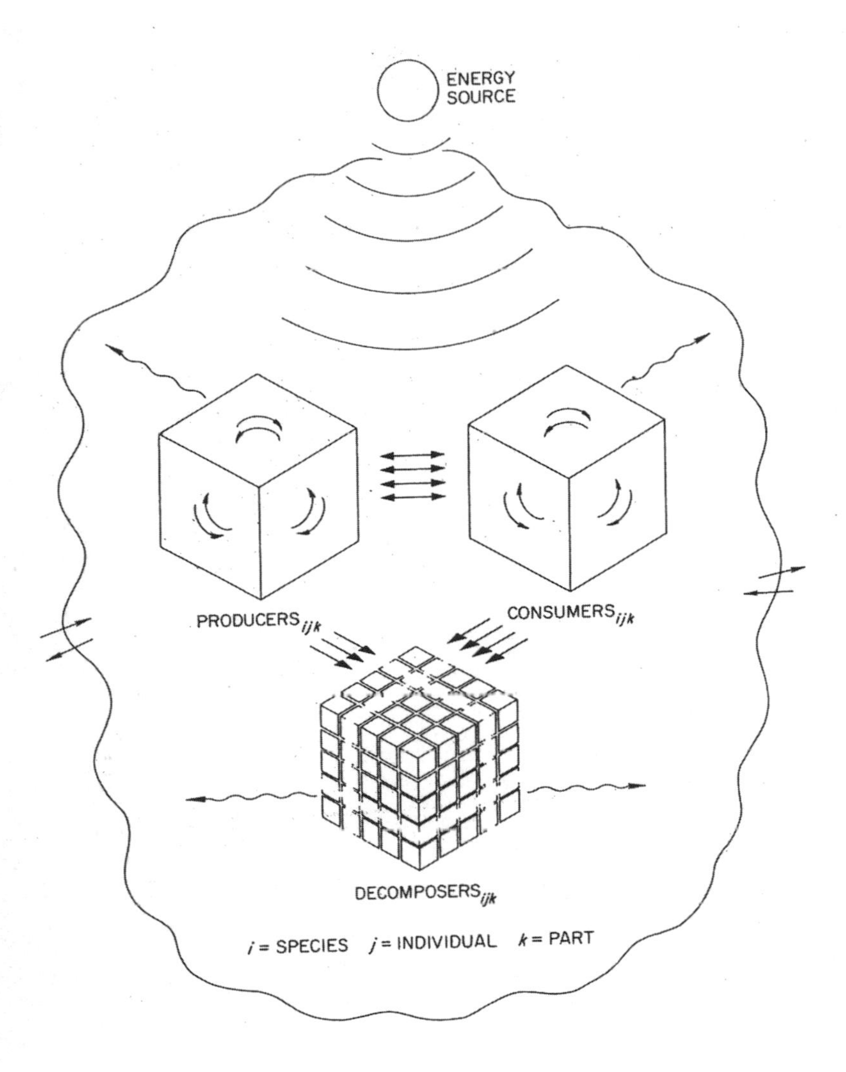

Fig. 8.2. Trophic levels of the ecosystem. George van Dyne's representation of the ecosystem, adapted to pseudoalgebraic computer languages, depicts each trophic level as a three-dimensional matrix. Straight lines represent matter, and wavy lines represent energy. From George Van Dyne, "Ecosystems, Systems Ecology, and Systems Ecologists," Health Physics Division, Oak Ridge National Laboratory, Oak Ridge, TN, June 1966. Courtesy of Oak Ridge National Laboratory, managed by UT-Battelle, LLC, for the U.S. Department of Energy.

The growing divorce between modern ecology and natural history meant that just at the time that ecology was being popularized in the environmental movement, it was also becoming an esoteric subject. Increasingly, any attempt to foster public understanding of ecology would require ecologists to translate their image of the machinery of nature into layman's terms. The layman was most likely to be captivated by old-fashioned natural history, by the fascinating stories of startling adaptations in plants and animals that became, and have remained, the stock in trade of nature programs on television. But the modern mood of systems ecology eschewed descriptive natural history and any suggestion that ecologists were nothing but bug-hunters or bird-watchers. Natural history was equated with lack of rigor, intellectual softness, and low status. In his introductory remarks at the head of a four-volume collection of articles on systems ecology and simulation published in the 1970s, Bernard Patten described systems ecology as an attempt to convert "soft" science into "hard" science, that is, to work toward the goal of prediction and not just description or explanation.[27] Prediction implied control: systems ecology was about learning to design and control ecosystems, and the systems ecologists knew they would meet resistance from colleagues who found those goals inappropriate.[28] Systems ecology brought together biology and engineering in a new synthesis. Humans were not just components of ecosystems, they were the designers and manipulators of systems.

For all its modernity, the new systems ecology retained many of the assumptions that had guided Frederic Clements's organismic concept of the plant community, even though plant ecologists were criticizing Clements's theories at this time. George Van Dyne, for instance, described systems as progressing toward a stable climax, whereas human intervention produced a different kind of equilibrium to which he applied Clements's term "disclimax" (for disturbed climax).[29] Models included human-dominated systems, but humans were still seen as disturbers of "natural" systems. The guiding assumptions of Clements's theory were translated easily into the new modeling language. Succession was thought to lead to higher diversity of life forms within the ecosystem, and hence to greater stability. The introduction of mathematical techniques tended to reinforce the presumption that systems tended toward equilibrium unless disturbed.

One might even conjecture that systems science operated subtly to rehabilitate the organismic perspective in the post-Clements world, making it less mystical by emphasizing the underlying machinery of nature, and thereby allowing its perpetuation beyond the demise of the complex-organism concept. Eugene

Odum could propose that the ecosystem was analogous to an organism in a state of homeostasis, with humans being analogous to pathogens, but what might seem a throwback to vitalism or mysticism could be met with the assurance that the ecosystem "organism" really functioned like a complicated machine.

The new methods did not so much change the traditional concerns and beliefs of ecology as introduce a new language for expressing those beliefs, refining the concepts, and analyzing the data. The question of man's place in nature was rephrased and adapted to the new environmental problems of the postwar period, but the underlying problem that had plagued George Perkins Marsh remained, as troubling as it ever was. As long as humans were so obviously endangering the planet and their own health, and as long as they held the power to shape the world through science and technology, it was difficult to see them as anything other than separate from nature and destroyers of equilibrium. The ecosystem concept brought humans into the picture, yet still considered them in some sense as operating contrary to nature's way, defying, to their peril, the wisdom of nature. The debate on this topic continued as ecologists, armed with new methods, new ideas about community structure and dynamics, and better data, revisited the problem of stability and its converse, the role of disturbance.

As ecologists worked on the development of general theories in the 1960s, a major concern was what characterized a stable population, community, or ecosystem. The interests of population and community ecologists linked up with those of ecosystem ecologists on this question. Population ecology, which had been advancing slowly from the 1920s, became a dominant field within ecology in the 1960s and led the development of general theory and the application of mathematical techniques to the study of species interactions. Population ecologists studied predator-prey and host-parasite relationships and competition within and between species, trying to understand how these population interactions affected the structure of communities. The work of Margalef and Evelyn Hutchinson, as well as innovative theoretical work by Hutchinson's student Robert MacArthur, focused interest on the connection between the structure of ecological communities, the complexity of food webs, and the diversity of organisms in the community.[30] In making this connection between population ecology and community ecology, the questions about stability and diversity that preoccupied ecosystem ecologists moved to center stage. By the end of the 1960s, the problem of stability seemed ready for fresh analysis and sharper definition.

How should one define *stability*, *diversity*, *regulation*, and *organization*? What did these concepts mean in natural systems, and what did they mean in human-dominated systems?

The question of ecological stability took on new urgency in the context of deep anxieties about the long-term viability of human-dominated systems. As Margalef argued, "Ecology should inspire a wiser management of nature: the feedback should work."[31] The new systems ecology, focusing attention on how the flow of energy affected the way the machinery of nature functioned, allowed one to assess how humans altered the energy flow and consequently made the machine work differently. Margalef pointed out that it was no longer possible to have a "natural" ecology that ignored human impact.

Heightened concern about overpopulation, pollution, and dwindling resources fed a growing and increasingly militant environmental movement. Ecologists expressed a new sense of urgency as they drew attention to looming environmental problems. Their rhetorical stance was feisty, though hardly bolder than Frederic Clements had been when he urged Americans to "meet change with change" back in the 1930s. Garrett Hardin in 1968 published a controversial analysis of common property resources, "The Tragedy of the Commons," and concluded that some degree of "mutual coercion, mutually agreed upon" was needed to prevent exploitation of resources and overpopulation.[32] Paul Ehrlich published his controversial book *The Population Bomb* in the same year, also calling for immediate measures to limit population growth.[33] Sales of Eugene Odum's textbook soared as students flocked to ecology courses. The magnitude of environmental problems kept growing, and ecology came increasingly to be linked to environmentalism and seen as the "subversive science" that Paul Sears had described in 1964.

The sense of impending crisis permeated discussions of long-standing ecological questions. A major symposium convened at Brookhaven National Laboratory in 1969 brought together ecologists and evolutionary biologists to discuss the various ways of analyzing stability in relation to diversity.[34] In between the more technical subjects, the papers and discussions reflected great concern about continued economic growth, overpopulation, and whether "civilized technological man" was able to create a society that was stable in its relationship to the environment.[35] Were people capable of adapting to "spaceship earth"?

In this age of social unrest, riots, and youthful rebellion, with the threat of nuclear war hanging over all, discussions drifted at times into apocalyptic visions. Crawford S. "Buzz" Holling, then at the University of British Columbia, delivered a paper on the relationship between stability in ecological and social systems, in which he commented on the slowness with which social institutions responded to such problems as overpopulation, a theme that he continually elaborated over the course of his scientific career. In the discussion someone asked whether Holling thought the world was going to become unstable. He answered that although he did not think human survival was at issue, there could occur "drastic socioeconomic collapse triggered by nuclear holocaust or social erosion of some kind."[36]

The mood of crisis was picked up also by Garrett Hardin, who gave the plenary address, entitled "Not Peace, but Ecology." He ended with a rousing call for a new Declaration of Independence to guide the crowded world, starting with the right of all people to use the media of the world. He envisioned "spirited and at times bloody warfare" against "misunderstandings, public apathy, vested interests, and widespread addiction to the religious opiate called 'Progress.'" Concluding, he asked, "Once the magnitude of the issues involved is understood, who would prefer peace?"[37] That the problem of stability should be a focus of concern in an age of instability and looming crisis is not surprising.

As these questions were being debated, ecology was about to benefit from a major infusion of funds in connection with the International Biological Program (IBP), which began in imitation of the International Geophysical Year of 1957. The idea of coordinating international scientific projects around certain key themes in biology was proposed in 1959, received a favorable response from scientists, and then became the subject of discussion and planning in international scientific organizations during the 1960s. In 1961 three main themes were identified: human genetics, conservation, and improvements in the use of natural resources. Planning committees worked to define the organization of projects during the 1960s; by the late 1960s the ecological analysis of large biomes had emerged as a major focus of attention. A biome was understood to be a region controlled by climate, in which certain kinds of plants or animals dominated.

American scientists had contributed to planning for the IBP almost from the start, although there was much ambivalence about whether the United States should join the program. Some scientists thought it might take funding away from other research, and others argued that the basic questions being posed in the IBP were not American problems.[38] In the end the National Science Foundation funded five major biome studies as part of the IBP from 1968 to 1974; these focused on grasslands, coniferous forests, deciduous forests, tundra, and deserts.[39] More than eighteen hundred scientists were involved in the biome programs, which provided scientists with an opportunity to explore the potential of systems ecology.[40] The information gained from these studies helped to document the various services that ecosystems performed in stabilizing and maintaining the environment, and hence it offered additional reasons for promoting conservation.

But the projects undertaken during the International Biological Program got mixed reviews from scientists. There were significant problems in the management of these "big science" projects, concerns about the quality of research produced, and intellectual problems arising from the way the ecosystem was viewed and studied. The idea of studying the totality of natural ecosystems was simply unworkable: the myriad individual research projects were never integrated into a general theory. Eugene Odum identified as a problem the lack of a central theme or a set of hypotheses that scientists would agree to test.[41] Frank Golley drily remarked that the IBP carried "bricolage to a new level of activity and scale."[42] Odum, as one of the planners behind the IBP projects, had hoped that the program would focus on landscapes that might not be confined to natural areas, establishing a new science of landscape ecology as a "pure science" contributing to landscape planning. But as Golley observed, the biome projects did not in the end promote landscape ecology and hence failed to capitalize on this opportunity to extend the reach of ecology. (Landscape ecology was developed later, in the 1980s).

A related criticism dealt with the failure to connect the scientific goal of constructing an ecosystem model with the needs of the people who might be using ecological knowledge to solve specific problems. Kenneth Watt, by then on the faculty of the University of California at Davis, critiqued the biome ecosystem models as the IBP projects drew to a close, making pertinent observations about the relationship between the modeling enterprise and the decision-making process.[43] His comments were consistent with his early background in systems analysis, where the purpose of modeling was to aid decision making. As systems ecologists tried to develop more complex models, they moved further away from the decision-making process. Watt argued that trying to develop large, realistic models of ecosystems was inefficient and even pointless. Rather, the form of the model had to be related to the kind of decision that needed to be made and also to the decision maker, who often was not the scientist. Decision makers could not cope with cumbersome models whose assumptions were hard to grasp.

The whole strategy of modeling, Watt explained, should be adapted to the context in which the model would be used. A cabinet official and a mayor needed to know different things and therefore needed different models to guide their decisions. If ecological research and modeling was about predicting and controlling, and if the model was to lead to a decision of some kind, then the question of how to steer between "unrealistic oversimplicity" and "unwieldy and untestable complexity" became crucial. Watt returned to the problem of communication: ecologists needed to be guided not just by academic interest in studying nature, but by the need to apply the results of their work, which meant making it understandable to nonexperts.

The application of systems science included expert testimony in court actions concerning environmental damage. Orie Loucks, then at the Institute for Environmental Studies at the University of Wisconsin, Madison, discussed in the early 1970s how systems science was being used in court actions.[44] Examples of such actions included hearings on the impact of DDT in Wisconsin waters, the impact of radionuclides and thermal pollution on fisheries, and the flooding of river basins in Florida. Scientists were called upon to discuss various aspects of the ecological questions concerned.

Loucks, drawing on his experiences in two environmental court actions, remarked that systems science provided a way of viewing the environmental system in an integrated or holistic manner. In a case involving whether DDT should be considered a pollutant, for instance, it had to be demonstrated that DDT was causing deterioration in the ecosystem as a whole. Loucks's job as scientific expert was to pull together the technical testimony given by other people and to explain how the entire ecosystem was affected by this pesticide. That strategy worked, and the court ruled that DDT was an environmental pollutant as defined by Wisconsin law.

Loucks also suggested that the systems approach was effective in part because the flow charts and equations provided a structured way of looking at systems and assessing the impact of any perturbation. Much of the scientific testimony could not be cross-examined; as Loucks remarked, "It is always difficult to cross-examine mathematical relations." He viewed the outcry following Rachel Carson's indictment of pesticide misuse, *Silent Spring*, as having been "partly emotional, and clearly based on scattered, piecemeal scientific evidence."[45] To be effective in the courtroom, the evidence had to be brought together and presented in a consistent way, dressed in the garb of the most up-to-date scientific method, systems analysis. Loucks thought that systems science could and should have an important role in environmental court cases, along with other scientific fields in ecology, biochemistry, and statistics. He also suggested that hearings of this kind created opportunities for interaction between disciplines and between scientists and laymen, creating a sense of community and forcing scientists out of the isolation of their specialized fields of research.[46]

In his assessment of the use of systems analysis in environmental litigation, Loucks noted that the weakest link was at the level of system definition and description. The ecosystem was a concept or perspective defined by the existence of cycles of matter and flows of energy, but as a real object the system could be defined in many different ways. Frederic Clements had thought it was possible to identify boundary zones separating different communities, and he went so far as to coin a term, *ecotone*, for these boundaries. Sometimes boundaries were clear, as on the shores of a lake. But in other environments when ecologists looked for boundaries, they had trouble seeing them. A careful definition of the ecosystem was a necessary step in testing hypotheses concerning how succession proceeded and in clarifying the meaning of concepts of stability, homeostasis, or equilibrium.

The relationship between how ecologists defined the ecosystem and how they probed the meaning of these concepts was particularly well illustrated in a long-term analysis of a forest ecosystem led by Herbert Bormann, a plant ecologist, and Gene Likens, a freshwater ecologist. The ecosystem was identified not by the dominant plants and animals in the system, but as a watershed, that is, the total area drained by a given body of water. Bormann and Likens began using small watersheds to study ecosystem processes in the 1960s and over a couple of decades built up a highly successful model that allowed them to investigate some of the assumptions that had been guiding studies of ecological succession for decades. Stephen Bocking, in a history of their research, has noted that the study of experimental watersheds went back to 1909, when the Forest Service set up experimental watersheds to assess land management practices and their effects on stream flow. Some of these watersheds were located in the Hubbard Brook Experimental Forest in New Hampshire, and they in turn became the sites of Bormann and Likens's research program.

The Experimental Forest, with mixed deciduous and coniferous trees, covered about three thousand hectares of rugged land in the central part of the state. The entire forest had been cut once, starting in the early twentieth century, and after that had not been cut. Some of the small watersheds were located on bedrock, preventing deep seepage, and therefore were relatively watertight. Bormann and Likens realized that they could use such watersheds to get fairly precise measures of nutrients going into and coming out of the ecosystem. They could also impose experimental treatments, for example removing vegetation from entire watershed ecosystems and comparing the responses of experimental systems to undisturbed ones. This allowed them to investigate and test the assumptions about ecological succession that were current in the literature on ecosystem ecology. Their research, reported in a 1974 *Science* article and their 1979 book *Pattern and Process in a Forested Ecosystem*, was extremely important in providing evidence of acid precipitation in North America. They concluded that acid rain added stress to the forest ecosystem, although they did not have enough information to conclude that it was having a role in the decline of forest growth that had been seen in the United States since 1950.[47]

Ecosystem development, they explained, had been seen as a kind of battle between the development of organization in the ecosystem and destructive forces that diminished ecosystems.[48] Ecosystems were thought to achieve increased control of the physical environment as they developed. To use Eugene Odum's concept of ecological homeostasis, the ecosystem, like an organism, protected itself against outside disturbances. Succession therefore proceeded in an orderly way and culminated in a stabilized ecosystem, the climax. Continuing the organismic analogy, Odum inquired whether mature ecosystems might, like aging individuals, develop unbalanced metabolisms and become vulnerable to disease or perturbation. The overall "strategy" of development, to use Odum's word, was directed toward achieving as large and diverse an organic structure as was possible.[49] This was the controversial idea that Bormann and Likens challenged.

Bormann and Likens tackled this model of ecosystem development by performing a simple experiment: they clear-cut the forest and observed the successional processes that followed. They divided successional development in their experimental system into four phases, which they called reorganization (the first decade or two), aggradation (accumulation of biomass, lasting over a century), transition (when biomass declines), and steady state (when biomass fluctuates about a mean). They studied the aggradation phase, beginning about fifteen years after clear-cutting, the most intensively; their treatment of the steady state was more theoretical. The idea of the steady state was suggested "with some trepidation," because they believed there was change even at that level. Moreover, the steady state could possibly be "a period of quiet antecedent to a cyclic revolution." Steady state, which occurred when the total biomass fluctuated about a mean, in fact was best seen as a "shifting mosaic," meaning that the forest was made up of irregular patches of vegetation at different ages.[50]

The idea that the ecosystem was a shifting mosaic of patches picked up on a theme developed in 1947 by British plant ecologist Alexander S. Watt.[51] In his presidential address to the British Ecological Society, Watt discussed the relationship of pattern and process, which ecologists explored using both analytic and synthetic methods. He pointed out that ecologists moved back and forth between a synthetic or ecosystem approach and analysis of the component parts of ecosystems. The analytical extreme was illustrated by Henry Gleason's individualistic approach, which cut the community into small pieces. Watt's address examined the relationship between these two perspectives in the study of plant communities.

Studying plants individually was not feasible, so Watt instead chose aggregates of individuals that formed patches. The patches existed in a mosaic that formed the community, and the dynamic relationship between patches created an overall pattern of orderly change in the community. However, Watt also realized that the "inherent tendency to orderliness" could be disrupted by obstacles that upset the order.[52] The task of the ecologist was to understand both what created order and what upset that order.

Watt's article established a model for later studies of heterogeneity, or "patchiness." Bormann and Likens's concept of a "shifting mosaic steady state," characterized by irregular oscillations about a mean, was similar to Watt's model of succession. At any point within the system there was change, but in the system as a whole the forces of aggradation and decomposition were roughly balanced, producing a steady state. Bormann and Likens realized that the concept of the steady state was only an approximation. In the strict sense there was no absolute steady state, for even a system thought to be in a steady state experienced slow, long-term change.[53]

Bormann and Likens used their study to challenge the received wisdom on ecosystem development, in particular the idea, which they took from Eugene Odum, that diversity increased as development proceeded.[54] Odum had also suggested that as systems became more biologically controlled, species diversity should increase. Bormann and Likens concluded that the period of maximum stability, understood as maximum control over destabilizing forces, was not the main feature of the steady state condition. In fact the closest control was in the aggradation phase, when biomass was accumulating. Yet they also found that species richness, which could be used as a measure of diversity, was lowest during the mid-aggradation phase, even though biotic control of metabolism and biogeochemistry was at the maximum during that phase. Species richness actually peaked in the early stage of development after clear-cutting, when the system was least stable.

Another important question concerned the impact of outside disturbance on the development of ecosystems. Traditional models of ecosystem development depicted succession as an orderly progression, as though there was an inherent tendency toward order.[55] However, ecosystems were subject to random disturbances or perturbations. Some occurred within the system and might be considered part of its normal development, as when a tree fell down. Other disturbances originated outside the system and would not be considered part of normal development. Human impacts were in this category of external or exogenous causes of disturbance. But ecologists had begun to question the idea that exogenous agents were necessarily separate from normal developmental processes. This research focused in particular on the role of fire, which held special significance in that it was produced both through impersonal causes and through human action.

Orie Loucks, for instance, had developed a model showing that in forests of southern Wisconsin, periodic fires had perturbed the system so that the vegetation went through cycles of growth.[56] Moreover, this perturbation actually functioned to create a high level of species diversity as well as high productivity in periodic "waves." He concluded that the periodic, random perturbations were beneficial and that in suppressing fires in recent decades, humans had actually caused the "greatest upset of the ecosystem of all time." Other ecologists had also criticized concepts of the steady state, suggesting that exogenous disturbances, including not just fire but even hurricanes, might destroy and restart forests at irregular intervals. Despite these criticisms, Bormann and Likens did not dispense with the idea of development toward a steady state, which their research suggested was still a useful concept. They felt that the importance of catastrophic recycling had been overemphasized and that in the forested areas they studied it probably did not occur frequently and was therefore not a deterrent to the development of a steady state.

Bormann and Likens recognized that part of the difficulty they had in discussing concepts like stability stemmed from the vagueness of ordinary language. In physics, terms are often defined mathematically by relating them to quantities that can be measured. A term such as *force*, which might be ambiguous in common usage, can be defined as "mass times acceleration," referring to two measurable variables. Definitions of this kind can be standardized and everyone can use terms the same way. In ecology, consensus about how terms should be defined and how they could be related to measurable quantities was hard to achieve, given that ecologists worked in diverse ecological contexts.

Ecologists used many terms in a large and metaphorical sense, much as Darwin had done in 1859 when he used "struggle for existence" to denote a wide variety of ecological relationships, including situations in which organisms were not engaged in a direct struggle with other organisms. The problem of ambiguous language continued to trouble ecologists as the science developed. Stuart Pimm suggested in 1984 that careful definition of ecological quantities would allow for a more systematic and rigorous approach to the analysis of these problems.[57] He identified five meanings of the term *stability* and then differentiated the meanings with a set of subsidiary terms, adding *resilience*, *persistence*, *resistance*, and *variability* to the original term *stability*. Pimm followed up his critique with a book-length study that examined the various meanings of stability and explored their relationship to community structure.[58] He noted that field ecologists were working on these problems separately from mathematical ecologists and that good communication between these groups was lacking. He also identified a "sect" who thought that disagreements about these concepts stemmed from the fact that people were looking at different parts of a complicated field. This comment echoed Margalef's observation that the way ecologists developed their ideas and approaches depended a great deal on the kinds of landscapes and species they studied.

Whatever the cause of the disagreements, the fact was that the same words were used in very different senses, and this lack of standardization continued even as new terms were coined and new definitions proposed. C. S. Holling, for instance, commented that Pimm's use of the term *resilience* was virtually opposite to his own understanding of the concept.[59] Pimm's definition of *resilience* coincided with Holling's concept of *stability*, namely a return to an equilibrium state following perturbation. Holling had used *resilience* to talk about conditions far from equilibrium.

These ecological terms were also metaphors—ecology was rife with metaphors taken from physics—which if anything compounded the problems that ecologists had faced decades earlier when organismic metaphors were introduced. By the 1980s the language of physics dominated, starting with the ecosystem concept and moving on to metaphors that described concepts of ecological resilience, perturbation, equilibrium, and nonequilibrium. But without a means of quantifying these definitions, and without agreement about what the words meant, the plethora of new terms could just as easily spark disagreement in ecology as solve the linguistic ambiguities that beset the science. In addition, multiplication of terms produced a more abstract, jargon-laden depiction of ecological processes, removing ecological concepts even further from the layman's grasp.

The metaphorical language common to ecology tended to reinforce the assumption that nature was something lying "out there" that was studied more or less objectively by scientists whose aim was to discover what nature was "really like." Their disagreements concerned how that natural world should be characterized and especially how to characterize equilibrium states and the ability of systems to perpetuate themselves. Humans exploited, interfered with, or disturbed nature. The goal of ecology was to learn nature's ways so as to minimize the effects of that disturbance. As we have seen, the dramatic evidence of human disturbance in the postwar decades fostered this view of humans as out of balance with and fundamentally alienated from the natural world.

There were dissenting voices from this negative vision, and the most eloquent one was that of René Dubos, who in 1980 in a short book for a general audience, *The Wooing of Earth*, took up the problem of human creativity in relation to environmental problems. Dubos, then professor emeritus at Rockefeller University, came to ecology from a background in medical science and had a deep interest in humanized environments, an interest he traced to his childhood in the farm country north of Paris. When this book was published, toward the end of his life, he was a well-known author of many books and articles for a general audience. He had also been a prominent supporter of Rachel Carson.

Dubos started with the assumption that human work, aided by the creative imagination, could transform the world in positive ways and bring out its potential, could in fact improve upon nature. The "humanization" of the earth had certainly involved the destruction of wilderness, going back to ancient times. No doubt, Dubos mused, the old civilizations around the Mediterranean had impoverished the land as their populations grew. But Dubos denied that these actions were all negative. The ancient Greeks, in cutting down the forests, had brought out the beauty of the underlying landscape and had enabled different species of plants and animals to flourish. These changes in turn affected the human psyche, perhaps encouraging the philosophical and poetic imagination.

"Ecology becomes a more complex but far more interesting science," he argued, "when human aspirations are regarded as an integral part of the landscape."[60]

He was fully aware that the human impact on the earth had been enormously destructive. He did not fail to note the modern threats of nuclear warfare, over-population, waste of energy and resources, pollution, and environmental degradation, all themes that had been in the forefront of the environmental literature for many years. Yet despite these grave concerns, he sought a way to incorporate the human perspective into the discussion of how such problems might be corrected. The concept of ecological homeostasis, which had been such a central feature of ecological thought, was another major point. He noted places where nature exhibited remarkable powers of self-healing: "Even Bikini and Eniwetok, pulverized and irradiated by fifty-nine nuclear blasts between 1946 and 1958, are said to be reacquiring a complex biota, despite the destruction of their topsoil."[61]

But the ability of the land to heal itself did not mean that environmental recovery was confined to reestablishing an original state of equilibrium. Dubos pointed out that random events influenced the evolution of ecosystems in unpredictable ways. Nature did not simply bounce back once people stepped aside. Sometimes new adaptive responses of a creative nature could be made, and in those cases human design could play an important role. This was the theme he wanted to explore.

The idea of ecological equilibrium, he argued, did not imply the conclusion that many people adopted, which was that "nature knew best."[62] Dubos said that "natural" systems showed clumsy and wasteful "solutions" to ecological problems. Many animal populations crashed at regular intervals, which meant suffering on a large scale. Accumulations of vast quantities of fossil fuels and piles of guano on islands were nothing if not failures in recycling. Humans had created artificial but stable ecosystems every bit as admirable as natural systems, with high levels of diversity replacing the wilderness areas that had been destroyed. Dry areas had been made into prosperous agricultural lands, and in some countries, notably the Netherlands, large areas had been reclaimed from the sea, an impressive process of landscape evolution that started in the early Christian era. Without romanticizing human management of the environment, Dubos drew attention to the many positive effects of human ingenuity, including the conversion of desolate lands into productive regions so lovely that their artificiality had been forgotten by the public. Artificial landscapes mellowed as they evolved over time, acquiring "the poetic nobility of natural creations."[63]

Dubos was trying to develop a logic that would accept the inevitability of human manipulation of nature while recognizing human responsibility toward the earth and other people and anticipating the long-term consequences of human actions. He summed up his management philosophy in the dictum "Think globally and act locally." Doing that required not just scientific knowledge but also ethical principles to guide human action, principles that ecology alone could not provide. What brought an ethical component into environmental problems was that ecological relationships were influenced by the presence of humans. Our actions determined the direction and extent of environmental changes. Therefore science had to be "supplemented by humanistic value judgments concerning the effect of our choices and actions on the quality of the relationship between humankind and Earth, in the future as well as the present."[64] This was one of the greatest challenges faced by modern society.

How could a humanistic perspective help us reorient our values, incorporating the knowledge available through science but not losing sight of the positive aspects of human involvement with the world? In the urban context, considered in chapter 9, balancing the need to expand scientific knowledge and awareness with the human perspective on environmental problems became especially challenging.

CHAPTER NINE

New Frontiers

As ecology evolved over the course of the twentieth century, its social role evolved with it. By the 1990s, with the human population topping 5 billion, it became evident that humans were having an impact on the entire planet. Figuring out how to respond required not just more ecological science but better communication of an ecological perspective to the general public. Whereas ecologists had often sought out relatively pristine or wilderness areas for study, in part to provide a benchmark by which to judge human influence on the environment, it seemed a logical next step to bring ecological studies into populated areas in a more deliberate manner. The "new frontier" of ecology was the city. Thus we come full circle, ending close to where we began our story, in a city.

In the United States, discussion of the need to incorporate human ecology into general ecology went back at least to the 1920s but did not significantly affect the evolution of ecology as a biological discipline. The idea of including human populations in ecosystem analysis was clearly part of the original vision of ecosystem ecology, but the bulk of ecological study was directed toward regions that were sparsely populated. Hence bringing ecology into the city was a relatively novel thought. In the urban setting it is difficult to evade the question raised by René Dubos: How "human-centered" should ecology be as it attempts to address the environmental problems of today? (see chapter 8). In the city the problems of general ecology, which involve the study of ecological systems, the services they provide, and the mechanisms that sustain them, become more closely connected to the problems of human ecology, which involve the study of what sustains human societies in a state of reasonable health and happiness. Connecting how ecosystems work with an understanding of how human societies function, both in the past and in the present, is an extremely complex problem. It challenges ecologists to envision their discipline differently and in particular to open themselves to new relationships with other disciplines and with the general public.

Ecologists, when entering new environments, have always had to adapt and to shape their research questions to the demands of those environments. As Ramón Margalef pointed out in the 1960s, the ways in which ecologists study and think about nature reflect the *genii loci*, the special characteristics of the landscapes and species that they study. The unique features of the urban environment should suggest a point of view and a plan of attack that are appropriate to that environment and to the study of the human population that dominates it. But ecology, having had a century to develop as a biological discipline, had established methods and traditions that defined its point of view and its disciplinary boundaries. Ecologists' sense of their boundaries and purpose, and their need to operate within established institutional structures, constrained their freedom to operate, just at the time that they were being challenged to adapt their vision to new environments and problems. There was a tension between the desire to maintain disciplinary traditions, to operate within existing traditions and continue along well-trodden paths, and the need to evolve new approaches in response to the demands of different environments and social contexts.

Modern urban studies exemplify the problems faced by the discipline of ecology as a whole at the end of the twentieth century. Ecology involves understanding how nature works; right from the start of the discipline, biologists have been anxious to establish the principle that basic research deserves support. But ecology must also be a guide in helping us to solve environmental problems. It must show us what we need to do to adapt successfully to lands that are changing rapidly under our influence. In order to teach us these lessons, it must be able to link human happiness and welfare to the survival of ecological systems, where happiness and welfare refer not only to material benefits but also to deeper spiritual values that we wish to foster. In the urban setting one cannot avoid thinking about how ecological system health relates to human health in its broadest physical, social, and spiritual sense. One must constantly consider how to apply ecological knowledge, how to bridge the gap between the academy and the public, and how to communicate scientific findings in a meaningful way.

This challenge entails the ability to shift one's perspective, as suggested in Dubos's comments, so that a human-centered viewpoint can be properly articulated and appreciated along with traditionally less human-centered forms of analysis. I suggest that this is a challenge for the whole discipline of ecology and is by no means confined to research in urban settings. However, it is easier when studying remote environments to ignore the human-centered perspective or to interpret expressions of a human-centered view as being a capitulation to anti-environmental attitudes, as though it were an argument against the preservation of wilderness. In the urban environment the question of how human-centered ecology should be jumps to center stage. Hence urban studies provide opportunities to view these problems under a magnifying glass and to derive some lessons that might apply to the whole of ecology.

To set the stage, we return to the New York Botanical Garden, the starting point for our story of the history of American ecology. In 1980 the Garden held a conference on the future of the plant sciences, focusing on the role of botanical gardens and arboreta. René Dubos delivered the keynote address. He challenged the Garden to think about its mission in the modern world and involve itself more directly in solving contemporary problems. "You wish to be what you have always been, and you think you cannot go beyond what you are," he said. "I assure you that this is the best formula for suicide for a scientific institution."[1] He argued that damaged ecosystems required more human intervention to restore them, and such intervention required more knowledge of plant ecology and the properties of plants. He asked scientists to pay attention to human needs and to use their science to meet those needs. Dubos's comments stimulated the Garden's director, James McNaughton Hester, to review the Garden's mission. From his review came several recommendations, among them a decision to create the Institute of Ecosystem Studies to promote ecological research and education.

The Institute of Ecosystem Studies owed its existence to the same industrial wealth that had made the creation of the New York Botanical Garden possible a century earlier. The first steps were taken in Nathaniel Britton's day, by Mary Flagler, a young heir to the Standard Oil fortune. She was the granddaughter of Henry Morrison Flagler. In partnership with John D. Rockefeller and Samuel

Andrews, Henry Flagler helped to found Standard Oil Corporation, which in the 1870s became the leading oil refining company in America and the richest industrial company in the world.[2] Flagler moved to New York City when the company set up its headquarters there in 1877. In the 1880s Flagler turned his attention to development in Florida, recognizing its potential as a tourist destination. He built a railroad to link the towns on the east coast, promoted agricultural development, erected luxury hotels, and made other civic improvements that enabled towns such as St. Augustine, Daytona, Palm Beach, and Miami to grow into major resorts. He had plans to drain and develop a section of the Everglades when he died in 1913. His company continued with development plans, unsuccessfully as it turned out, until the land was sold to become part of the Everglades National Park, dedicated in 1947.

Henry Flagler's heir Mary Flagler married Melbert Cary, a publisher, in 1923, and over time the money generated by industrial development was steered toward its opposite, wildlife preservation. During the depression the couple bought fourteen farms and other properties near the town of Millbrook in New York state. The property became their weekend retreat from the city. Melbert died in 1941, but Mary continued to look after the property. Over the years the fields and pastures, nearly two thousand acres of land, were allowed to revert to forest, gradually turning into a wildlife refuge.

Mary Flagler Cary's will provided for a trust to preserve the property after her death in 1967. In 1971 the trustees chose the New York Botanical Garden to be custodian of the property, and the Garden developed an arboretum on it. When the Garden in the early 1980s embarked on its quest for a new mission and decided to expand research in ecology, it sought a scientist of international reputation to direct the Cary Arboretum and create a new ecology center there. The choice was Gene Likens, whose pioneering long-term ecosystem studies of the Hubbard Brook Experimental Forest had yielded the first evidence of acid precipitation in the United States. In 1983, under Likens's directorship, the Millbrook property became the home of the Institute of Ecosystem Studies, a division of the New York Botanical Garden.[3] The Institute's mission was to provide a research environment conducive to creative thought and without the distractions of red tape. In 1985 the Institute began a series of international workshops exploring themes on the frontiers of ecological science, with particular attention to ecosystem research and the value of long-term studies. Affiliations with several universities—Cornell, Rutgers, the University of Connecticut, and Yale—enabled the Institute to train graduate students, and the Institute also ran an educational program for the local population. It was fitting that the matured science of ecology, having benefited in its infancy from the Garden's existence, should return to its place of origin. But following a pattern of creation and spin-off that we have already observed at the Botanical Garden, the Institute formally separated from the Garden in 1993, a decade after its creation, and became an independent not-for-profit corporation with a core research staff of more than twenty scientists.

At the time the Institute acquired its independence, ecologists around the world were confronting evidence that humans were altering the entire planet.[4] The population of the earth, having just passed 5 billion, had doubled since the "Marsh festival" at Princeton in 1955. In half of a human lifetime it could double again. In the words of human ecologist Stephen Boyden, "for the first time in the history of life on Earth, a single species is now producing progressive ecological change at the level of the planet as a whole."[5] What was the price of our success as colonizers? The time was ripe for a comprehensive look at what changes humans had wrought during the modern industrial age. In 1987 an international group of scientists convened at Clark University in Worcester, Massachusetts, to revisit the themes that George Perkins Marsh had laid out in the nineteenth century, examine the major human impacts on the earth in the past three centuries, and bring new scientific methods to bear on their analysis.[6] The proceedings of the conference were published in 1990 under the title *The Earth as Transformed by Human Action*, echoing the later editions of Marsh's book *Man and Nature*, which had been retitled *The Earth as Modified by Human Action.*

Robert Adams, writing the foreword to the conference volume, noted the disparity between the "massive environmental transformation in which we are now almost routinely engaged" and the limited means available to monitor, ameliorate, or counteract those changes.[7] Martin Holdgate added a postscript in which he envisioned a "sea change" in late-twentieth-century ecology, one that would entail a reinterpretation of the "human relationship with nature through the drawing together of many strands of culture, experience, and thought." Ecology itself, he imagined, would no longer look askance at applied scientists and those who worked in human ecology. It would begin to recognize that care for the environment "had to be built into the fundamental policies of nations—their agriculture, industry, energy, and commerce."[8]

Environmental concerns, Holdgate argued, would have to be woven into the process of economic development of nations and pervade national thinking in order to achieve "sustainable" living, that is, a style of life less wasteful and damaging both to the environment and to human health. *Sustainability* referred to the idea of meeting present needs without compromising the ability of future generations to meet their needs. He observed that new collaborations between ecologists and social scientists would be necessary if the goal of enhancing human life in a sustainable way was to be achieved. Communication of scientific research to the public was important, but conversely, scientists had to understand the public better, to know how people perceived their world.

Also in the late 1980s the Ecological Society of America was beginning an evaluation called the Sustainable Biosphere Initiative, which was to define research priorities in the late twentieth century. A committee of eminent ecologists published a report in 1991 calling for bold initiatives and stressing the need to focus research so as to aid environmental decision making.[9] Monitoring the earth's climate revealed significant global climate change, and an alarming rise in species extinctions stimulated urgent demands to protect biodiversity. The committee identified these problems as major priorities for research. They were linked closely to the overarching goal of creating sustainable ecological systems, which meant wiser management of resources and improved conservation measures.

The theme of interdisciplinary cooperation ran throughout the report. Large-scale and long-term experiments, use of remote sensing technologies, and growth of large-scale data sets held out the promise for better synthesis of ecological research as well as closer cooperation between scientific disciplines. Creating sustainable systems required changes in human behavior and new environmental policies, and the report called for collaboration between ecologists and a wide range of social scientists, policymakers, and planners. The writers hoped that ecologists could play a key role in fostering new interdisciplinary approaches to the environment. Better interactions with applied scientists (in forestry, agriculture, and resource management) were also needed. However, the report concluded that neither the funding nor the infrastructure in the United States was sufficient to address the research needs that the scientists had identified. Moreover, the report recognized that achieving success depended on communicating information to citizens and especially to political leaders.

In 1997 ecologists Peter Vitousek, Harold Mooney, Jane Lubchenco, and Jerry Melillo published in *Science* an overview of the extensive human alteration of the earth, which followed up on the themes addressed in the Sustainable Biosphere Initiative.[10] Their article introduced a special series of reports on human-dominated ecosystems, discussing how to use science to create management strategies to achieve sustainable agriculture, forestry, and fisheries and to restore degraded lands.[11] The essays not only dealt with ecological principles but also drew attention to the need to understand human behavior, social systems, and economics, an endeavor that suggested more interaction between ecologists, applied scientists, social scientists, and economists. The introductory essay emphasized the way human activities were producing global climatic change and loss of biological diversity. The conclusion that humans dominated the planet, as George Perkins Marsh had begun to discern nearly 150 years earlier, now was inescapable. And the solution foreseen by Marsh, namely that human impact had to be met by better engineering and management, was equally inescapable. Maintaining "wild" ecosystems, the authors concluded, with the word *wild* now in quotation marks, would require increasing human involvement.

Both the Clark University symposium and the Sustainable Biosphere Initiative placed great value on funding long-term, continuous studies of different regions. Ever since the recognition of succession as a basic ecological process, ecology in the United States has been concerned with changes occurring in landscapes over periods of time measured in decades and, where possible, centuries. Ecologists such as Frederic Clements, as well as his colleagues at the Desert Laboratory in Tucson, had from the start realized that some patterns of change could be discerned only over several decades. The creation of a knowledge base formed from long-term observations that tracked changes in the land was important in enabling ecology to evolve into a discipline (see chapter 4). However, many ecological studies did not adopt longer-term historical perspectives, so it was still necessary in the late twentieth century to make a case for the benefits of long-term studies.

Historical ecology had by no means disappeared in the postwar decades, as indicated by the historical approach taken by geographer Carl Sauer and like-minded colleagues in the 1950s (see chapter 6). A modern extension of this discussion can be seen in a 1994 book by Gordon G. Whitney that lays out the history of environmental change in North America over four centuries.[12] Whitney explicitly built on the work of people such as James Malin, Carl Sauer, George Perkins Marsh, and others who had studied the long-term transformation of the North American landscape. But as ecology neared its centennial as a discipline, most research projects in academic ecology were both short-term and on a small scale. In part such choices reflected pragmatic decisions about what was feasible, given the constraints of the academic environment and the inherent difficulties of studying a highly complex world. Whatever its causes, the bias toward short-term studies existed despite the fact that one of the earliest accomplishments of first-generation ecologists was to establish the custom of studying a region continuously over long periods of time.

Only in the 1970s, following the International Biological Program, did the National Science Foundation (NSF) turn its attention to funding long-term ecological studies. From 1977 to 1979 the NSF sponsored three workshops to explore the feasibility of collaborative long-term research projects. In 1980 it initiated a pilot project in long-term ecological research, selecting six proposals for studies across the mainland United States. During the 1980s twelve sites were added, but two later withdrew from the program. Among those added was the Hubbard Brook Experimental Forest in New Hampshire, where Herbert Bormann, Gene Likens, and colleagues had been engaged in studies of forest succession since 1962. Two sites in Antarctica were added in the early 1990s, and one of the original six sites (in South Carolina) withdrew. The NSF network of long-term ecological research projects, known as LTERs, now involved more than one thousand scientists and students. Growing concern about environmental problems around the globe in the 1990s propelled the development of national and international ecological networks.[13]

The LTER sites were located in relatively pristine areas, that is, in protected natural areas, although most included human-manipulated ecosystems as well as natural ones. They represented a variety of ecosystems and climatic types ranging from tropical rain forest (Puerto Rico) to polar desert (Antarctica). Research at each site was defined by certain core questions, so that despite the variation among sites, the kinds of data generated would be comparable. In 1994 the NSF started to fund cross-site comparisons within the LTER network as well as comparisons with sites outside the network. The core questions dealt with basic ecological issues, including patterns of primary production (capture of energy), movement of nutrients through the system, and the effects of disturbances on systems. Addressing the core questions was a requirement of each project, not an option.

Just under half of the world's population (including about three-quarters of the U.S. population) lived in urban areas; therefore the understanding of global processes also required deeper understanding of how cities and their surrounding regions worked ecologically. By the year 2000 there were more than three hundred cities with populations greater than 1 million and sixteen megacities larger than 10 million people. These cities had an enormous footprint; that is, they consumed resources from a much larger area than they occupied and were having a global impact on biogeochemical cycles and climate change.[14] In 1997 the LTER program brought human impact on the environment squarely into the picture by funding two urban studies, one in Phoenix, Arizona, and the other in Baltimore, Maryland.

An ecological approach to human-dominated environments had been identified sporadically as a fruitful research subject from the 1920s on.[15] Ellsworth Huntington had tried to define ecology in a way that would include attention to public health problems but had met with little success (see chapter 5). In proposing the term *ecosystem* in 1935, Arthur Tansley had assumed that it could be applied to human-dominated systems as well as more pristine areas. When the concept of the ecosystem entered ecology, it seemed that this new perspective might provide a way to look more closely at human impact on the environment and also to bring ecologists and social scientists together. In 1950 at the annual meeting of the American Association for the Advancement of Science, the section on social science focused on interdisciplinary work, including discussion of urban ecology. Francis Evans of the Institute of Human Biology at the University of Michigan suggested that Tansley's concept of the ecosystem could be fruitfully applied to the city and its hinterland and that the urban area was a natural laboratory for the study of how biological and social causes interacted in human evolution.[16] At the "Marsh festival" in 1955, Lewis Mumford hoped that the future would see more attention paid to the natural history of urbanization so that progress could be made toward creating "an urban and industrial pattern that will be stable, self-sustaining, and self-renewing."[17] Several of the papers in that symposium dealt with human ecology.

Edgar Anderson, director of the Missouri Botanical Garden, urged biologists to come out of mountaintops and jungles and concentrate more on human activities and human-created landscapes.[18] In the 1950s he started taking his botany classes to the dump heaps, alleys, and vacant lots of St. Louis to stimulate students to think differently and creatively about city living.[19] He argued that quiet contemplation of specimens of *Homo sapiens* was just as enlightening as the intelligent observation of nature and would in fact lead to deeper understanding of natural history.[20] Doing ecological work in the city was as convenient as in the country, and Anderson believed that it was also necessary, for "naturalists who will not face resolutely the fact that man is a part of nature cannot become integrated human beings."[21] He argued that ecologists who focused on remote landscapes to the detriment of urban environments could not help to heal society's ills. In the city, where nature appeared to be out of balance, it was easier to analyze the interplay of forces that produced the natural balances that ecologists valued elsewhere. Cities needed to be studied so that people would accept them instead of running from them and so they would be inspired to "mold them into the kind of communities in which a gregarious animal like man can be increasingly effective." Anderson's insight into the importance of solving ecological problems by studying human-dominated landscapes also made him open to dialogue with social scientists.[22]

Such arguments were controversial because they could be taken to imply complacency about the loss of wilderness areas, or to suggest that humans, being tamers of wilderness, did not need wilderness at all. Ecologist Daniel McKinley, without naming his targets, took aim at "people watchers" and commented that "city-watching and people-watching leave you with no norm. You cannot measure the health of the city against itself." The danger, he argued, was not that naturalists failed to see that humans were part of nature, but that the majority of humans failed to grasp that "nature is a part of man." The danger was that people would totally lose contact with nature, would not understand the value of biological variety, and would not be motivated to preserve it. He had no truck with trendy movements that purported to put humans into nature or that argued that humans were "natural," while not insisting on the importance of preserving wilderness areas. Although supportive of human ecology, McKinley saw a problem in putting too much emphasis on urban ecology if it overshadowed the need to study ecologically richer, and to him more interesting, wilderness areas.[23]

However, interest in urban ecology in the 1960s and 1970s was not necessarily a symptom of the devaluation of nature, as McKinley feared. It was a response to the decline of the American urban environment and an attempt to find ways to revitalize cities. In this quest human geographers began to adopt a more dynamic view of the city landscape, drawing on social theory to interpret the city environment.[24] Ecologists also began to rethink their approach to the urban environment. The Institute of Ecology, an organization of applied ecology formed in the late 1960s, set up eight task groups to consider urban ecology in 1972. Their recommendations were published in 1974 as *The Urban Ecosystem: A Holistic Approach*, which mapped out several areas of research. As the title suggests, the report adopted an ecosystem framework. It called for a "new conception of the urban ecosystem," which meant an extended definition that included natural ecosystems as well as built environments.[25] The urban ecosystem had both natural and social control mechanisms that had to be understood in relation to each other. Human aspirations and how humans attained their goals had to be considered in assessing these relationships.

By the same token, people had to be educated to alter their behavior and lifestyle, so the report called for research on the motives and values of urban dwellers. These were problems of adaptation: how well humans were adapted to city living, what problems confronted them, and what might beautify the city environment and promote social stability.[26] These were also some of the dominant concerns of the time. The thrust of the report was to find ways to motivate people to rescue declining cities and build vibrant urban communities in ways that were ecologically sound. The report identified clear communication to the public as a priority.[27]

Concurrently, on the international scene, UNESCO launched a new program called Man and the Biosphere in the 1970s, with the idea of addressing some of the gaps in the International Biological Program, which had ended in 1974. European and Russian scientists had faulted the studies of the International Biological Program for their lack of practicality. Most of those studies had focused on natural areas, although one study located in Belgium included an ecological analysis of Brussels. Man and the Biosphere devoted attention to systems in which humans were an integral part, including agricultural systems and cities. Scientists conducted urban ecosystem projects in Hong Kong, Tokyo, Sydney, and Rome.[28] Thus there was by the 1970s growing interest worldwide in urban ecology and in the study of large cities.

The two urban LTER projects located in Phoenix and Baltimore, though not entirely novel in advancing the cause of urban ecology, differed from their predecessors in that they evolved directly from the mainstream ecology of the existing LTERs, which studied relatively pristine systems. While working within the boundaries of the LTER program, ecologists entered into a new environment and a new set of social interactions. New relationships would be demanded between academic ecologists, applied scientists, and social scientists, as well as between ecologists and the public. Given the expectation that ecologists would be addressing the people who were trying to solve the problems of each urban environment, one likely outcome was that ecological research would begin to look less academic and more practical, that is, more attentive to the unique problems of each location. The scientists did want to be responsive to the local settings and to improve the quality of life of the people living in both cities. They did want political leaders and managers to recognize that their studies had something tangible to contribute to the cities.

Yet the studies had to be comparable to the other long-term ecosystem projects. The researchers did not have the freedom to reconstruct urban ecology from the ground up, as it were, following the dictates of their particular locations. In addition, the scientists had to worry about seeming to adopt an advocacy role, something that might taint their value as "objective" researchers. They saw their role not as setting policy, but as providing data to policymakers and managers. So they had to adjust to and fit into the new urban environment while maintaining a critical distance and not "going native."

The urban LTERs operated under different, and at times opposing, selective pressures. First of all, they had to conform to existing practices in ecology; but, given the novel environment, they also had to be innovative enough to justify continued funding of research in that environment. In addition, their studies had to be seen as research projects, contributing to basic research in ecology. Yet the scientists could not afford to appear too detached from practical problems for fear of losing the support of civic leaders, local activists, and the public.

The ecologists involved in both urban studies adopted a bold stance in outlining their ambitions, at the same time emphasizing how well they fitted into the existing LTER framework. The city environment became the "new frontier," and even the "last frontier," of American ecology, although one might equate the "last frontier" more with outer space than the inner city.[29] The projects aimed high on both scientific and social fronts. They hoped to contribute to the advance of basic ecology, turn urban mind-sets toward the environment by educational reforms (including curricular changes in schools), help solve particular environmental problems in the regions studied, and also create models for studies of urban ecosystems elsewhere. These goals were communicated in a consistently upbeat series of pitches in scientific journals that explained the importance of the new frontier, albeit in rather unspecific terms. As stated in an overview of both projects in 2000, "Acknowledging the central human component leads to an emphasis on new quantitative methods, new approaches to modeling, new ways to account for risk and value, the need to understand environmental justice, and the importance of working within a globally interacting network."[30] Well before the projects had run long enough to generate many results, publications rolled out to spread the word about the advantages that would accrue from these studies.

In these urban studies ecologists would continue to ask the traditional questions concerning such matters as the flow of energy, the cycling of material, and the impact of disturbance. By using the same core questions as in other ecosystem studies, the urban studies could link up to other long-term ecological research projects in the network. Additional questions had to do with how ecosystem processes were affected by the people living on the land. What about flows of information and the role of institutions and organizations in shaping people's perceptions and constraining their actions? How did people use the land, domesticate animals, consume resources, and produce wastes? What was the connection between socioeconomic status and these activities? What were the feedbacks between human and ecological systems? The projects as initially outlined were extremely ambitious in setting out a broad range of general questions. The difficult problem would be to figure out which of these questions could be addressed concretely in the environments selected.

Studies would ideally involve interdisciplinary interaction. People from various disciplines were supposed to learn to speak each other's language and shape their research questions in response to the perspectives of other disciplines.[31] Yet the scientists were aware that in the culture of criticism that defined scientific practice, selling interdisciplinary research and getting it through the peer-review process would be tough.[32] Not only was interdisciplinary research hard, but it was a hard sell. Linking the physical and biological data to the social data in a meaningful way would prove to be one of the largest challenges of the projects, a task that would require continual evaluation.

The methodological approach adopted in the ecosystem studies was the "hierarchical patch dynamic" analysis that had been developed in ecology in the 1980s.[33] Patch dynamics is a way to study how complex systems change over time. It divides the land into patches of various kinds and sizes, depending on the questions being asked and the scale of the inquiry, and then analyzes the dynamic relationships between the patches. The approach is meant to link processes occurring on a small scale with the features of ecosystems as viewed on a larger scale.

The patch-dynamic approach was related to the "shifting mosaic" concept used by Herbert Bormann and Gene Likens (see chapter 8). The idea that an ecosystem is heterogeneous, or patchy, and that there is a complicated dynamic relationship between these patches, can be compatible with the belief that systems can achieve a steady state. However, the patch-dynamic method was intended to be a marked departure from the Hubbard Brook approach, in that there was no requirement that the system be in equilibrium. It was a method designed specifically for the study of disturbance. Those who used patch dynamics deliberately played down the idea of ecological equilibrium and instead emphasized the dynamic properties of the system and the role of disturbance in creating the observed patterns.

This method was felt to be well suited to the urban environment.[34] The underlying idea was that the city was not to be treated as a black box but something whose fine details, on the level of forests, streams, parks, houses, neighborhoods, and roadways, would be probed and properly understood. The city was a composite of "patches" of different size and structure, each patch interacting with other patches, all adding up to a complex mosaic that could be modeled. The patch-dynamic approach departed from the "macroscopic" view of ecosystems advocated by Tom Odum (see chapter 7) and instead sought to put the city's neighborhoods under the ecologist's microscope.

The city environment was heterogeneous relative to more pristine environments, and the goal was to capture the details of its dynamic, patchy structure, then relate that spatial heterogeneity to changes in critical resources that might be ecological (energy, nitrogen) or social and economic (information, labor, capital). Patches are hierarchically organized, with groups of patches on a smaller scale forming larger patches with different characteristics on a broader scale. The application of the patch-dynamic approach was made technically feasible by inexpensive computer technologies, which allowed for the use of Geographic Information Systems (GIS) and the development of GIS databases that could facilitate modeling.[35]

It was obvious that "patchiness" or heterogeneity could be perceived quite differently depending on one's disciplinary background or how one used the land. The ecologists noted, for example, that "the civil engineer and the urban recreationist will have different views of the boundaries of a watershed. The first may see a 'sewershed'; the second, a visually unified landscape that is engaging on a morning jog."[36] From the ecologists' perspective, these differences implied that different methods of classifying heterogeneity were needed. Classification methods had to incorporate several dimensions, both ecological and social. The desire to take account of multiple ways of interpreting the landscape translated into the idea that a basic task of such projects was to devise new, multidimensional methods of *description*, *classification*, and *mapping*. The challenge was to understand the causal connections between these dimensions or data sets, that is, to link one layer of the map to another.

Studies set in a human-dominated environment pose many conceptual and practical problems. Understanding the subtle nature of human effects demands a historical approach and encourages taking a long view of landscape history. Archaeological evidence becomes important, and historical records have to be plumbed to chart the patterns of land use. On even longer time scales, pollen from core samples can be analyzed to track changes in vegetation cover when humans occupy the land. With a longer view, the line between "natural" environments and "cultural" landscapes becomes increasingly blurred, and it becomes more difficult to make judgments about how landscapes ought to look.

Charles Redman, an archaeologist working on the Phoenix LTER, questioned whether it was even possible to reconstruct an environment unaffected by humans. The participants at the "Marsh festival" in 1955 had recognized the very same problems many years earlier (see chapter 6). As Redman pointed out, it is not meaningful to talk about "nature" as an absolute concept, for humans have been important modifiers of the landscape from ancient times. Understanding their influences can lead to a positive view of human accomplishments. "Seeing the world through the eyes of a human, it is a good world, because we have acted to make it so."[37] This was not to deny that humans had created environmental problems and might now be on a collision course with disaster. But adopting a long view of human action left one with an appreciation of the human ability to solve problems. What needed to be grasped, Redman argued, was the way human values and human goals influenced how environments were perceived and judged. Defining the "ideal" or "best" environment depended on the viewpoint of the observer. At the very least, he suggested, such judgments should not assume that the "best" environment was one in pristine condition, untouched by human hands, for such environments simply did not exist. It was highly unrealistic to adopt a romantic view of the pristine wilderness.

Like Dubos, Redman believed that judgments about landscapes and land uses always had to return to the question of human values and objectives rather than grasping for an image of what nature "intended." Instead of worrying about whether an environment was natural or unnatural, one should ask what objectives a society had for a given environment and what actions needed to be taken to reach those objectives. These were of course economic, political, and cultural issues, with many opposing interests at work. The purpose of ecology was to weigh in with scientific data that showed the ramifications of human actions on a large scale and over the long term, ideally so that problems could be fixed.

These discussions about what we mean by "nature" are of interest because they suggest the related question raised by Ellsworth Huntington and reiterated by Carl Sauer, Edgar Anderson, and René Dubos: How "human-centered" should ecological research be? Using the Baltimore LTER as my main source of information, I argue that although urban LTERs study human-dominated environments, and although they hope to educate and improve the lives of the citizenry, in crucial ways this approach to urban ecology is not human-centered. Instead, it conforms to the traditions of ecosystem research on more pristine systems.

This observation is made not to fault the urban projects but rather to draw attention to the way institutional and disciplinary constraints affect science. The ecologists were, first of all, entering a new environment already inhabited by members of other disciplines whose job involved studying human ecology. To the extent that the scientific territory was already "claimed" by other disciplines, primarily those that constitute the public health field, ecologists had to work out what their niche was in a competitive environment. In addition, the urban studies operated under terms set largely by their link to the other LTERs, and these constraints prevented them from adapting freely to the particular human problems of each city. They evolved from a biological discipline and from a history of ecosystem analysis that emphasized certain kinds of ecological problems, mainly those dealing with flows of energy and material through the system.

These were legitimate questions, important for solving environmental problems within the larger regions. But they were questions that could be explored without the kind of direct "people watching" that Anderson had in mind when he recommended five decades ago that ecologists pay more attention to *Homo sapiens*. The decision not to adopt an explicitly human-centered approach, which was taken deliberately, in turn had implications for how ecologists related to people in other disciplines, or to people who might envision the city in different ways. The decision also had implications for how the science-public interface was imagined to operate. These implications form the basis for a discussion of ecology's prospects in the conclusion.

The Baltimore Ecosystem Study was a collaborative project, but from the earliest stages the lead role was played by the Institute of Ecosystem Studies. The director of the Baltimore study, Steward T. A. Pickett, was a senior scientist at the Institute, and several other investigators and students were connected with the Institute. Pickett was trained in plant ecology and was originally interested in understanding ecological succession from an evolutionary standpoint. He had applied the method of patch dynamics to problems in ecological succession, finding that the method was a useful foil to models based on equilibrium theories and that it allowed one to focus on the role of disturbance in ecological systems.[38] He joined the Institute as a visiting scientist in 1985 and then became a permanent member of the scientific staff.

Pickett had been thinking about urban ecology well before the LTER. In 1990 Pickett and Mark McDonnell, also at the Institute, argued that urbanization was a "massive, unplanned experiment" that could benefit from systematic ecological study.[39] Drawing on ecosystem studies in the New York City region, they explored ways of studying changes along the urban-rural gradient. The urban-rural gradient is not studied as an example of a successional series but is of interest because of the changes that can be observed along the gradient. Pickett and McDonnell noted that this gradient created opportunities for new kinds of experimental manipulations.[40]

The Institute's regular program of conferences, known as Cary conferences, began to home in on problems in human ecology. In 1991 a Cary conference titled "Humans as Components of Ecosystems" looked at various kinds of subtle human ecological effects.[41] McDonnell and Pickett edited the published volume that resulted from that conference. In 1992 a workshop at the Institute, "Integrated Regional Models," again took up the problem of humans and their environment, exploring the possibility of interaction between biological, physical, and social sciences and assessing various modeling strategies to determine what approaches might prove most successful.[42] Ecologists at the Institute of Ecosystem Studies were deeply involved in discussions about how to incorporate human effects into ecosystem analysis at the time that the LTER program was being extended to the urban environment.

These discussions, which opened up a range of questions that needed testing, took concrete form in a proposal to the National Science Foundation for what became the Baltimore Ecosystem Study.[43] For the study, the institute collaborated with several institutions: two campuses of the University of Maryland, Johns Hopkins University, and especially the USDA Forest Service. In addition, the ecosystem study involved various partnerships with other institutions, including some international collaborators. A link to economics was provided through the Institute for Ecological Economics, directed by Robert Costanza at the University of Maryland, College Park. Costanza had for years been trying to develop links between ecology and economics and had developed a landscape model based on the Patuxent River watershed that was relevant to the Baltimore study.[44] In 2002 Costanza's group moved to Vermont, but they remained connected with the Baltimore project.

What is perhaps most important for an urban ecosystem study, and what distinguishes it from other kinds of ecosystem studies, is that data from historical records can provide a picture of how actions taken long ago have left their mark on landscapes over the centuries. Grace Brush, a plant ecologist at Johns Hopkins University, provided the deep historical perspective for this project.[45] Her research involved reconstructing the history of the Chesapeake system over several millennia, using data derived from core samples of sediments. Changes in plant pollen, for example, revealed shifts in land use as the region was colonized. Different land uses, as well as disturbances such as hurricanes, affect the rate of sedimentation. By tracking changes in sediment, Brush could investigate the long history of the impact of human populations around the Chesapeake Bay. The most dramatic changes could be seen not on the land itself, but on the Chesapeake estuary, which lost many bottom-dwelling organisms as the forest cover disappeared from the surrounding land.

In the shorter term, the history of the city of Baltimore—its pattern of economic development, its settlement, and its growth of neighborhoods—was all highly relevant to understanding the present form of the watershed ecosystem. Chris Boone, a geographer from Ohio University, added a historical and geographic perspective, drawing on past records to chart the development of Baltimore and its infrastructure. A sensitivity to historical context was crucial to one of the general goals of the research: to test whether knowledge derived from the study of more pristine systems was applicable to urban environments. Scientists engaged in historical analysis were better armed to challenge prevailing wisdom in ecology.

The most important antecedent for the Baltimore project was the Hubbard Brook ecosystem study by Gene Likens, Herbert Bormann, and their colleagues, which was also part of the LTER network. Baltimore contains four small watersheds, defined by three streams and a region around the downtown inner harbor. Each stream watershed represents a gradient from rural or suburban regions at the headwaters to an increasingly urban environment as one moves downstream. The watersheds are part of the larger regional watershed flowing into the Chesapeake Bay, one of the most productive estuaries in the world until recently. The scientists selected the Gwynns Falls watershed in the western part of metropolitan Baltimore to begin their study. This watershed defined the ecosystem. Although the sampling sites and much of the experimental work were confined to this watershed, covering about one-third of the metropolitan area, the intention was to extend the study eventually to the whole metropolitan region of Baltimore.

By defining the watershed as an ecosystem, the study conformed to the Hubbard Brook and similar kinds of ecosystem studies, even though it was departing from the Hubbard Brook study in its use of patch dynamics. But the decision to define the ecosystem by the boundaries of the watershed was not just a reflection of Hubbard Brook's prestige: it also reflected ecological outreach projects already under way in Baltimore. In 1984 the Parks and People Foundation was created to help in the revitalization of Baltimore's green spaces. Through a project called the Urban Resources Initiative, which involved partnership with the USDA Forest Service, Parks and People researched ways to improve cities through their green spaces, for instance by converting abandoned land into community gardens. In 1994 the Forest Service created a project called Revitalizing Baltimore, which included community forestry and watershed restoration in all the watersheds of Baltimore. The Parks and People Foundation managed the project, much of which concentrated on a group of connected parks located in the Gwynns Falls watershed. Designed by the Olmsted brothers in 1904, this large urban wilderness area had suffered from years of neglect and had a somewhat unsavory reputation, being a common dumping ground for murder victims. Revitalizing Baltimore hoped to promote other uses by improving the parks with new trails and educational programs and in general encouraging citizen appreciation of nature. With these improvements, the parks were much more welcoming, in effect becoming more parklike and less of a wilderness.

Two scientists who had been connected with these projects, William Burch and Morgan Grove, had key roles in the later Baltimore Ecosystem Study. Burch, from the Yale University School of Forestry and Environmental Studies, had been on the advisory council of the Institute of Ecology workshop that produced the 1974 report on urban ecology, discussed earlier. Grove, a sociologist with the Forest Service, was one of the central participants in the LTER from the start, especially in its educational outreach programs. Here again Hubbard Brook provided an important model for the development of the social and cultural dimensions of the project. Bormann and Likens had asked, "Does the vegetation structure of the watershed affect its hydrologic cycles?" Grove and Burch in 1997 posed a parallel question, reframed to fit the Urban Resources Initiative and the LTER: "Do sociocultural patterns and processes affect a watershed's hydrologic cycles?"[46] Such parallel questions pointed to what scientists fully expected, for instance that the extensive paving characteristic of cities would indeed affect hydrologic cycles.

There were advantages to using a proven long-term study as a springboard, but there were other reasons as well for taking a watershed approach to the city. Baltimore's nascent "green" movements, aided by such initiatives as Revitalizing Baltimore, were being organized around the three stream watersheds at about the same time that the Baltimore Ecosystem Study was getting started. The other two—the Jones Falls and Herring Run watersheds—both were acquiring vocal community groups that were working to enhance the quality of the environment along the course of the streams, with expectations of improving and stabilizing the adjacent communities. The "green" movements in Baltimore, many of them informal groups of concerned citizens, were to some extent organized around these landscape features. There were connections between the activist groups, but each one focused its environmental concerns on the communities within a particular watershed that were most likely to be affected by urban renewal along the streams. The development of a watershed-based "green" culture therefore nicely complemented the organization and goals of the ecosystem study.

The fact that urban revitalization was already targeting the Gwynns Falls watershed made this choice for the Baltimore study logical, but one obvious problem was that the human population was not in any sense bounded by or limited by the watershed. Could a definition of the ecosystem devised for a natural area really be used in a city, if the intention was to integrate humans into ecology? Should not the ecosystem have been defined by the distribution of its dominant human population instead?

The apparent disjunction between the watershed model and human population distribution and movement was partly compensated for by the method of patch analysis, which allowed the watershed to be broken down into smaller pieces for detailed study. This method can be applied to any size area, and the restriction of the study to one watershed was as convenient a starting point as any. Moreover, even within the one watershed there was an obvious gradient from rural to urban as one moved downstream. Examining changes along this gradient could give insight into the character and function of the whole city, because the one watershed was like a cross-sectional "slice" of Baltimore city and the surrounding suburban county.

The use of the small watershed model also meshed well with existing ways of addressing environmental problems in the wider region. A large proportion of scientific research in the Baltimore study focused on forests and the ecosystem services provided by forests, especially in maintaining water quality. In Baltimore interest in water quality was connected to broader concern about nitrogen-loading in Chesapeake Bay. Scientists had long been worried about the amount of nitrogen that was entering the bay. Nitrogen pollution is a result of fossil fuel burning, fertilizer use, and sewage effluent. Changes in water quality were having a devastating impact on underwater vegetation and animal populations. Forests absorbed nitrogen and helped to protect the bay from this source of pollution.

The problems of the Chesapeake watershed were familiar to scientists involved in the Hubbard Brook Ecosystem Study, who were part of a network of ecologists making a systematic evaluation of nitrogen pollution in New England and New York.[47] One of the conclusions of this analysis was that nitrogen problems had to be assessed on a watershed-by-watershed basis, because the response of watersheds to management protocols depended on the site.[48] In the light of these larger concerns about watershed management, studying urban watersheds made sense as part of this larger assessment.

A related problem was the role of forests in maintaining water quality in the reservoirs of Baltimore.[49] A growing deer population, pushed into greater contact with humans by urban sprawl, was consuming the understory in the forests surrounding the reservoirs. The loss of saplings meant the forest would not regenerate itself when existing trees died, and a deforested buffer zone was likely to be less effective in maintaining water quality. Here was an example of an environment, the buffer around reservoirs, that had to be managed in order to preserve a "natural" vegetation cover that would ensure the continuation of the forest or, in other words, maintain something like a state of equilibrium around the reservoirs. Human values, in this case the need for fresh water, dictated what states of nature were preferable.

Yet another area of scientific interest was the ecology of the riparian ecosystem, the regions bordering on streams, and the effect of urbanization on this ecosystem.[50] Historically, Gwynns Falls had been a waste-disposal system for the industries along the stream. Industrial and agricultural activity gradually ceased in the twentieth century, but urban development continued. One of the most important effects of urbanization is to lower the water table, producing a condition known as hydrologic drought. This condition alters soil chemistry, which in turn changes the action of microbes in the soil. Microbial actions affect the processes that produce and consume nitrates and hence are important in controlling nitrogen pollution. Progressive drying of the habitat as one moves from rural to urban areas also affects the diversity of vegetation. These shifts could have repercussions on ecosystem services in the regions bordering the streams. Such ecological concerns shaped the questions and methods of the Baltimore LTER and created a context that made the watershed approach meaningful.

Public outreach and education in the LTER was closely focused on connecting urban revitalization to environmental restoration. Trees, green spaces, parks, and gardens were signs of urban renewal and hence of higher environmental quality. Poor areas had fewer resources for these landscape embellishments and hence could be expected to have poor-quality environments. Improvement of parks, creation of gardens in schoolyards, and development of community green spaces were all part of the educational outreach of the LTER. From the standpoint of the ecosystem study, outreach was directed toward transforming people's attitudes by enhancing their experience of nature. Income level affected the ability of communities to improve their neighborhoods in line with "green" principles, and it affected their access to green environments. To the extent that the LTERs might improve poor people's access to green spaces, therefore, the projects were seen by the scientists as ways of promoting social justice.

Scientists imagined that the process would work in the following way. People drawn to parks would experience nature and become aware of the value of this resource. They would thus be motivated to demand maintenance of the resource. In making their demands they would start to "function as a regulatory feedback mechanism in the ecosystem, in the same way that a complex web of interactions maintains stability (resistance and resilience) in unmanaged forest ecosystems."[51] In describing these goals, scientists cited Bormann and Likens' studies of the Hubbard Brook watershed, underscoring the way that study was still serving as a model. The quotation also illustrates a theme explored in chapter 8, namely that the ecosystem concept focuses one's attention on the feedback conditions that ensure system stability. Even though the methodology of hierarchical patch dynamics did not assume a state of equilibrium, the concept of equilibrium (or stability) was still important in thinking about what the ecosystem's condition should be. Humans, described here in abstract and idealized terms, were supposed to ensure that their ecosystems were as well regulated and therefore as stable (in an ecological sense) as a natural ecosystem (such as the Hubbard Brook ecosystem).

A traditional ecological idea, that humans should learn to imitate nature's ways and that stability was a sign of health in systems, was still relevant even though patch dynamics was described as heralding a shift from a classical equilibrium paradigm to nonequilibrium concepts in ecology. But equilibrium concepts and ideas of order and harmony were not banished, despite claims of a paradigm shift: the new viewpoint did not imagine a world that was eternally changing and incapable of stability. Indeed, the new perspective was almost Heraclitean in finding order within constant change: harmony was embedded in patterns of fluctuation and ecological persistence reflected "order within disorder."[52] The new methodology paid more attention to heterogeneity in the environment, but ecologists were still concerned with what produced order in systems and what enabled them to persist. The nonequilibrium perspective was not so much a paradigm "shift" as an attempt to analyze ecosystem dynamics in greater detail. In the urban ecosystem, the goal was to get humans into the feedback "loops" so that the ecosystems would be maintained and stabilized rather than degraded.

Achieving that goal could not be the sole responsibility of ecologists, though. Social scientists involved in the project had to think through the problems of human behavior and what motivated people to change their behavior. But social science also depicts people in abstract terms; that is, it generalizes about populations but does not focus on individuals in communities in a fine-grained way. The product of scientific study might appear in the form of statistics, graphs, charts, or maps, rather than as individual stories of people's lives, which after all would not pass muster in peer review, for they would not conform to scientific standards. Inevitably there is distance between the academic study of populations and the people who are the subjects of the study.

The gap between the academy and "the people" was revealed most clearly near the start of the project in presentations by Bryant Smith, project coordinator with the Parks and People Foundation, at Baltimore Ecosystem Study (BES) meetings held in 2000 and 2001.[53] With funding from the BES, Smith supervised a photography project by a group of young people at an inner-city community center that served a predominantly poor community. The center was not located in the ecosystem, but its directors had made a connection to the BES as soon as they learned about it in 1998, hoping that the partnership would provide educational resources for the neighborhood kids.[54]

Under Smith's direction the teenagers embarked on a neighborhood digital photography project that was designed to help them explore problems in their area. They were encouraged to talk to their subjects and get a better sense of why things happen and how people are prevented from taking positive steps to solve problems. There was a scientific component to the study in that the students used global positioning systems to develop a GIS map of neighborhood characteristics. They learned to collect and document data and to use information technologies. The project, which included a photography exhibit, was a study of the social processes and deep problems of the inner city, in particular problems connected to drugs, homelessness, and teenage pregnancy.

This was an exercise in listening to what ordinary people had to say about their environment and an opportunity to reflect on what environmental issues were of concern to Baltimoreans. I imagine the project as the kind of urban study that Edgar Anderson had in mind back in the 1950s, when he began taking his botany classes to the seedier parts of St. Louis and when he advocated "people watching" to capture people's lives and understand their points of view. Smith pointed out that the students saw their environment as consisting of other people; that is, their photographs were not of landscapes but of people. The problems the students encountered, the ones that mattered to them, were problems that directly affected people's lives. As Smith said, in these stark urban environments the neighborhood was the people.

Smith's project benefited from connection with the BES, but in crucial respects its outcome contrasted markedly with that of the approach exemplified in the rest of the ecosystem study. His poignant observations about how the students perceived their environment, and what "environment" meant to them, were very different from the style and tone of the rest of the Baltimore study. In the inner-city project, the students had freedom to define the subject matter. In the other presentations at the same meetings, science rather than people-watching was the primary goal. The result was that the "human face" of the ecosystem was obscured elsewhere in the BES. Smith's project stood out because of its direct human focus.

The contrast can be seen by comparing Smith's project to another educational project that went on at the same time, in the more formal setting of a Baltimore high school. Formal education through the schools meant teaching students about ecosystem services provided by forests and streams and how different land uses affected those services. Ultimately the challenge was to show students how to connect ecology to their own lives. The vehicle for making this connection was the problem of water quality, especially the quality of drinking water in Baltimore's reservoirs, which the Forest Service was also concerned about. At a magnet high school in Baltimore, students developed long-term projects on the natural history of the city, focusing on land use surrounding water reservoirs.[55]

Their teacher, Karen Hinson, worked closely with researchers of the Baltimore Ecosystem Study, and the student projects echoed the scientific questions being pursued in the ecosystem research. Students enjoyed the projects and their scores on national tests improved. They developed skills in collecting data, linking disciplines, and seeing the relevance of history to current conditions. Hinson felt the students became less myopic as they learned how to connect local processes with the larger world. The project results were very impressive; as an educational experiment this was an undoubted success.

But one surprising thing was missing from the projects, Hinson noted.[56] She commented that her students somehow managed to miss the people in this story. The dynamic human interactions that occurred daily were absent from their reports; they did not see people as the heart of the city, although the projects included interviews with people. As I listened to Hinson's assessment of her students' accomplishments and blind spots, her remarks did not strike me as surprising, for in a basic sense the Baltimore Ecosystem Study was also lacking a human face, even though it was all about human influences on the environment.

In fact, the scientific reports at the annual meeting of the BES covered a variety of topics that largely conformed to one's expectations of an ecological meeting. Historical perspectives were included, and urban geography, economics, and sociology were part of the offerings, but they were relatively minor compared to reports on hydrology and ecology. One learned about nitrogen fluxes, heavy metal concentrations in soil, changes in diversity of plant species, and the spread of invasive species. Information about new sensing technologies was reported. Microclimate, the impact of deer populations on forests, and the role of fallen logs in purifying water flowing downstream were all discussed. But one heard less often about the people living in the ecosystem.

The LTER scientists wanted to view people "not as disturbances to the ecosystem but as important drivers of and limitations to it."[57] An important goal was to change the way humans thought about the environment. People made decisions that affected how the urban ecosystem worked, and one of the purposes of this research was to improve the quality of the environment, and the quality of life of citizens, by altering how these decisions were made. But the means of achieving this goal was to "fit" people into the ecosystem framework, as part of the machinery of feedback mechanisms that would stabilize the system. That meant, paradoxically, not adopting a strongly human-centered perspective. The health and well-being of the ecosystem was the primary objective. The result was in fact to obscure the human face in the ecosystem project.

I have devoted so much attention to this urban study, despite its being a relatively new project, because it captures and foregrounds the challenges that ecology as a whole confronts. This challenge is to understand, and continually address and reassess, the relationship between general ecology and human ecology. We should think about how our studies of nature are bound up with conversations we have always had about ourselves, what we value, and how we should behave in a changing world. In the pristine wilderness it is possible to study plants, animals, water, air, and soil and contribute to biological knowledge without worrying too much about how to solve human problems. Yet we know that there are no truly pristine environments left and that human impact has reached global proportions. Ecological knowledge has to be applied to find solutions to environmental problems, and ecologists must find ways to communicate their lessons and show people how to approach every kind of problem ecologically, that is, to realize how actions in one place have repercussions down the line. But to fulfill this social obligation, ecologists have to see the human face in the ecosystems of the world.

Even if the urban studies were to end abruptly, the lessons of their early stages are important. In the city the hard questions about how to apply general knowledge, assess human impacts, create an ecology with a human face, and deal on a practical level with people cannot be evaded. The difficulties that urban ecologists face are magnified versions of problems that all ecologists need to address. In the concluding section we will consider a few reasons why the gap between the scientific approach of traditional ecology and the humanistic perspective that has been expressed by various writers should be bridged in order to achieve what ecologists have always wanted: legitimacy for this challenging subject, which is now a scientific discipline but also, as Frederic Clements pointed out in the 1930s, a point of view and plan of attack.

CONCLUSION

Expanding the Dialogue

I view the history of ecology as an ecological problem in itself. Ecologists not only study how plants and animals are adapted to environments. They themselves must adapt to new demands as societies evolve and continually transform the environment. In responding to such transformations a century ago, biologists devised new research strategies and began to accumulate a record of ecological change over time. Their research established traditions in theory and practice, points of view and habits of thought. This foundation, which eventually came to define ecology, channeled the future course of the science. As it matured, ecology became increasingly bound by its traditions, yet it was also challenged to keep responding and adapting to new demands. Survival depended on adaptability. In short, the evolution of ecology occurs in the rather complicated ecological context of a changing landscape and a changing society and is constrained by a disciplinary culture.

This book tells part of the story of how ecologists have struggled for a place in American society, or, as ecologists themselves might say, how they have tried to define their niche. An architectural niche is fixed, a recessed space in a wall that holds an object, but an ecological niche is in motion. It is a multidimensional concept that in the biological application includes everything animals or plants do throughout their lives. The niche is shaped by interactions with an environment and with other organisms. It can evolve. In the disciplinary sense, finding one's niche means creating a role and performing various functions professionally and socially in dynamic interaction with other disciplines and in interaction with various groups of people.

More than a century ago, scientists with ambitious plans for the development of American biology created the conditions that enabled ecology to become a distinct discipline. The definition of the field has always been problematic, however—the niche evolves continually—and periodically ecologists have rethought the nature and purpose of the discipline in response to changing environments and new social demands. Now, when looming environmental problems require ever more sophisticated ecological insight, ecologists recognize the need to refresh their vision of the science and its role.[1] The ability of biologists to respond flexibly to new demands depends not just on vision and leadership, but also on the culture of science and on the traditions that develop over time. Traditions established over several decades channel research along certain paths and set limits to the adaptive response. The perennial challenge is to think creatively about how to overcome those constraints.

We have seen that as ecologists entered the "new frontier," the city, their research expectations were so constrained by tradition that it was difficult to find the "human face" in the ecosystem, despite its being dominated by humans. Losing sight of the human face means that in crucial ways these studies were not human-centered, which seems paradoxical for a study of urban ecology. Does this paradox need to be resolved? I believe it does, because if ecologists want to achieve their goals of shifting people's perceptions and motivating people to behave differently, then they must connect with their target audience. Adopting a human-centered perspective would help to make the most effective connection to the human population, which is both the source of environmental problems and the means by which the problems may be solved.

In the urban environment environmental problems are inseparable from the myriad conditions that make cities good or bad places to live: access to good education, public transportation, employment, health care, recreation—all the things that together produce vibrant and healthy city environments. Thus the social goal has to go beyond educating people about how ecosystems function, which can be somewhat passive in nature, a form of sermonizing that may not alter behavior. The goal must be to give people the feeling that they can take active control over their environment, understand what is wrong with it, and act to shape their future. As Lester Ward wrote more than a century ago, what one wants is to achieve a synergistic interaction between people and environment. As an entry into these problems, it would be more effective to begin with a human-centered viewpoint and from there move outward to the more abstract viewpoints of ecosystem ecology.

Seeing that the city environment consists of people, as compared with seeing it as a built environment on top of a watershed, is the first step toward understanding that questions about the physical and biotic environment (nutrient loading, species diversity, and so forth) are not separate from social questions affecting the health of cities. The social goal should be to make people better activists in general, to give them the ability to effect change on all levels. People are motivated to act when they can relate problems to their needs and interests and when they see that their actions can have results.

How might we extend the lessons of modern urban ecology to the discipline of ecology as a whole? One relevant question is whether the goals of ecology can be achieved without adopting new perspectives. A stated goal of the urban long-term ecological research projects (LTERs) is to connect with people through educational programs and various forms of outreach, such as park revitalization. Urban ecologists hope to alter people's perceptions of city ecosystems and thereby to improve those environments and reduce the huge gaps between the haves and the have-nots. They want to help people "learn things they can actually discover and apply in their homes and neighborhoods." They want a form of education that is "geared to fostering citizenship, empowerment, and relevancy, and one that blurs the distinction between school learning and real-world, scientific learning." They see ecological understanding as a "key right for all people."[2] They imagine that such knowledge would be eye-opening, enabling people to look beyond what they know and see the larger picture:

> People accustomed to thinking about crime in the city will also think about trash and air pollution; people thinking about urban birds will consider issues of poverty and how history has shaped the urban landscape; people thinking about childhood asthma will consider the functioning of the information dissemination system, and the distribution of environmental burdens; people thinking about whether to spray mosquitoes for West Nile Virus control will automatically think of potential impacts on the poor and infirm, on adjacent aquatic ecosystems, and on nontarget organisms.[3]

The hope, as Grace Brush explained to me, is that understanding the ecological trade-offs between different forms of economic development—a bushel of corn or a bucket of chicken for a basket of crabs—would at least enter into political calculations. Furthermore, people need to understand that ecosystems possibly cannot be "fixed": once altered, they may never return to what they were, and something valued will be lost forever.

In many respects these goals echo the aims of the first scientists who built the New York Botanical Garden and faced the prospect of training an unruly population to become nature-lovers. Back then the goal was to teach people to see the Garden as private property, there for people's recreation but not to be defaced. They had to be taught how to observe and how to modify their behavior to prevent damage. This experience was meant to carry forward to their experiences in the larger world. The strategy could be and was criticized for placing too much emphasis on individual reform and not enough on the larger economic forces of development that were of greater destructive potential than a population of flower-pickers.

Today the goals include making people aware of these broader economic forces, yet the emphasis on individual reform is still the mainstay of the educational agenda. The modern ecologist is more willing to admit that scientific experts cannot know what is important to people, so the experts need to be educated as well. But the faith remains that the cumulative effect of many reformed individuals will have significant impact on a region's ecological condition. Whereas a century ago it mattered to teach people to see nature as a garden, now ecologists want to entice them into gardens and teach them to see the world as an ecosystem, of which they are a part.

The critical question is whether such goals of individual reform and environmental improvement can be achieved by relatively conservative strategies, that is, by following established traditions and maintaining traditional disciplinary boundaries. Should we modify existing models of research by tacking on a human dimension as a kind of extra layer or extra box in the flow chart? Or do we need a more basic reinterpretation of what ecology is about?

The Baltimore study adopted the reasonable strategy of building on past models of ecosystem ecology, extending these successful models to the urban environment and bringing in people from other disciplines to participate in various ways in the study. That strategy, which is conservative and safe, is a prudent one, given the culture of science and the difficulties of conducting interdisciplinary work. It is influenced by the policies of funding agencies like the National

Science Foundation and by the mandates of government agencies like the Forest Service. It grows out of the history of ecology and the expectations that ecologists have for what is acceptable scientific method. It is influenced by the separation between "pure" and "applied" science, which can create barriers between those doing basic research and those doing practical work. Valuable research is being done. The question is whether a conservative strategy goes far enough in identifying and solving the kinds of challenges that one meets in our overcrowded world.

When one thinks of the city of Baltimore, for instance, many health problems come to mind: for example, lead poisoning among children and high rates of asthma and diabetes. Social problems stem from poverty, a vigorous drug trade, and a failing school system. Infrastructure problems include an old sewer system in need of repair and the lack of efficient public transportation. All of these are environmental problems, and all might enter into a study of the urban ecosystem. These issues are not entirely absent from the Baltimore study, but they are in the background, and some are not mentioned at all. They are not all deemed to be within the "ecology niche"; instead they are assigned to the "public health niche" or the "resource management niche" or are left to the politicians and educators. The result is that environmental issues are addressed in a fragmentary way because no single discipline takes on the task of synthesizing or integrating the knowledge being produced by all the disciplines. In addition, within ecology itself the loss of the ability to frame discussions in a human-centered way may not even be recognized as a problem. Yet unless a human-centered perspective can be built into environmental discussions, it will be difficult to convey to ordinary people, and also to politicians, the links between environmental reforms and other issues that affect human health and societal health. This failure also creates a barrier to the teamwork that is needed to solve environmental problems.

Of course one would not suggest that it is up to ecologists to solve all the problems of the modern city; they cannot even address all these problems. But ecologists could think about how these various "niches" relate to each other, and keep in mind the need for open dialogue with people in other disciplines who deal with such problems using their own methods. Teaming up need not imply interdisciplinarity, the buzzword of the modern age that is often set up as an ideal but is seldom realized—and for very good reasons, given the difficulties of forging syntheses between disciplines. Interdisciplinary work is not well rewarded in the academic world and is not well funded. Every discipline has its own vigorously defended approach and turf. Cultural change to overcome these obstacles cannot be achieved by a handful of researchers working on one project. It must be pushed forward by the larger community of scientists, who need to find ways to get the ear of politicians holding the purse strings. Then again, interdisciplinarity may not always be a viable objective, especially when there is no single goal to which all research is aiming.

Although promoting interdisciplinarity as a key step in changing the culture of ecology has been a major theme of recent discussions, connecting with and learning from other disciplines may be achievable without major cultural transformation in science. If ecology itself is not sufficiently human-centered, there are other disciplines that are, and these can be brought into the discussion and juxtaposed with ecology, not in order to create an interdisciplinary hybrid, but with the idea of showing the public how ecological questions permeate other disciplines and how several disciplines may be converging on similar conclusions. In the urban environment one can see, for example, that there should be links between ecology, public health, and human geography and that urban planners or landscape architects should also adopt an ecological perspective in their work. These other fields take different approaches to the problems of human behavior and social organization. They have evolved different ways to frame problems in a human-centered way, within the boundaries of their professions. Bringing together these voices to present a comprehensive picture of environmental problems and possible solutions is well worth the effort. The Baltimore study is taking steps in this direction, but institutional and funding constraints practically require that the steps be taken tentatively.

What is true of the urban environment can extend to ecology more broadly. Communication and cooperation between ecology, public health, and geography, three parallel disciplines that are sometimes in a competitive relationship, is the prerequisite for effective integration of knowledge from these fields. The benefits to ecology would be to suggest ways to apply a more human-centered approach to the analysis of ecological problems, which ultimately will help ecologists get their message across to the public better. A human-centered point of view does not have to imply one group of people imposing their values on another; it can mean a willingness to sit down together and engage in dialogue. If adaptive change is necessary, then the urban environment, which is the meeting place for many different disciplines and perspectives and the source of many environmental problems, seems an ideal location in which to think through the challenges and needs of modern ecology, from the ground up. And it is the ideal place to foment change.

But surely these problems are too important to be left to the academic world: the entire community must be involved. Arguably a more human-centered perspective needs to be at work in the urban setting, a perspective that puts people front and center and thinks through the problems of cities in a way that is sensitive to human needs. As an example of such a human-centered approach, I will reach back four decades to the work of Jane Jacobs and her classic 1961 study of the American city, *The Death and Life of Great American Cities.* Jacobs saw cities as problems in "organized complexity" where dozens of quantities could vary simultaneously but in subtly interconnected ways. Cities, she believed, represented many kinds of problems of organized complexity.[4]

Take, for instance, a city park and its uses, one of her examples.[5] The park's use depends in part on its design. But it also depends on who is nearby to use the park, and that in turn depends on how the city around the park is used. Each surrounding part of the city might influence the park, but these parts do not operate in isolation; each portion of the city interacts to form a complicated set of influences. Change any one part, and new sets of influences come into play. Such interactions all had to be studied. Jacobs saw cities as dynamic, evolving places, and her approach could certainly be characterized as ecological for that reason. But it was ecological thinking in the way that Edgar Anderson's or René Dubos's thinking was ecological, with insights derived from sensitive people-watching, attentive to the things that made people function better in some environments than in others.

Jacobs's book, which is still in print, stimulated readers to think of cities in ecological terms, to understand the complex relationships that were operating and thereby to avoid oversimplifying things and producing maladaptive planning formulas. We can contrast her idea of "organized complexity" with Pickett and McDonnell's view of urbanization as a "massive, unplanned experiment."[6] Organized complexity does not suggest something unplanned, but something that has been partly planned and yet is evolving in response to many influences and interactions. The city is a blend of planned elements; ecological interactions shape its development in ways that cannot easily be foreseen. But their complexity does not mean they cannot be understood or studied in a rigorous way.

I mention Jacobs to illustrate one form of ecological understanding applied to cities that has always been firmly human-centered and concrete and because of that has been successful in communicating its core message. Rachel Carson's writing had similar power, also because it was human-centered. Though recognizing that humans could destroy habitat and exhaust resources, Jacobs also knew, like Dubos, that humans had traits that led to the preservation, cultivation, and admiration of nature. She asked questions designed to accentuate the positive in human nature. She appreciated "lively, diverse, intense cities" where people communicated, built, and invented things needed to combat their problems. To Jacobs a city sidewalk, its bordering uses, and its users, are "active participants in the drama of civilization versus barbarism in cities."[7] The well-used sidewalk could be a means to assure public safety, a place for people to connect and communicate, and a place for children to play under the watchful eyes of adults.

Things could go well in cities, or they could go badly wrong. Those judgments about success and failure were made in relation to the human condition and the potential of human life. Jacobs had her eye firmly on the people, on how they lived and interacted and what made city environments work from the point of view of the people who lived there. As Edgar Anderson also argued, a city environment has to be a good place for people to live. Solving environmental problems associated with cities must go hand in hand with solving the human problems that make cities unpleasant and unsafe for people.

Ecologists view urban environments from a far less human-centered perspective. When an ecologist looks at paved areas, whether sidewalk or street, he or she sees an impermeable surface that almost always has negative connotations. It allows the rainfall to run off the land too quickly, drying out the city and perhaps polluting the waters into which the runoff flows. For the ecologist, paved areas elicit research questions like the following: "Does the water draining from an area move more rapidly into storm sewers because of intact curbs, or is it more likely to infiltrate into the soil along unguarded road verges? How do these structures affect loading of the water with toxins or potentially disease-causing organisms?"[8]

The modern ecological literature, as illustrated in the LTERs, depicts the urban environment quite differently from the way Jacobs writes about the city. Indeed one has to look hard to find even a passing citation to Jacobs's work in the LTER literature.[9] The difference is an understandable result of the evolution of ecology, and there is nothing wrong with the questions ecologists are asking. Yet Jacobs was also concerned about solving environmental problems. The difference lies not so much in the general goals as in the point of view taken. Both perspectives are necessary to attain the goal of persuading people to think differently and to create better environments.

The need to see things from the human viewpoint has in fact influenced urban renewal projects that are linked to the ecosystem studies. After all, when it was realized that something should be done to bring more people into the Gwynns Falls park system in Baltimore, the critical step was to build a wide asphalt walkway through the park so that people could walk or bicycle more easily, despite the fact that ecologists view paving as harmful to the environment. But without that paving, the park would be a forbidding environment and would not function well as an educational resource. This simple example illustrates that human needs must be understood and addressed if humans are expected to alter their relationship to the land. This basic lesson could easily be extended and developed in the broader framework of ecological discussions: How does ecology help people to live better? How do we establish compromises between our needs and larger environmental concerns?

As we saw in chapter 9, ecologists involved in modern urban ecosystem studies do acknowledge that people will see the environment differently depending on their disciplinary orientation or on how they use the landscape. A civil engineer and a morning jogger will look at the same landscape very differently. Ecologists, however, take this fact to mean that they must create complicated multidimensional classifications of the city environment.[10] The different perspectives are seen to create challenges to producing a scientific description or map of the landscape. I am suggesting that acknowledging these perspectives has different significance. Unless one understands what people need from their environment and how to satisfy their desire for safe and healthy environments, it will be difficult to relate abstract and esoteric scientific arguments to ordinary lives and produce fundamental change in how we exploit our environment.

Moving across scientific disciplines is certainly controversial, and bridging the "two cultures" of non-human-centered and human-centered discourses, as Rachel Carson and Dubos hoped to do, is even more so. Those writers who take a highly people-centered approach may appear subjective, emotional, and not scientific at all. The perception that arguments not couched in graphs, maps, and charts are not rigorous or are lacking in scientific method is not accurate. In Jacobs's most recent book, *Dark Age Ahead*, she emphasizes the importance of basing decisions on data, of carrying out experiments and evaluating outcomes.[11] In fact she faults scientific disciplines for failing to test their cherished assumptions and therefore perpetuating mistaken policies that have done harm to city environments. Being "human-centered" does not imply that rigor is not respected.

Knowing when and how to frame environmental problems in a human-centered way is particularly important if one of the goals of ecological research is to promote social justice.[12] Fully recognizing the link between environmentalism and social justice is arguably the key to successful resolution of many of our looming worldwide environmental crises. In 2004 the Norwegian Nobel Committee awarded the Peace Prize to the Kenyan activist and environmentalist Wangari Maathai, in recognition of her leadership in Africa's Green Belt movement and in promoting the causes of democracy and women's rights. Her acceptance speech stressed the importance of linking environmentalism to the broader defense of human rights, peace, and good government: "The Norwegian Nobel Committee has challenged the world to broaden the understanding of peace: there can be no peace without equitable development; and there can be no development without sustainable management of the environment in a democratic and peaceful state."[13] Whether one considers nitrogen loading in a small Baltimore stream or the expanding deserts of Africa, the solutions demand both science and the ability to see the human face in the landscape. Science has to connect to citizens, so that ordinary people will understand the value of the science and its significance in their lives. Environmental justice can then be understood in its fullest sense, as applying not just to parks, community gardens, and experiences of nature, but to all the problems that beset the poor in the city.

Once again, it is not necessarily the ecologist's responsibility to achieve this integration of knowledge: others bear equal responsibility for synthesizing and communicating knowledge. But if ecologists see their science as an integrative discipline, as Eugene Odum argued a quarter of a century ago, and if they wish to have a larger voice in the modern world, then the problem of integration becomes one that they need to tackle. If we take seriously the environmental problems that scientists warn us about, then it is clear we must be prepared over the next generation to make radical changes in our lives and expectations, to adapt ourselves to these changes as best we can. Ecology also must reinvent itself, as it has done in the past, and adapt to these challenges. Learning from adjacent disciplines and other perspectives is an obvious approach. Being outspoken about the need for cultural change within ecology may be needed as well.

Notes

INTRODUCTION: The Struggle for Place

1. Philip Pauly and Mark Barrow examine the professionalization of biology mainly from the zoological side, although Pauly discusses the botanical work of federal government scientists. Their books complement my story in the first half of this book. Mark V. Barrow Jr., *A Passion for Birds: American Ornithology after Audubon* (Princeton, NJ: Princeton University Press, 1998); Philip J. Pauly, *Biologists and the Promise of American Life, from Meriwether Lewis to Alfred Kinsey* (Princeton, NJ: Princeton University Press, 2000).

2. Samuel P. Hays, *Conservation and the Gospel of Efficiency; The Progressive Conservation Movement, 1890–1920* (Cambridge, MA: Harvard University Press, 1959).

3. Malcolm Nicolson, "Humboldtian Plant Geography after Humboldt: The Link to Ecology," *British Journal for the History of Science* 29 (1996): 289–310; Eugene Cittadino, *Nature as the Laboratory: Darwinian Plant Ecology in the German Empire, 1880–1900* (Cambridge: Cambridge University Press, 1990).

4. Robert Kohler, *Landscapes and Labscapes: Exploring the Lab-Field Border in Biology* (Chicago: University of Chicago Press, 2002), pp. 88–89.

5. An analysis of American exceptionalism in relation to the emergence of the social sciences can be found in Dorothy Ross, *The Origins of American Social Science* (Cambridge: Cambridge University Press, 1991).

6. George Perkins Marsh, *Man and Nature; or, Physical Geography as Modified by Human Action* (New York: Scribner, 1864; repr., Cambridge, MA: Harvard University Press, 1965). The 1965 edition has an introduction and annotations by David Lowenthal; page references are to the 1965 edition.

7. For Marsh's place in the history of conservation, see Arthur A. Ekirch, *Man and Nature in America* (New York: Columbia University Press, 1963).

8. David Lowenthal, *George Perkins Marsh, Versatile Vermonter* (New York: Columbia University Press, 1958); rev. ed., *George Perkins Marsh, Prophet of Conservation* (Seattle: University of Washington Press, 2000).

9. Marsh, *Man and Nature*, p. 280.

10. Ibid., p. 42.

11. David Lowenthal, introduction to Marsh, *Man and Nature*, p. xxv.

12. Marsh, *Man and Nature*, p. 464.

13. Ibid., p. xxiv, from a letter written by Marsh to Charles Scribner in 1863.

14. Ibid., pp. 41, 37.

15. Ibid., p. 45.

16. Lowenthal, introduction to Marsh, *Man and Nature*, p. xxvi.

17. All references here are to a later printing, Nathaniel Southgate Shaler, *Nature and Man in America* (New York: Scribner's, 1895).

18. Ibid., pp. 6–8.

19. Mary P. Winsor, *Reading the Shape of Nature: Comparative Zoology at the Agassiz Museum* (Chicago: University of Chicago Press, 1991), pp. 41–42.

20. Shaler, *Nature and Man in America*, p. 19.

21. Ibid., p. 110. The contests between organic armies are mainly discussed in chapter 4.

22. Ibid., p. vii.

23. Ibid., p. 168.

24. Ibid., p. 281.

25. Ibid., p. 282.

26. Nathaniel Southgate Shaler, ed., *The United States of America*, 2 vols. (New York: Appleton, 1894), quotes from the introductory chapter by Shaler, 1:49.

27. Ibid., 1:12–13.

28. Shaler, *Nature and Man in America*, p. 283.

29. David N. Livingstone, *Nathaniel Southgate Shaler and the Culture of American Science* (Tuscaloosa: University of Alabama Press, 1987), chap. 7.

30. Ibid., p. 212.

31. On Ward as an example of the ideology of American exceptionalism, see Ross, *Origins of American Social Science*, pp. 91–94.

32. Lester F. Ward, "The Local Distribution of Plants and the Theory of Adaptation," *Popular Science Monthly* 9 (1876): 676–84.

33. Ibid., p. 676.

34. Ibid., p. 679.

35. Ibid., p. 680.

36. Charles Darwin, *On the Origin of Species: A Facsimile of the First Edition* (Cambridge, MA: Harvard University Press, 1964), p. 73.

37. Ward, "Local Distribution of Plants," pp. 681–82.

38. Ibid., pp. 682–83.

39. Lester F. Ward, *Applied Sociology: A Treatise on the Conscious Improvement of Society by Society* (Boston: Ginn, 1906), pp. 124, 123, quotes p. 123. This definition of life was enunciated in 1865 by Claude Bernard, *An Introduction to the Study of Experimental Medicine*, trans. Henry C. Greene (New York: Dover, 1957).

40. Ward, *Applied Sociology*, p. 128.

41. Ibid., p. 131.

ONE: Entrepreneurs of Science

1. Anecdote told by Charles F. Cox, "Advantages of the Alliance to the Scientific Societies," in *The Scientific Alliance of New York: Addresses Delivered at the First Joint Meeting* (New York: Scientific Alliance of New York, 1893), pp. 9–17, quote p. 13.

2. Mary P. Winsor, *Reading the Shape of Nature: Comparative Zoology at the Agassiz Museum* (Chicago: University of Chicago Press, 1991), chaps. 1–2; A. Hunter Dupree, *Asa Gray: American Botanist, Friend of Darwin* (Baltimore: Johns Hopkins Press, 1959).

3. For an overview of the changes in American biology, with particular attention to zoology, see Philip J. Pauly, *Biologists and the Promise of American Life, from Meriwether*

Lewis to Alfred Kinsey (Princeton, NJ: Princeton University Press, 2000); Mark V. Barrow Jr., *A Passion for Birds: American Ornithology after Audubon* (Princeton, NJ: Princeton University Press, 1998); Ronald Rainger, Keith R. Benson, and Jane Maienschein, eds., *The American Development of Biology* (Philadelphia: University of Pennsylvania Press, 1988); Jane Maienschein, *Transforming Traditions in American Biology, 1880–1915* (Baltimore: Johns Hopkins University Press, 1991).

4. Andrew Denny Rodgers III, *American Botany, 1873–1892: Decades of Transition* (Princeton, NJ: Princeton University Press, 1944), p. 227.

5. Mark V. Barrow Jr., "The Specimen Dealer: Entrepreneurial Natural History in America's Gilded Age," *Journal of the History of Biology* 33 (2000): 493–534.

6. Janet Browne, *Charles Darwin: The Power of Place* (New York: Knopf, 2002), pp. 172–73.

7. John Merle Coulter, "The Future of Systematic Botany," *Botanical Gazette* 16 (1891): 243–54, quotes p. 247.

8. Malcolm Nicolson, "Humboldtian Plant Geography after Humboldt: The Link to Ecology," *British Journal for the History of Science* 29 (1996): 289–310; Malcolm Nicolson, "Alexander von Humboldt, Humboldtian Science, and the Origins of the Study of Vegetation," *History of Science* 25 (1987): 167–94.

9. See, for example, Keith R. Benson, "The Darwinian Legacy in the Pacific Northwest: Seattle's Young Naturalists' Society, P. Brooks Randolph, and Conchology," in *Darwin's Laboratory: Evolutionary Theory and Natural History in the Pacific*, ed. Roy MacLeod and Philip F. Rehbock (Honolulu: University of Hawaii Press, 1994), pp. 212–38.

10. For a fuller study of Britton's work, see Peter P. Mickulas, "Giving, Getting, and Growing: Philanthropy, Science, and the New York Botanical Garden, 1888–1929" (Ph.D. diss., Rutgers University, 2002).

11. Douglas Sloan, "Science in New York City, 1867–1907," *Isis* 71 (1980): 35–76; Simon Baatz, *Knowledge, Culture, and Science in the Metropolis: The New York Academy of Sciences, 1817–1970* (New York: New York Academy of Sciences, 1990); Scientific Alliance of New York, *Proceedings of the Second Joint Meeting, in Memory of John Strong Newberry* (New York: Press of L. S. Foster, 1893).

12. Andrew Denny Rodgers III, *John Torrey, A Story of North American Botany* (New York: Hafner, 1965).

13. Ann Blum, *Picturing Nature: American Nineteenth-Century Zoological Illustration* (Princeton, NJ: Princeton University Press, 1993), p. 237.

14. Arthur Hollick, "Torrey Botanical Club Reminiscences," *Bulletin of the Torrey Botanical Club* 17 (1918): 29–30.

15. Rodgers, *John Torrey*.

16. Henry H. Rusby, "An Historical Sketch of the Development of Botany in New York City," *Torreya* 6 (1906): 101–11, 133–45. Rusby published the same article also in *Plant World* 9 (1906): 153–61, 186–90.

17. On Low's role in the development of Columbia University, see Thomas Bender, *New York Intellect: A History of Intellectual Life in New York City, from 1750 to the Beginnings of Our Own Time* (Baltimore: Johns Hopkins University Press, 1987), pp. 279–84.

18. John Hendley Barnhart, "Historical Sketch of the Torrey Botanical Club," *Bulletin of the Torrey Botanical Club* 17 (1918): 12–21. A history of the Torrey Botanical Society by Patrick L. Cooney, dated November 7, 2000, is on the Web site NY-NJ-CT Botany Online at http://nynjctbotany.org/. See the link "History of the Torrey Botanical Society."

19. Hollick, "Torrey Botanical Club Reminiscences," pp. 29–30.

20. Marshall A. Howe, "Arthur Hollick, February 6, 1857–March 11, 1933," *Bulletin of the Torrey Botanical Club* 60 (1933): 537–53.

21. Pamphlet entitled *Public Discussion on Socialism, held at the New Theater, Leicester, on the Evenings of Tuesday and Wednesday, April 14th–15th, 1840, between Mr. John Brindley and Mr. Hollick, Socialist Missionary* (Leicester, Eng.: Price, 1840).

22. Frederick Hollick, *The Marriage Guide; or, Natural History of Generation: A Private Instructor for Married Persons and Those about to Marry, both Male and Female* (New York: American News, 1860). Other works by Hollick were *The Origin of Life: A Popular Treatise on the Philosophy and Physiology of Reproduction in Plants and Animals, Including the Details of Human Generation with a Full Description of the Male and Female Organs* (New York: T. W. Strong; G. W. Cottrell, 1845); *The Nerves and the Nervous* (New York: American News, 1873); *The Matron's Manual of Midwifery; and the Diseases of Women during Pregnancy and in Childbed* (New York: T. W. Strong, 1848).

23. Marcia Myers Bonta, *Women in the Field: America's Pioneering Women Naturalists* (College Station: Texas A&M University Press, 1991), pp. 124–31.

24. Margaret W. Rossiter, *Women Scientists in America: Struggles and Strategies to 1940* (Baltimore: Johns Hopkins University Press, 1982), p. 83.

25. William C. Steere, "Elizabeth Gertrude Knight Britton," in *Notable American Women, 1607–1950*, ed. Edward T. James (Cambridge, MA: Harvard University Press, 1971), 1:243–44.

26. Henry H. Rusby, "Nathaniel Lord Britton," *Science* 80 (1934): 108–11, quote p. 109.

27. Rusby, "Historical Sketch," p. 111.

28. George Thurber to Asa Gray, August 16, 1880, quoted in Susan M. Rossi-Wilcox, "Henry Hurd Rusby: A Biographical Sketch and Selectively Annotated Bibliography," *Harvard Papers in Botany* 4 (1993): 1–30, p. 2. See also Rodgers, *American Botany*, pp. 317–18.

29. Rossi-Wilcox, "Henry Hurd Rusby," p. 3.

30. Fanchon Hart, "Rusby, Henry Hurd," *Dictionary of American Biography*, suppl. 2, 1957, pp. 590–92.

31. George A. Bender, "Henry Hurd Rusby, Scientific Explorer, Societal Crusader, Scholastic Innovator," *Pharmacy in History* 23 (1981): 71–85, esp. p. 73.

32. William Lewis Herndon and Lardner Gibbon, *Exploration of the Valley of the Amazon*, 2 parts (Washington, DC: Robert Armstrong, 1854). Part 1 was by Herndon, and part 2 by Gibbon.

33. Rossi-Wilcox, "Henry Hurd Rusby," p. 5.

34. Henry H. Rusby, *Jungle Memories* (New York: Whittlesey House, 1933).

35. Diary of Elizabeth Britton, entry dated September 14, 1888, Archives of LuEsther T. Mertz Library, New York Botanical Garden, Bronx, New York.

36. Janet Browne, "Biogeography and Empire," in *Cultures of Natural History*, ed. N. Jardine, J. A. Secord, and E. C. Spary (Cambridge: Cambridge University Press, 1996), pp. 305–21, cite p. 313.

37. Donal P. McCracken, *Gardens of Empire: Botanical Institutions of the Victorian British Empire* (London: Leicester University Press, 1997), pp. 150–52.

38. Dupree, *Asa Gray*, p. 402.

39. Ibid., pp. 342–50.

40. Rusby, "Historical Sketch," p. 138.

41. Rusby, "Nathaniel Lord Britton," p. 109.

42. Addison Brown, "The Elgin Botanical Garden, Its Later History, and Relation to Columbia College and the Vermont Land Controversy," *Bulletin New York Botanical Garden* 5 (1909): 319–72. On early-nineteenth-century American botanical gardens, see Therese O'Malley, "'Your garden must be a museum to you': Early American Botanic Gardens," in *Art and Science in America: Issues of Representation*, ed. Amy R. W. Meyers (San Marino, CA: Huntington Library, 1998), pp. 35–59.

43. "A Great Garden Needed," *New York Herald*, November 26, 1888, p. 8; "Our Botanical Weakness," *New York Herald*, November 27, 1888, p. 11.

44. "A Great Garden Needed," p. 8.

45. "Nichols on Botany," *New York Herald*, December 2, 1888, p. 11. Hollick's letter is quoted in this article.

46. Addison Brown, *Judge Addison Brown, Autobiographical Notes for His Children* (Berryville, VA: Virginia Book, 1972), p. 125.

47. Harold E. Hammond, *A Commoner's Judge: The Life and Times of Charles Patrick Daly* (Boston: Christopher, 1954), p. 377.

48. "Public Botanic Garden," *New York Times*, January 13, 1889, p. 5.

49. Nathaniel Britton, "Botanical Gardens," *Bulletin New York Botanical Garden* 1 (1896): 62–77.

50. McCracken, *Gardens of Empire*, p. 78.

51. Rusby, "Nathaniel Lord Britton," p. 110.

52. Hammond, *A Commoner's Judge*.

53. Ibid., pp. 375–78.

54. "Botanical Gardens for This City," *New York Times*, February 8, 1889, p. 8; "For a Public Botanic Garden," *New York Times*, February 24, 1889, p. 5.

55. "An American Kew Proposed," *New York Times*, March 8, 1891, p. 13.

56. "The New York Botanic Garden," *Harper's Weekly*, May 23, 1891.

57. Brown, *Judge Addison Brown*, pp. 126–27.

58. On the history of the Alliance and Britton's role in it, see Baatz, *Knowledge, Culture, and Science*, pp. 147–64.

59. Cox, "Advantages of the Alliance," pp. 9–17.

60. Ibid., pp. 13–14.

61. Seth Low, "Advantages to New York City of the Alliance of the Scientific Societies," in *The Scientific Alliance of New York: Addresses Delivered at the First Joint Meeting* (New York: Scientific Alliance of New York, 1893), pp. 5–9.

62. Addison Brown, "The Need of Endowment for Scientific Research and Publication," in *The Scientific Alliance of New York*, pp. 18–41.

63. Ibid., p. 41.

64. Nathaniel L. Britton, "Kind of Building Required by the Scientific Alliance," in *The Scientific Alliance of New York: Addresses Delivered at the First Joint Meeting* (New York: Scientific Alliance of New York, 1893), pp. 61–64, quote p. 64.

65. Baatz, *Knowledge, Culture, and Science*, pp. 155–56.

66. Brown, *Judge Addison Brown*, pp. 128–30. The *Sun* was owned by Charles A. Dana, whose son Paul was on the Park Board of Commissioners. Paul Dana was also editor of the *Sun* during this time.

67. *New York Times*, August 8, 1895, p. 5, col. 3.

68. "No Better Site Found; the Choice of Bronx Park for a Botanical Garden," *New York Times*, August 18, 1895, p. 2, col. 2.

69. The committee's report was summarized in an article, "Might Mar Bronx Park: Change in Botanical Garden Plan Advised," *New York Sun*, June 15, 1897, p. 7.

70. Ibid.

71. See the following articles in the *New York Sun:* "The Botanic Garden," June 16, 1897, p. 6; "Bronx Garden Plans Go," June 22, 1897, p. 7; "The Botanical Garden Plans," June 23, 1897, p. 6 (mentions "well-reputed gentlemen"); "The Outrage in Bronx Park," July 23, 1897, p. 6; "The Better Element—A Study," August 3, 1897, p. 6.

TWO: A Botanical Revolution

1. Henry A. Gleason, "The Scientific Work of Nathaniel Lord Britton," *Proceedings American Philosophical Society* 104 (1960): 205–26, cite p. 209.

2. For a summary of de Candolle's ideas, with commentary by Asa Gray, see Asa Gray, "Botanical Nomenclature," *Scientific Papers of Asa Gray* (Boston: Houghton, Mifflin, 1889), 1:358–83.

3. E. E. Sterns, "The Nomenclature Question and How to Settle It," *Bulletin of the Torrey Botanical Club* 15 (1888): 230–35, quote p. 230.

4. Mark V. Barrow Jr. *A Passion for Birds: American Ornithology after Audubon* (Princeton, NJ: Princeton University Press, 1998), chap. 4. Barrow discusses the nomenclature controversy as it affected ornithology.

5. Quoted by Lester Ward, "The Nomenclature Question," *Bulletin of the Torrey Botanical Club* 22 (1895): 313. Ward was quoting from Frederick Coville, "A Reply to Dr. Robinson's Criticism of the 'List of Pteridophyta and Spermatophyta of Northeastern America,'" *Botanical Gazette* 20 (1895): 163. The source for the quotation was not given by Coville.

6. The story of the nomenclature debate, with an emphasis on the role of Nebraska botanist Charles Bessey, is recounted in Richard A. Overfield, *Science with Practice: Charles E. Bessey and the Maturing of American Botany* (Ames: Iowa State University Press, 1993), chap. 5. Further details of the controversy are in Andrew Denny Rodgers III, *John Merle Coulter, Missionary in Science* (Princeton, NJ: Princeton University Press, 1944); and Andrew Denny Rodgers III, *American Botany, 1873–1892: Decades of Transition* (Princeton, NJ: Princeton University Press, 1944).

7. Asa Gray to N. L. Britton, November 27, 1887, in *Letters of Asa Gray*, ed. Jane L. Gray (Boston: Houghton Mifflin, 1893), 2:813–14.

8. A. Hunter Dupree, *Asa Gray: American Botanist, Friend of Darwin* (Baltimore: Johns Hopkins Press, 1959), pp. 395–402. The controversy between Greene and Gray is also discussed in Rodgers, *American Botany*, chap. 11.

9. Dupree, *Asa Gray*, p. 134.

10. Robert P. McIntosh, "Edward Lee Greene, the Man," in *Landmarks of Botanical History: Edward Lee Greene*, ed. Frank N. Egerton (Stanford, CA: Stanford University Press, 1983), 1:18–53, cite p. 34.

11. McIntosh, "Edward Lee Greene," p. 38.

12. Frederic T. Bioletti, "Reminiscences of an Amateur Botanist," *Scientific Monthly* 29 (1929): 333–39.

13. Ibid., pp. 334–36.

14. Egerton, *Landmarks of Botanical History*. This is a two-volume collection of his-

torical essays by Greene. Ancient botanical history is discussed in volume 1; see the essays "The Rhizotomi" and "Theophrastus of Eresus."

15. Ibid., p. 127.

16. McIntosh, "Edward Lee Greene," p. 39.

17. On Greene and Brandegee, see Barbara Ertter, "People, Plants, and Politics: The Development of Institution-Based Botany in California, 1853–1906," in *Cultures and Institutions of Natural History: Essays in the History and Philosophy of Science*, ed. Michael T. Ghiselin and Alan E. Leviton, vol. 25 in Memoirs of the California Academy of Sciences (San Francisco: California Academy of Sciences, 2000), pp. 203–48; Barbara Ertter, "The Flowering of Natural History Institutions in California," *Proceedings of the California Academy of Sciences* 55, suppl. 1 (2004): 58–87.

18. McIntosh, "Edward Lee Greene," p. 39.

19. Ibid., p. 38.

20. Dupree, *Asa Gray*, p. 398.

21. McIntosh, "Edward Lee Greene," p. 34.

22. James Britten, "Recent Tendencies in American Botanical Nomenclature," *Journal of Botany* 26 (1888): 257–62.

23. N. L. Britton, letter to editor, *Journal of Botany* 26 (1988): 292–95.

24. Charles Sprague Sargent, *The Silva of North America* (1890; New York: Peter Smith, 1947), 1:viii.

25. Charles Bessey to Elizabeth Britton, April 29, 1896, in Elizabeth Gertrude Britton Correspondence, Charles E. Bessey file, Archives of the LuEsther T. Mertz Library, New York Botanical Garden, Bronx, New York.

26. Sterns, "The Nomenclature Question," p. 230.

27. H. H. Rusby, "Note on Nomenclature," *Bulletin of the Torrey Botanical Club* 15 (1888): 218–19.

28. Diary of Elizabeth Britton, entry dated August 24, 1888, Archives of the LuEsther T. Mertz Library.

29. Sterns, "The Nomenclature Question," p. 235.

30. Anonymous editorial, *Botanical Gazette* 17 (1892): 164–65.

31. See Simon Schama, *Landscape and Memory* (New York: Vintage Books, 1995), pp. 185–201, on the symbolism of the Big Trees.

32. Sereno Watson, "On Nomenclature," *Botanical Gazette* 17 (1892): 169–70; N. L. Britton, "The Plea of Expediency," *Botanical Gazette* 17 (1892): 252–53.

33. Gleason, "Scientific Work of Britton."

34. This name appears not to have been valid; in 1893 Britton described it as a new genus without recognizing Greene's priority. See Susan Rossi-Wilcox, "Henry Hurd Rusby: A Biographical Sketch and Selectively Annotated Bibliography," *Harvard Papers in Botany* 4 (1993): 1–30, esp. p. 3, p. 11n4.

35. Committee members were H. H. Rusby, N. L. Britton, J. M. Coulter, F. V. Coville, L. M. Underwood, L. F. Ward, and W. A. Kellerman.

36. George B. Sudworth, "On Legitimate Authorship of Certain Binomials with Other Notes on Nomenclature," *Bulletin of the Torrey Botanical Club* 20 (1893): 40–46.

37. Charles E. Bessey, "The New Check-List of Plants," *American Naturalist* 29 (1895): 349–51, quote p. 350.

38. Ibid., p. 350.

39. Anonymous editorial, *Botanical Gazette* 17 (1892): 297–98.

40. Lucien M. Underwood, "The Nomenclature Question at Genoa," *Bulletin of the Torrey Botanical Club* 19 (1892): 325–30, quote p. 327.

41. Anonymous editorial, *Botanical Gazette* 17 (1892): 22–23.

42. On the politics surrounding the formation of the Botanical Society, see Overfield, *Science with Practice*, pp. 115–20.

43. Otto Kuntze, "On a New Code of Nomenclature," *Botanical Gazette* 19 (1894): 126. This is an extract from a private letter, probably to Coulter, that Kuntze asked to be printed in this journal.

44. Ibid.

45. Rogers McVaugh, "Edward Lee Greene: An Appraisal of His Contribution to Botany," in Egerton, *Landmarks of Botanical History*, pp. 54–84, Greene quote p. 76.

46. Rodgers, *John Merle Coulter*, quote p. 137, no source given.

47. Coville, "A Reply to Dr. Robinson's Criticism," pp. 162–67.

48. B. L. Robinson, "The Nomenclature Question: On the Application of 'Once a synonym always a synonym' to binomials," *Botanical Gazette* 20 (1895): 261–63.

49. Ward, "The Nomenclature Question," pp. 308–29.

50. Ibid., pp. 325–26.

51. Ibid., pp. 320–21.

52. Lester Frank Ward, *Dynamic Sociology; or, Applied Social Science*, 2nd ed., 2 vols. (New York: Appleton, 1897). The first edition was published in 1883.

53. Ward, "The Nomenclature Question," p. 319.

54. "American Code of Botanical Nomenclature," *Bulletin of the Torrey Botanical Club* 34 (1907): 167–78. The authors listed on p. 168 were members of the Nomenclature Commission of the Botanical Club of the American Association for the Advancement of Science.

55. Addison Brown, *Judge Addison Brown: Autobiographical Notes for His Children* (Berryville, VA: Virginia Book, 1972), pp. 189–90.

56. Gleason, "Scientific Work of Britton," pp. 215–16.

THREE: Big Science

1. Janet Browne, *Charles Darwin: The Power of Place* (New York: Knopf, 2002), pp. 172–73.

2. Edwin G. Burrows and Mike Wallace, *Gotham: A History of New York City to 1898* (New York: Oxford University Press, 1999), pp. 906–16.

3. For more detail on the construction and design of the Garden and its displays, see Peter Mickulas, "A Scientific Garden in a Public Space: The New York Botanical Garden, 1900–10," *Studies in the History of Gardens and Designed Landscapes* 22 (2002): 193–213.

4. Frans Stafleu, "History of the New York Botanical Garden," chap. 4 file, p. 37, typescript, Archives of the LuEsther T. Mertz Library, New York Botanical Garden, Bronx, New York.

5. The scope of the Garden was described in Daniel T. MacDougal, "Botanic Gardens. I. Origin and General Organization," *Popular Science Monthly* 50 (1897): 172–86; and Daniel T. MacDougal, "The New York Botanical Garden," *Popular Science Monthly* 57 (1900): 171–78. The annual reports published in the Garden's *Bulletin* describe the building plans, plantings, and acquisitions in detail.

6. Henry A. Gleason, "Thumbnail Sketches of Botanists," typescript, Henry A. Gleason Papers, box 4, series 3, pp. 31–42 on Britton, Archives of the LuEsther T. Mertz Library, New York Botanical Garden, Bronx, New York.

7. Henry A. Gleason, "The Scientific Work of Nathaniel Lord Britton," *Proceedings American Philosophical Society* 104 (1960): 205–26, cite p. 205.

8. Carlton C. Curtis, "A Biographical Sketch of Lucien Marcus Underwood," *Bulletin of the Torrey Botanical Club* 35 (1908): 1–12.

9. Greater New York included the city of New York, Brooklyn, Staten Island, and Queens.

10. Henry Gleason, "The Short and Simple Annals of Henry A. Gleason," vol. 1, pp. 169–75; and "Thumbnail Sketches of Botanists," pp. 199–200, both in Henry A. Gleason Papers, Archives of the LuEsther T. Mertz Library.

11. Ibid.

12. Statistical data are taken from the annual *Bulletin* of the New York Botanical Garden, from the reports of the director and of the various other departments and managers. The figures for 1900–1909 come from the report for the year 1910, in *Bulletin of the New York Botanical Garden* 7 (1909–11): 319.

13. Andrew Denny Rodgers III, *American Botany, 1873–1892: Decades of Transition* (Princeton, NJ: Princeton University Press, 1944), pp. 227–28.

14. *Johns Hopkins University Circulars*, vol. 19, no. 142 (December 1899): 2–3; no. 144 (April 1900): 28–29.

15. On the Humboldtian tradition, see Malcolm Nicolson, "Humboldtian Plant Geography after Humboldt: The Link to Ecology," *British Journal for History of Science* 29 (1996): 289–310; see Eugene Cittadino, *Nature as the Laboratory: Darwinian Plant Ecology in the German Empire, 1880–1900* (Cambridge: Cambridge University Press, 1990); William Coleman, "Evolution into Ecology? The Strategy of Warming's Ecological Plant Geography," *Journal of the History of Biology* 19 (1986): 181–96; Joel B. Hagen, "Ecologists and Taxonomists: Divergent Traditions in Twentieth-Century Plant Geography," *Journal of the History of Biology* 19 (1986): 197–214.

16. Eugene Cittadino, "Ecology and the Professionalization of Botany in America, 1890–1905," *Studies in History of Biology* 4 (1980): 171–98.

17. John Merle Coulter, "The Future of Systematic Botany," *Botanical Gazette* 16 (1891): 243–54.

18. N. L. Britton, "Our Conception of 'Species' as Modified by the Doctrine of Evolution," *Transactions of the New York Academy of Sciences* 13 (1894): 132–35.

19. Ibid., p. 132.

20. Lucien Underwood, "The Last Quarter: A Reminiscence and an Outlook," *Science* 12 (1900): 161–70. Further scholarly study to investigate the relationship between these ideals and actual practice in systematics would be highly desirable.

21. Joseph Charles Arthur, "Development of Vegetable Physiology," *Botanical Gazette* 20 (1895): 381–402, quote p. 382.

22. Ibid., pp. 389, 394.

23. Daniel T. MacDougal, "Botanic Gardens. II. Tübingen and Its Botanists," *Popular Science Monthly* 50 (1897): 312–23.

24. William Coleman, "Cell, Nucleus, and Inheritance: An Historical Study," *Proceedings American Philosophical Society* 109 (1965): 124–58.

25. Hugo de Vries, *Die Mutationstheorie*, 2 vols. (Leipzig: Veit, 1901, 1903). The Eng-

lish translation is Hugo de Vries, *The Mutation Theory: Experiments and Observations on the Origin of Species in the Vegetable Kingdom,* 2 vols., trans. J. B. Farmer and A. D. Darbishire (Chicago: Open Court, 1909).

26. On the place of de Vries's work in the Darwinian tradition, see Lindley Darden, "Reasoning in Scientific Change: Charles Darwin, Hugo de Vries, and the Discovery of Segregation," *Studies in History and Philosophy of Science* 7 (1976): 127–69. On the Dutch context of de Vries's work, see Bert Theunissen, "Knowledge Is Power: Hugo de Vries on Science, Heredity, and Social Progress," *British Journal of the History of Science* 27 (1994): 291–311. See also Peter J. Bowler, "Hugo de Vries and Thomas Hunt Morgan: The Mutation Theory and the Spirit of Darwinism," *Annals of Science* 35 (1978): 55–73.

27. D. T. MacDougal, "Mutation in Plants," *American Naturalist* 37 (1903): 737–70, quote p. 745.

28. Janet Browne, *Charles Darwin: The Power of Place* (New York: Knopf, 2002). Darwin published a book on insectivorous plants in 1875 and, aided by his son Francis, a book on the power of movement of plants in 1880.

29. Soraya de Chadarevian, "Laboratory Science versus Country-House Experiments: The Controversy between Julius Sachs and Charles Darwin," *British Journal for the History of Science* 29 (1996): 17–41.

30. Wilhelm Pfeffer, *The Physiology of Plants: A Treatise upon the Metabolism and Sources of Energy in Plants,* 2 vols., 2nd ed., rev., trans. Alfred J. Ewart (Oxford: Clarendon Press, 1900–1903).

31. Hugo de Vries, *Species and Varieties: Their Origin by Mutation* (Chicago: Open Court, 1905), p. 12.

32. D. T. MacDougal, "Discontinuous Variation and the Origin of Species," *Science* 21 (1905): 540–43, quote p. 543.

33. D. T. MacDougal, "The Origin of Species by Mutation," *Torreya* 2 (1902): 65–68, 81–84, 97–101.

34. MacDougal, "Mutation in Plants," pp. 737–70.

35. MacDougal to C. B. Davenport, July 19, 1904, Charles B. Davenport Papers, B:D27, D. T. MacDougal file 2, American Philosophical Society, Philadelphia.

36. D. T. MacDougal, "Alterations in Heredity Induced by Ovarial Treatments," *Botanical Gazette* 51 (1911): 241–57.

37. Charles Stuart Gager, "Effects of the Rays of Radium on Plants," *Memoirs of the New York Botanical Garden,* vol. 4 (New York, 1908); D. T. MacDougal, "The Direct Influence of Environment," in *Fifty Years of Darwinism: Modern Aspects of Evolution* (New York: Henry Holt, 1909), pp. 114–42, quote p. 129. MacDougal's views about specific characters on the chromosomes were taken from Edmund Beecher Wilson's research at Columbia University.

38. Peter J. Bowler, *The Eclipse of Darwinism: Anti-Darwinian Evolution Theories in the Decades around 1900* (Baltimore: Johns Hopkins University Press, 1983).

39. D. T. MacDougal, "Heredity and the Origin of Species," *Monist* 16 (1906): 32–64.

40. *New York Times,* December 24, 1905, first magazine section, pp. 1–2.

41. Philip J. Pauly, *Controlling Life: Jacques Loeb and the Engineering Ideal in Biology* (New York: Oxford University Press, 1987), pp. 100–105.

42. MacDougal, "Origin of Species by Mutation."

43. Eugene Cittadino describes this campaign in "Ecology and Botany in America."

44. Lucien M. Underwood, "The Department of Botany," *Columbia University Quarterly* 4 (June 1903): 278–92, quote p. 292.

45. Lucien Underwood, "The Royal Botanic Gardens at Kew," *Science* 10 (1899): 65–75, quote p. 70.

46. D. E. Boufford and S. A. Spongberg, "Eastern Asian–Eastern North American Phytogeographical Relationships: A History from the Time of Linnaeus to the Twentieth Century," *Annals Missouri Botanical Garden* 70 (1983): 423–39.

47. Underwood, "Royal Botanic Gardens at Kew," p. 73.

48. Underwood, "The Last Quarter," p. 170.

49. Underwood, "Royal Botanic Gardens at Kew," p. 75.

50. Ibid., pp. 72, 75.

51. Underwood, "The Last Quarter," pp. 169, 170.

52. Simon Baatz, *Knowledge, Culture, and Science in the Metropolis: The New York Academy of Sciences, 1817–1970* (New York: New York Academy of Sciences, 1990), pp. 202–16.

53. Gleason, "Thumbnail Sketches of Botanists," p. 39 on Britton.

54. Ibid.

55. Henry A. Gleason, "The Scientific Work of Nathaniel Lord Britton," *Proceedings American Philosophical Society* 104 (1960): 205–26, cite p. 220.

56. Ibid., p. 220.

57. Philip Pauly makes a related observation about Spencer Baird's insistence that collectors gather unusual as well as common animals and note their exact locations, in *Biologists and the Promise of American Life, from Meriwether Lewis to Alfred Kinsey* (Princeton, NJ: Princeton University Press, 2000), p. 46.

58. A related issue was concern about invasive species. Philip Pauly analyzes how botanical research in the federal government was affected by "ecological imperialism" after 1898, and he relates these biological discussions to concern about human "aliens," in ibid., chap. 3.

59. George A. Bender, "Henry Hurd Rusby, Scientific Explorer, Societal Crusader, Scholastic Innovator," *Pharmacy in History* 23 (1981): 71–85.

60. Henry H. Rusby, "Recent Botanical Collecting in the Republic of Colombia," *Memoirs Torrey Botanical Club* 17 (1918): 39–47.

61. Editorial, *Botanical Gazette* 23 (1897): 200.

62. Eugen Warming, "On the Vegetation of Tropical America," *Botanical Gazette* 27 (1899): 1–18; Coleman, "Evolution into Ecology?"

63. William Fawcett, "The Public Gardens and Plantations of Jamaica," *Botanical Gazette* 24 (1897): 345–69.

64. David Fairchild, letter to *Botanical Gazette* 27 (1899): 320–22.

65. Lucien Underwood, report in *Journal New York Botanical Garden* 4 (July 1903): 109–19.

66. Forrest Shreve, "A Collecting Trip at Cinchona," *Torreya* 6 (1906): 81–84.

67. N. L. Britton, report in *Journal New York Botanical Garden* 5 (January 1904).

68. Gleason, "Scientific Work of Britton," p. 205.

69. Ibid., p. 222.

70. Ibid.

71. Charles E. Bessey, "The Taxonomic Aspect of the Species Question," *American Naturalist* 42 (1908): 218–24, quote p. 222.

72. Ibid.

73. Henry C. Cowles, "An Ecological Aspect of the Conception of Species," *American Naturalist* 42 (1908): 265–71, quote p. 271.

74. Nathaniel L. Britton, review of *Flora of the District of Columbia, Torreya* 19 (1919): 244–46, quote p. 246.

75. Stuart McCook, "The Agricultural Awakening of Latin America: Science, Development, and Nature, 1900–1930" (Ph.D. diss., Princeton University, 1996), Pittier quote pp. 200–201.

76. Report of the director, *Bulletin New York Botanical Garden* 1 (January 1897): 48.

77. On Central Park's evolution, see Roy Rosenzweig and Elizabeth Blackmar, *The Park and the People: A History of Central Park* (Ithaca, NY: Cornell University Press, 1992), chap. 12.

78. C. Stuart Gager, "Elizabeth Britton and the Movement for the Preservation of Native American Wild Flowers," *Journal New York Botanical Garden* 41 (1940): 137–42.

79. Helen Horowitz, "Animal and Man in the New York Zoological Park," *New York History* 56 (1975): 426–55, quotes p. 445.

80. Gager, "Elizabeth Britton and Wild Flowers," p. 137.

81. Ibid., p. 138.

82. Ibid.

83. Elizabeth G. Britton, "Vanishing Wild Flowers," *Torreya* 1 (1901): 85–94, quotes pp. 85, 86.

84. Ibid., p. 92.

85. One sees this ideology at work in other areas of conservation at this time, for example, in the conflict between scientists and fishermen over conservation of oysters in the Chesapeake Bay. See Christine Keiner, "Scientists, Oystermen, and Maryland Oyster Conservation Politics, 1880–1969: A Study of Two Cultures" (Ph.D. diss., Johns Hopkins University, 2000).

86. The modern literature on the history of conservation movements distinguishes between movements for "conservation" and "preservation" as having different goals and motives. I have not followed this distinction and instead use the terms *conservation* and *preservation* interchangeably, because this usage more accurately reflects how the terms were employed in the early twentieth century.

87. Mark V. Barrow Jr., *A Passion for Birds: American Ornithology after Audubon* (Princeton, NJ: Princeton University Press, 1998), esp. chaps. 5 and 6 on the Audubon Society, conservation, and professionalization within ornithology.

88. Daniel MacDougal, "Report of Expedition to Arizona and Sonora," *Journal New York Botanical Garden* 3 (1902): 98–99.

89. G. Gordon Copp, Stokes prize essay published in *Journal New York Botanical Garden* 7 (1906): 26–29.

90. F. H. Knowlton, "Suggestions for the Preservation of Our Native Plants," *Journal New York Botanical Garden* 3 (1902): 41–47.

91. Mary Perle Anderson wrote about the nature-study movement and also won the first prize for the Stokes fund competition, in *Journal New York Botanical Garden* 10 (1909).

92. "Mrs. Britton Honored in Dedication of Plaque by New York Bird and Tree Club in Wild Flower Garden," *Journal of New York Botanical Garden* 41 (1940): 129–37, quote p. 130.

FOUR: Science in a Changing Land

1. The relationship between ecology and agricultural science is not well understood. For this period we need a study along the lines of the analysis of agricultural genetics by Barbara A. Kimmelman, "A Progressive Era Discipline: Genetics at American Agricultural Colleges and Experiment Stations, 1900–1920" (Ph.D. diss., University of Pennsylvania, 1987).

2. Robert E. Kohler, *Landscapes and Labscapes: Exploring the Lab-Field Border in Biology* (Chicago: University of Chicago Press, 2002), reference to ecology as "schizoid" on p. 75, "lurching" on p. 137, and "double bind" on p. 187.

3. "Report of the Advisory Committee on Botany," *Carnegie Institution of Washington Yearbook* 1 (1902): 3–12.

4. Ibid., p. 6.

5. F. V. Coville, *Botany of the Death Valley Expedition*, (Washington, DC: Government Printing Office, 1893), p. 18; Andrew Denny Rodgers III, *John Merle Coulter, Missionary in Science* (Princeton, NJ: Princeton University Press, 1944), pp. 225–27.

6. Samuel Hays, *Conservation and the Gospel of Efficiency: The Progressive Conservation Movement, 1890–1920* (Cambridge, MA: Harvard University Press, 1959). The concepts underlying the "gospel of efficiency" are discussed on pp. 122–27.

7. The consequences of the act are discussed in Donald Worster, *Rivers of Empire: Water, Aridity, and the Growth of the American West* (New York: Pantheon, 1985).

8. Hays, *Conservation and Gospel of Efficiency*, pp. 55–56.

9. "Report of Advisory Committee," p. 5.

10. Quote in William T. Hornaday, *Camp-Fires on Desert and Lava* (New York: Scribner's, 1914), p. 4; Eugene Cittadino, "Ecology and the Professionalization of Botany in America, 1890–1905," *Studies in History of Biology* 4 (1980): 171–98, esp. pp. 182–84.

11. Hornaday, *Camp-Fires on Desert and Lava*, p. 14.

12. F. V. Coville and D. T. MacDougal, *Desert Botanical Laboratory of the Carnegie Institution* (Washington, DC: Carnegie Institution of Washington, 1903), pp. 12–17.

13. Ibid., p. 14.

14. M. Mitchell, "The Founding of the University of Arizona, 1885–1894," *Arizona and the West* 27 (1985): 5–36.

15. Hornaday, *Camp-Fires on Desert and Lava*, pp. 18–19.

16. Godfrey Sykes, *A Westerly Trend . . . Being a Veracious Chronicle of More than Sixty Years of Wanderings, Mainly in Search of Space and Sunshine* (Tucson: University of Arizona Press, 1984).

17. Judith C. Wilder, "The Years of a Desert Laboratory," *Journal of Arizona History* 8 (1967): 179–99.

18. Ibid.; Godfrey Sykes, *The Colorado Delta*, published jointly by the Carnegie Institution of Washington and American Geographical Society (Baltimore: Lord Baltimore Press, 1937).

19. Ellsworth Huntington, "The Desert Laboratory," *Harper's Monthly Magazine* 122 (1911): 651–62, quote p. 655.

20. For descriptions of the Desert Laboratory and assessments of its role in the development of ecology, see especially Janice E. Bowers, "A Debt to the Future: Scientific Achievements of the Desert Laboratory, Tumamoc Hill, Tucson, Arizona," *Desert Plants*

10 (1990): 9–12, 35–47; Janice E. Bowers, *A Sense of Place: The Life and Work of Forrest Shreve* (Tucson: University of Arizona Press, 1988); Robert P. McIntosh, "Pioneer Support for Ecology," *BioScience* 33 (1983): 107–12; Ray Bowers, "Mr. Carnegie's Plant Biologists: The Ancestry of Carnegie Institution's Department of Plant Biology," pamphlet (Washington, DC: Carnegie Institution of Washington, 1992); William G. McGinnies, *Discovering the Desert: Legacy of the Carnegie Desert Botanical Laboratory* (Tucson: University of Arizona Press, 1981); Rodgers, *John Merle Coulter*, pp. 227–33.

21. D. T. MacDougal, *Botanical Features of North American Deserts* (Washington, DC: Carnegie Institution of Washington, 1908), pp. 109–11.

22. MacDougal to Tucson Chamber of Commerce, March 19, 1907, MacDougal Papers, Arizona Historical Society, Tucson, box 7, file 41.

23. Peter Dreyer, *A Gardener Touched with Genius: The Life of Luther Burbank* (Berkeley: University of California Press, 1985).

24. Nathan Reingold, "National Science Policy in a Private Foundation: The Carnegie Institution of Washington," in *The Organization of Knowledge in Modern America, 1860–1920*, ed. Alexandra Oleson and John Voss (Baltimore: Johns Hopkins University Press, 1979), pp. 313–41.

25. Hugo de Vries, "A Visit to Luther Burbank," *Popular Science Monthly* 67 (1902): 329–47; Sharon E. Kingsland, "The Battling Botanist: Daniel Trembly MacDougal, Mutation Theory, and the Rise of Experimental Evolutionary Biology in America, 1900–1912," *Isis* 82 (1991): 479–509.

26. Robert S. Woodward, Record of Minutes of Meeting of Carnegie Institution of Washington, Dec. 8, 1908, Carnegie Institution of Washington Archives, Washington, DC, p. 731.

27. McGinnies, *Discovering the Desert*, pp. 10–12.

28. Francis Darwin, presidential address, *Report of the 78th Meeting of the British Association for the Advancement of Science*, Dublin, September 1908 (London: John Murray, 1909), pp. 3–27.

29. "New Wonders of Science in Dealing with Plants," *New York Times*, October 4, 1908, part 5, p. 8. For further discussion of Francis Darwin's theories, with a contribution by MacDougal, see "Do Plants Really Possess the Power of Thinking?" *New York Times*, February 28, 1909.

30. D. T. MacDougal, "Alterations in Heredity Induced by Ovarial Treatments," *Botanical Gazette* 51 (1911): 241–57.

31. MacDougal to Charles F. Cox, March 31, 1909, MacDougal Papers, box 9, folder 50.

32. D. T. MacDougal, "Organic Response," *American Naturalist* 45 (1911): 5–40, quote p. 40.

33. Garland Allen, *Thomas Hunt Morgan: The Man and His Science* (Princeton, NJ: Princeton University Press, 1978), p. 124n63.

34. For a review of genetic and cytological research on *Oenothera* between 1909 and 1913 and its relation to Mendelism, see R. Ruggles Gates, "Recent Papers on *Oenothera* Mutations," *New Phytologist* 12 (1913): 290–302.

35. McGinnies, *Discovering the Desert*, pp. 9–10; Bowers, *A Sense of Place*, pp. 44, 69.

36. Quoted in Bowers, *A Sense of Place*, pp. 69–70.

37. D. T. MacDougal, "Department of Botanical Research," *Carnegie Institution of Washington Yearbook* 15 (1916): 95.

38. William P. Blake, "Sketch of the Region at the Head of the Gulf of California,"

introductory essay in *The Imperial Valley and the Salton Sink*, by H. T. Cory (San Francisco: John J. Newbegin, 1915), pp. 1–35.

39. Worster, *Rivers of Empire.*

40. George Kennan, *The Salton Sea: An Account of Harriman's Fight with the Colorado River* (New York: Macmillan, 1917); Cory, *The Imperial Valley and the Salton Sink.*

41. Kennan, *The Salton Sea;* William E. DeBuys, *Salt Dreams: Land and Water in Low-Down California* (Albuquerque: University of New Mexico Press, 1999).

42. D. T. MacDougal, William Phipps Blake, Godfrey Sykes, E. E. Free, W. H. Ross, A. E. Vinson, George J. Peirce, Melvin A. Brannon, J. Claude Jones, and S. B. Parish, *The Salton Sea: A Study of the Geography, the Geology, the Floristics, and the Ecology of a Desert Basin* (Washington, DC: Carnegie Institution of Washington, 1914), pp. 179–82.

43. MacDougal et al., *The Salton Sea;* D. T. MacDougal, "The Salton Sea," *American Journal of Science* 39 (1915): 231–50; Sykes, *The Colorado Delta.*

44. Jeffrey P. Cohn, "Saving the Salton Sea," *BioScience* 50 (2000): 295–301.

45. Volney M. Spalding, "Present Problems in Plant Ecology: Problems of Local Distribution in Arid Regions," *Annual Report of the Smithsonian Institution for Year Ending June 30, 1909* (Washington, DC: Government Printing Office, 1910), pp. 453–63, quote p. 462.

46. Francis E. Lloyd, "A Botanical Laboratory in the Desert," *Popular Science Monthly* 66 (1905): 239–42.

47. Burton L. Livingston, "Present Problems of Physiological Plant Ecology," *Plant World* 12 (1909): 41–46.

48. Edith Shreve was prevented from being a paid staff member of the laboratory by antinepotism rules of the Carnegie Institution. On her career, see Janice E. Bowers, "A Career of Her Own: Edith Shreve at the Desert Laboratory," *Desert Plants* 8 (1986): 23–29. On women scientists in the Southwest, including the women at the Desert Laboratory, see George E. Webb, "A Woman's Place Is in the Lab: Arizona's Women Research Scientists, 1910–1950," *Journal of Arizona History* 34 (1993): 45–64.

49. Goebel's major treatise on causal morphology, *Organography of Plants*, first was published in German in 1898 and was translated into English in two volumes published by Clarendon Press in 1900 and 1905. B. L. Livingston, "The Experimental Morphology of Plants," *Plant World* 11 (1908): 151–54; F. E. Lloyd, review of *Organography of Plants*, by Karl Goebel, *Torreya* 5 (1905): 167–69.

50. D. T. MacDougal, "By Caravan through the Libyan Desert," *Harper's Magazine* 127 (1913): 489–500.

51. D. T. MacDougal, "The Measurement of Environic Factors and Their Biologic Effects," *Popular Science Monthly* 84 (1914): 417–33, quote p. 419. This article was based on a lecture given to the trustees of the Carnegie Institution in December 1913.

52. Ibid.

53. MacDougal, "Department of Botanical Research," p. 86; D. T. MacDougal, "Can We Grow Our Own Rubber? Guayule, a Native American Rubber, Is Being Cultivated on a Large Scale in California," *Scientific American* 139 (1928): 16–19.

54. W. L. Bray, "Desert Plants," *American Naturalist* 44 (1910): 443–48, quote p. 448.

55. Jane Maienschein, "Pattern and Process in Early Studies of Arizona's San Francisco Peaks," *BioScience* 44 (1994): 479–85.

56. Bowers, *A Sense of Place;* Burton L. Livingston and Forrest Shreve, *The Distribution of Vegetation in the United States, as Related to Climatic Conditions* (Washington, DC: Carnegie Institution of Washington, 1921).

57. Forrest Shreve, "The Rate of Establishment of the Giant Cactus," *Plant World* 13 (1910): 235–40.

58. Ellsworth Huntington, "The Greenest of Deserts," *Harper's Monthly Magazine* 123 (1911): 50–58.

59. Hornaday, *Camp-Fires on Desert and Lava*, pp. viii–ix.

60. Ibid., pp. 8–9.

61. Ibid., pp. 15, 16.

62. Ibid., pp. 292, 267.

63. Ibid., p. vii.

64. Robert L. Burgess, "The Ecological Society of America: Historical Data and Some Preliminary Analyses," reprinted in *History of American Ecology* (New York: Arno Press, 1977).

65. Bowers, *A Sense of Place*, pp. 142–45.

66. Herman Augustus Spoehr, *Essays on Science* (Stanford, CA: Stanford University Press, 1956), p. 8. The book contains a family history of Spoehr and several of his general essays.

67. Herman A. Spoehr, "The Instruments of Plant Biology," *Science* 70 (1929): 459–63, reprinted in Spoehr, *Essays on Science*, pp. 88–98.

68. Herman A. Spoehr, "Form, Forces, and Function in Plants," in *Cooperation in Research* (Washington, DC: Carnegie Institution of Washington, 1938), pp. 435–62. The section on pp. 445–46 is taken largely from Frederic Clements and Ralph W. Chaney, *Environment and Life in the Great Plains* (Washington, DC: Carnegie Institution of Washington, 1937), Supplementary Publications no. 24 (revised).

69. Herman A. Spoehr, "Division of Plant Biology," *Carnegie Institution of Washington Yearbook* 39 (1941): 146–47.

70. Ronald C. Tobey, *Saving the Prairies: The Life Cycle of the Founding School of American Plant Ecology, 1895–1955* (Berkeley: University of California Press, 1981), chap. 7.

FIVE: Visioning Ecology

1. Arthur G. Tansley, "International Phytogeographical Excursion in America, 1913," *New Phytologist* 12 (1913): 322–36; 13 (1914): 30–41, 83–92, 268–75, 325–33; quote 13 (1914): 333.

2. Tansley, *New Phytologist* 12 (1913): 327.

3. George Perkins Marsh, *The Earth as Modified by Human Action* (New York: Scribner's, 1907). This was a reprint of the revised edition of *Man and Nature*.

4. Frederick Jackson Turner, *The Frontier in American History* (New York: Holt, 1953). The first essay in this collection is the original statement of the frontier thesis made at the World's Fair in Chicago in 1893, at the meeting of the American Historical Association.

5. William Coleman, "Science and Symbol in the Turner Frontier Hypothesis," *American Historical Review* 72 (1966): 22–49.

6. Webb drew on the ideas of Henry Fairfield Osborn, a paleontologist at the American Museum of Natural History, concerning the question of whether humans originated on the plains or in the forest. See Walter P. Webb, *The Great Plains* (1931; repr., Lincoln: University of Nebraska Press, 1981), pp. 489–91.

7. Geoffrey J. Martin, *Ellsworth Huntington, His Life and Thought* (Hamden, CT: Archon Books, 1973).

8. Quoted in ibid., p. 66.

9. These included the Russian evolutionist Peter Kropotkin, who argued that evolutionary advance depended on cooperation between individuals rather than competition. For a discussion of Russian variations on Darwinian reasoning and the impact of the Russian landscape on evolutionary theory, see Daniel Todes, *Darwin without Malthus: The Struggle for Existence in Russian Evolutionary Thought* (New York: Oxford University Press, 1989).

10. Ellsworth Huntington, *The Climatic Factor, as Illustrated in Arid America* (Washington, DC: Carnegie Institution of Washington, 1914), p. 185.

11. Ellsworth Huntington, *Civilization and Climate* (New Haven, CT: Yale University Press, 1915). On the general enthusiasm for scientific analysis of efficiency in relation to human work in this period, see Anson Rabinbach, *The Human Motor: Energy, Fatigue, and the Origin of Modernity* (New York: Basic Books, 1990).

12. Huntington, *Civilization and Climate*, quotes pp. 293, 276.

13. Ibid., p. 294.

14. Ellsworth Huntington, "The Desert Laboratory," *Harper's Monthly Magazine* 122 (1911): 651–62, quotes pp. 661–62.

15. Ellsworth Huntington, *World-Power and Evolution* (New Haven, CT: Yale University Press, 1919).

16. Eugene Cittadino, "The Failed Promise of Human Ecology," in *Science and Nature: Essays in the History of the Environmental Sciences*, ed. Michael Shortland (Stanford-in-the-Vale, Eng.: British Society for the History of Science, 1993), pp. 251–83.

17. Ellsworth Huntington, "The Control of Pneumonia and Influenza by the Weather," *Ecology* 1 (1920): 6–23.

18. Ibid., p. 22.

19. Barrington Moore, "The Scope of Ecology," *Ecology* 1 (1920): 3–5.

20. Ibid., p. 5.

21. Stanislaus Novakovsky, "The Probable Effect of the Climate of the Russian Far East on Human Life and Activity," *Ecology* 3 (1922): 181–201; Stanislaus Novakovsky, "The Effect of Climate on the Efficiency of the Russian Far East," *Ecology* 3 (1922): 275–83; also Jacques Redway, "City Street Dust and Infectious Diseases," *Ecology* 3 (1922): 1–6; Clark Wissler, "The Relation of Nature to Man as Illustrated by the North American Indian," *Ecology* 5 (1924): 311–18.

22. Cittadino, "The Failed Promise of Human Ecology."

23. Sharon E. Kingsland, *Modeling Nature: Episodes in the History of Population Ecology*, 2nd ed. (Chicago: University of Chicago Press, 1995); Jan Lindström, Esa Ranta, Hanna Kokko, Per Lundberg, and Veijo Kaitala, "From Arctic Lemmings to Adaptive Dynamics: Charles Elton's Legacy in Population Ecology," *Biological Review* 76 (2001): 129–58.

24. *Reports of the Conferences on Cycles* (Washington, DC: Carnegie Institution of Washington, 1929).

25. "Matamek Conference on Biological Cycles," Labrador, 1931, typescript of the full proceedings, dated 1932, including a report by Ellsworth Huntington, in Special Collections, Milton Eisenhower Library, Johns Hopkins University. The location on the Matamek River in Labrador was the summer home of Copley Amory, a Bostonian who organized the conference because he had become interested in animal population fluctuations that he had observed over the years.

26. George Evelyn Hutchinson, "Kroeber on Culture and Huntington on Heredity and

Environment," in *The Itinerant Ivory Tower: Scientific and Literary Essays*, by George Evelyn Hutchinson (New Haven, CT: Yale University Press, 1953), pp. 70–86, quote p. 86.

27. Ronald C. Tobey, *Saving the Prairies: The Life Cycle of the Founding School of American Plant Ecology, 1895–1955* (Berkeley: University of California Press, 1981).

28. Cornelius L. Shear, *Field Work of the Division of Agrostology: A Review and Summary of the Work Done since the Organization of the Division, July 1, 1895* (Washington, DC: Government Printing Office, 1901).

29. Ibid., p. 24. Shear is quoting from a report by Jared G. Smith published in 1895.

30. Shear, *Field Work of the Division of Agrostology*, pp. 27, 37–38. The term *carrying capacity*, which Shear uses often as a general term for productivity of the land, later became a central ecological concept.

31. Ibid., p. 56.

32. Frederic E. Clements and Ralph W. Chaney, *Environment and Life in the Great Plains*, Supplementary Publication no. 24 (Washington, DC: Carnegie Institution, of Washington, 1937), p. 44. This is possibly a reference to the part Clements himself was responsible for, since Shear noted that his work with Clements in Colorado had not been published. Shear, *Field Work of the Division of Agrostology*, p. 29.

33. Clements and Chaney, *Environment and Life in the Great Plains*, p. 41. See also Webb, *The Great Plains*, pp. 376–78.

34. Aughey resigned from the university following a financial scandal in 1883 and later scientists considered him a charlatan.

35. Samuel P. Hays, *Conservation and the Gospel of Efficiency: The Progressive Conservation Movement, 1890–1920* (Cambridge, MA: Harvard University Press, 1959). See also Arthur A. Ekirch Jr., *Man and Nature in America* (New York: Columbia University Press, 1963).

36. Frederic E. Clements, *Research Methods in Ecology* (Lincoln, NE: University Publishing, 1905).

37. Ibid., pp. 145–60.

38. Stanley A. Cain, *Foundations of Plant Geography* (New York: Harper, 1944), pp. 21–22.

39. On the Clementsian research school during the Carnegie years, see Joel B. Hagen, "Clementsian Ecologists: The Internal Dynamics of a Research School," *Osiris* 8 (1993): 178–95.

40. Joel B. Hagen, "Organism and Environment: Frederic Clements's Vision of a Unified Physiological Ecology," in *The American Development of Biology*, ed. Ronald Rainger, Keith R. Benson, and Jane Maienschein (Philadelphia: University of Pennsylvania Press, 1988), pp. 257–77; Tobey, *Saving the Prairies.*

41. Joel B. Hagen, *An Entangled Bank: The Origins of Ecosystem Ecology* (New Brunswick, NJ: Rutgers University Press, 1992), pp. 22–24.

42. Frederic E. Clements, *Plant Succession: An Analysis of the Development of Vegetation* (Washington, DC: Carnegie Institution of Washington, 1916).

43. A subtle interpretation of Cowles's research in the context of the growth of the city of Chicago is given by Eugene Cittadino, "A 'Marvelous Cosmopolitan Preserve': The Dunes, Chicago, and the Dynamic Ecology of Henry Cowles," *Perspectives on Science* 1 (1993): 520–59.

44. Henry C. Cowles, "The Ecological Relations of the Vegetation on the Sand Dunes of Lake Michigan," *Botanical Gazette* 27 (1899): 95–117, 167–202, 281–308, 361–91. On

Cowles's research and its impact, see J. Ronald Engel, *Sacred Sands: The Struggle for Community in the Indiana Dunes* (Middletown, CT: Wesleyan University Press, 1983).

45. Henry C. Cowles, "The Causes of Vegetative Cycles," *Botanical Gazette* 51 (1911): 161–83, quote p. 161.

46. Cittadino, "A 'Marvelous Cosmopolitan Preserve'."

47. The study of competition was a lifelong concern, but the fullest development of his ideas on competition is in Frederic E. Clements, John E. Weaver, and Herbert C. Hanson, *Plant Competition: An Analysis of Community Functions* (Washington, DC: Carnegie Institution of Washington, 1929).

48. Ibid., p. 10.

49. Clements and Chaney, *Environment and Life in the Great Plains*, p. 35.

50. Hagen, "Organism and Environment"; Hagen, *An Entangled Bank*, pp. 22–23.

51. Clements, Weaver, and Hanson, *Plant Competition*, pp. 314–15.

52. Examples from the late nineteenth century include Stephen Alfred Forbes's idea of the lake as a microcosm, Anton Kerner Von Marilaun's description of plant communities as comparable to peaceful human societies, and Karl Möbius's articulation of the concept of an ecological community in his studies of the oyster. Often these ideas bore the impress of a teleological view of nature that had its roots in natural theology.

53. Clements, *Research Methods in Ecology*, p. 199.

54. Arthur G. Tansley, "The Classification of Vegetation and the Concept of Development," *Journal of Ecology* 8 (1920): 118–44.

55. Clements, Weaver, and Hanson, *Plant Competition*, p. 314.

56. Donald Worster, *Nature's Economy: A History of Ecological Ideas*, 2nd ed. (Cambridge: Cambridge University Press, 1994), pp. 215–17.

57. Clements and Chaney, *Environment and Life in the Great Plains*, p. 44.

58. Shear, *Field Work of the Division of Agrostology*, p. 56.

59. John E. Weaver and Frederic E. Clements, *Plant Ecology*, 2nd ed. (New York: McGraw-Hill, 1938), esp. pp. 86–88, 472. See also John E. Weaver and F. W. Albertson, *Grasslands of the Great Plains: Their Nature and Use* (Lincoln, NE: Johnsen, 1956), chaps. 1, 2. Weaver was Clements's doctoral student and remained part of his research group for several years. His later writings still drew on Clements's ideas, although he abandoned some features of Clementsian ecology.

60. Frederic E. Clements, "Experimental Ecology in the Public Service," *Ecology* 16 (1935): 342–63. For further analysis of the ideology underlying Clements's work and his school, see Tobey, *Saving the Prairies*, chap. 7.

61. Weaver and Albertson, *Grasslands of the Great Plains*, pp. 89–93; also Donald Worster, *Dust Bowl: The Southern Plains in the 1930s* (Oxford: Oxford University Press, 1979); see chap. 1 for description of dust storms and chap. 13 on the work of Clements and other ecologists.

62. Clements, "Experimental Ecology," p. 342.

63. Ibid., p. 345.

64. Clements and Chaney, *Environment and Life in the Great Plains*, p. 52.

65. Arthur G. Tansley, "Succession: The Concept and Its Value," in *Proceedings of International Congress of Plant Sciences, Ithaca, New York, August 16–23, 1926*, ed. B. M. Duggar (Menasha, WI: George Banta, 1929), 1:677–86, cite p. 686.

66. Henry A. Gleason, "Autobiography," undated typescript in the Henry A. Gleason

Records, series 1: Biographical Material, record group 4, Archives of the LuEsther T. Mertz Library, New York Botanical Garden, cite p. 37.

67. Paul B. Sears, *Deserts on the March* (Norman: University of Oklahoma Press, 1935; repr., Washington, DC: Island Press, 1988).

SIX: Science, History, and Progress

1. Ronald C. Tobey analyzes Clements's influence in *Saving the Prairies: The Life Cycle of the Founding School of American Plant Ecology, 1895–1955* (Berkeley: University of California Press, 1981). An assessment of Clements's place in American ecology toward the end of his life can be gleaned from Stanley A. Cain, *Foundations of Plant Geography* (New York: Harper, 1944).

2. Arthur G. Tansley, "The Classification of Vegetation and the Concept of Development," *Journal of Ecology* 8 (1920): 118–44, quote p. 122. Tansley was quoting from H. Gams and C. E. Du Rietz.

3. Janice Bowers, *A Sense of Place: The Life and Work of Forrest Shreve* (Tucson: University of Arizona Press, 1988), pp. 56–62.

4. Henry Allan Gleason, "Autobiography," n.d., quote p. 13. Typescript in Henry A. Gleason Records, series 1: Biographical Material, record group 4, Archives of the LuEsther T. Mertz Library of the New York Botanical Garden. The thesis was published as Henry Allan Gleason, "A Revision of the North American Veronieae," *Bulletin of the New York Botanical Garden* 4 (1906): 144–243.

5. His ecological work in Illinois is discussed in Henry A. Gleason, "The Vegetation of the Inland Sand Deposits of Illinois," *Bulletin of the Illinois State Laboratory of Natural History* 9 (1910): 23–174.

6. Malcolm Nicolson, "Henry Allan Gleason and the Individualistic Hypothesis," *Botanical Review* 56 (1990): 91–161.

7. Henry A. Gleason and Mel. T. Cook, "Plant Ecology of Porto Rico," *Scientific Survey of Porto Rico and the Virgin Islands*, vol. 7 (New York: New York Academy of Sciences, 1927).

8. Henry A. Gleason, "The Structure and Development of the Plant Association," *Bulletin of the Torrey Botanical Club* 44 (1917): 463–81.

9. Henry A. Gleason, "The Individualistic Concept of the Plant Association," *Bulletin of the Torrey Botanical Club* 53 (1926): 7–26.

10. Ibid., p. 23.

11. "I know of no other paper . . . ," Gleason, "Autobiography," p. 49; George E. Nichols, "Plant Associations and Their Classification," in *Proceedings of the International Congress of Plant Sciences, Ithaca, New York, 1926*, ed. Benjamin M. Duggar (Menasha, WI: George Banta, 1929), 1:629–41, "order out of chaos" p. 639; Gleason's response is on pp. 643–46. Joel Hagen discusses Gleason's critique, playing down the controversy, in *An Entangled Bank: The Origins of Ecosystem Ecology* (New Brunswick, NJ: Rutgers University Press, 1992), pp. 28–31. A discussion of the impact of Gleason's ideas, using Thomas Kuhn's theory of scientific revolution as paradigm shift, is in Michael G. Barbour, "Ecological Fragmentation in the Fifties," in *Uncommon Ground: Toward Reinventing Nature*, ed. William Cronon (New York: Norton, 1995), pp. 233–55.

12. Henry A. Gleason, "Further Views on the Succession-Concept," *Ecology* 8 (1927): 299–326.

13. Henry A. Gleason, "The Individualistic Concept of the Plant Association," *American Midland Naturalist* 21 (1939): 92–110.

14. James C. Malin, *The Grassland of North America: Prolegomena to Its History* (Ann Arbor, MI: Edwards Brothers, 1947), esp. chaps. 2, 10, 11; on Gleason's article, see p. 157. A good introduction to Malin's thought and impact, along with a collection of some of his more important articles, is in James C. Malin, *History and Ecology: Studies of the Grassland*, ed. Robert P. Swierenga (Lincoln: University of Nebraska Press, 1984).

15. Frederic E. Clements and Ralph W. Chaney, *Environment and Life in the Great Plains* (Washington, DC: Carnegie Institution of Washington, 1937), p. 47.

16. Ibid., pp. 51–52.

17. John Ernest Weaver, *North American Prairie* (Lincoln, NE: Johnsen, 1954); John E. Weaver and F. W. Albertson, *Grasslands of the Great Plain: Their Nature and Use* (Lincoln, NE: Johnsen, 1956).

18. James C. Malin, "Man, the State of Nature, and Climax: As Illustrated by Some Problems of the North American Grassland," *Scientific Monthly* 74 (1952): 29–37; reprinted in Malin, *History and Ecology*, pp. 112–25, quotes p. 123.

19. James C. Malin, "The Grassland of North America," in *Man's Role in Changing the Face of the Earth*, ed. William L. Thomas Jr. (Chicago: University of Chicago Press, 1956), pp. 350–66, quotes p. 362.

20. Eugene Cittadino, "The Failed Promise of Human Ecology," in *Science and Nature: Essays in the History of the Environmental Sciences*, ed. Michael Shortland (Stanford-in-the-Vale, Eng.: British Society for the History of Science, 1993), pp. 251–83; Stephen A. Forbes, "The Humanizing of Ecology," *Ecology* 3 (1922): 89–92; Charles C. Adams, "The Relation of General Ecology to Human Ecology," *Ecology* 16 (1935): 316–35.

21. The biographical information is from John Leighly, ed., *Land and Life: A Selection from the Writings of Carl Ortwin Sauer* (Berkeley: University of California Press, 1969), pp. 1–8; Martin S. Kenzer, ed., *Carl O. Sauer, a Tribute* (Corvallis: Oregon State University Press, 1987), esp. essays by Martin Kenzer, Anne Macpherson, and Kent Mathewson.

22. Carl O. Sauer, "Forward to Historical Geography," *Annals of the Association of American Geographers* 31 (1941): 1–24; reprinted in Leighly, *Land and Life*, pp. 351–79.

23. Carl O. Sauer, "The Formative Years of Ratzel in the United States," *Annals of the Association of American Geographers* 60 (1971): 245–54.

24. Ellen Churchill Semple, *Influences of Geographic Environment, on the Basis of Ratzel's System of Anthropo-geography* (New York: Holt, 1911).

25. See the essays in Leighley, *Land and Life*, esp. those in part 3, "Human Uses of the Organic World"; Carl O. Sauer, "Grassland Climax, Fire, and Man," *Journal of Range Management* 3 (1950): 16–21; Carl O. Sauer, *Agricultural Origins and Dispersal* (New York: American Geographical Society, 1952).

26. Sauer, "Grassland Climax, Fire, and Man," p. 21. On modern views of fire control, see Douglas Gantenbein, "Burning Questions," *Scientific American* 287, no. 5 (2002): 82–89.

27. Carl O. Sauer, "The Agency of Man on the Earth," in Thomas, *Man's Role*, pp. 49–69, quote p. 51.

28. George F. Carter, "Man in America: A Criticism of Scientific Thought," *Scientific Monthly* 73 (1951): 297–307. This culture, named Folsom culture after its location, is now thought to have arisen 12,900 years ago, replacing the Clovis culture, which arrived in North America about 13,800 years ago. The subject is still controversial. See Tim Flan-

nery, *The Eternal Frontier: An Ecological History of North America and Its Peoples* (New York: Atlantic Monthly Press, 2001).

29. Carl O. Sauer, "A Geographic Sketch of Early Man in America," *Geographical Review* 34 (1944): 529–73; reprinted in Leighly, *Land and Life*, pp. 197–245.

30. For a history of theories of large-mammal extinctions, with discussion of Sauer's work, see Kevin Francis, "'Death Enveloped All Nature in a Shroud': The Extinction of Pleistocene Mammals and the Persistence of Scientific Generalists" (Ph.D. diss., University of Minnesota–Twin Cities, 2002).

31. Carl O. Sauer, "Homestead and Community on the Middle Border," in Leighly, *Land and Life*, pp. 32–41; Sauer, "The Agency of Man on the Earth," pp. 49–69, quotes p. 66.

32. Thomas, *Man's Role*.

33. John T. Curtis, "The Modification of Mid-Latitude Grasslands and Forests by Man," in Thomas, *Man's Role*, pp. 721–36.

34. James S. Fralish, Robert P. McIntosh, and Orie L. Loucks, *John T. Curtis: Fifty Years of Wisconsin Plant Ecology* (Madison: Wisconsin Academy of Sciences, Arts, and Letters, 1993).

35. J. T. Curtis, comments reported in discussion in Thomas, *Man's Role*, p. 939.

36. Curtis, "Modification of Mid-Latitude Grasslands," p. 734.

37. Thomas, *Man's Role*, essays by Omer C. Stewart, "Fire as the First Great Force Employed by Man," pp. 115–33; Lewis Mumford, "The Natural History of Urbanization," pp. 382–98; Edgar Anderson, "Man as a Maker of New Plants and New Plant Communities," pp. 763–77.

38. "Discussion: Changes in Biological Communities," in Thomas, *Man's Role*, pp. 930–43.

39. Marston Bates, "Process," in Thomas, *Man's Role*, pp. 1136–40, quote p. 1137.

40. See the three discussion sessions, "Limits of the Earth: Materials and Ideas," "Man's Self-Transformation," and "The Unstable Equilibrium of Man in Nature," in Thomas, *Man's Role*, pp. 1071–128.

41. Ibid., p. 1087.

42. Charles Percy Snow, *The Two Cultures and the Scientific Revolution* (New York: Cambridge University Press, 1959); Bates, "Process," pp. 1136–40.

43. Lewis Mumford, "Prospect," in Thomas, *Man's Role*, pp. 1141–55, quote pp. 1150–51.

44. Carl Sauer, "Retrospect," in Thomas, *Man's Role*, pp. 1131–35, quote p. 1133.

45. James R. Hastings and Raymond M. Turner, *The Changing Mile: An Ecological Study of Vegetation Change with Time in the Lower Mile of an Arid and Semiarid Region* (Tucson: University of Arizona Press, 1965).

46. Ibid., p. 22.

47. Ibid., p. 288.

48. Raymond M. Turner, Robert H. Webb, Janice E. Bowers, and James Rodney Hastings, *The Changing Mile Revisited: An Ecological Study of Vegetation Change with Time in the Lower Mile of an Arid and Semiarid Region* (Tucson: University of Arizona Press, 2003).

SEVEN: A Subversive Science?

1. Marston Bates, "Process," in *Man's Role in Changing the Face of the Earth*, ed. William L. Thomas (Chicago: University of Chicago Press, 1956), pp. 1136–40, quote p. 1140.

2. Ralph E. Lapp, *Must We Hide?* (Cambridge, MA: Addison-Wesley, 1949), "atomic bomb geography," pp. 143–45; "cities of the past," p. 85; "doughnut city," p. 163.

3. The critic was David Bradley, in his review of Must We Hide? by Ralph E. Lapp, *Saturday Review*, June 18, 1949, pp. 13–14. Positive reviews were published in *U.S. News and World Report*, April 29, 1949, pp. 11–13; and *New York Times Book Review*, July 3, 1949, p. 1.

4. United States Atomic Energy Commission, *Atomic Energy and the Life Sciences* (Washington, DC: Government Printing Office, 1949).

5. Ibid., pp. 54–55.

6. Frederick P. Cowan, "Everyday Radiation," *Physics Today* 5, no. 10 (1952): 10–16, quote p. 14.

7. Richard Rhodes, *Dark Sun: The Making of the Hydrogen Bomb* (New York: Simon and Schuster, 1995), pp. 542–43.

8. Ralph E. Lapp, *The Voyage of the Lucky Dragon* (London: F. Muller, 1958).

9. "Radioactive Cloud Floats over East," *Baltimore Sun*, March 11, 1955, p. 1.

10. "Brown Rain Stirs Case of Jitters Here," *Baltimore Sun*, March 12, 1955, p. 26.

11. Willard F. Libby, "Radioactive Fallout and Radioactive Strontium," *Science* 123 (1956): 657–60, quote p. 660.

12. John C. Bugher, "Effects of Fission Material on Air, Soil, and Living Species," in Thomas, *Man's Role*, pp. 831–48.

13. *Proceedings of the International Conference on the Peaceful Uses of Atomic Energy, held in Geneva, 8 August–20 August 1955*, 16 vols. (New York: United Nations, 1956).

14. John N. Wolfe, editorial introduction to special issue on ecology, *BioScience* 14 (1964): 9–10.

15. Arthur G. Tansley, "The Use and Abuse of Vegetational Concepts and Terms," *Ecology* 16 (1935): 284–307, quote p. 291. The debate has been analyzed in its British and South African context by Peder Anker, *Imperial Ecology: Environmental Order in the British Empire, 1895–1945* (Cambridge, MA: Harvard University Press, 2001), chap. 4.

16. Tansley, "Use and Abuse," p. 303.

17. Frederic Clements, John E. Weaver, and Herbert C. Hanson, *Plant Competition: An Analysis of Community Functions* (Washington, DC: Carnegie Institution of Washington, 1929), p. 314.

18. Frank B. Golley, *A History of the Ecosystem Concept in Ecology: More Than the Sum of the Parts* (New Haven, CT: Yale University Press, 1993), pp. 56–59.

19. G. Evelyn Hutchinson, "Circular Causal Systems in Ecology," *Annals New York Academy of Sciences* 50 (1948): 221–46.

20. Norbert Wiener, *Cybernetics, or Control and Communication in the Animal and the Machine* (New York: Wiley, 1950).

21. For example, Hutchinson discussed Lindeman's work in "Circular Causal Systems." See Raymond Lindeman, "The Trophic-Dynamic Aspect of Ecology," *Ecology* 23 (1942): 399–418.

22. Joel B. Hagen, *An Entangled Bank: The Origins of Ecosystem Ecology* (New

Brunswick, NJ: Rutgers University Press, 1992), p. 96; Robert E. Cook, "Raymond Lindeman and the Trophic-Dynamic Concept in Ecology," *Science* 198 (1977): 22–26.

23. Howard T. Odum, "The Stability of the World Strontium Cycle," *Science* 114 (1951): 407–11.

24. Eugene P. Odum, *Fundamentals of Ecology* (Philadelphia: Saunders, 1953). The book went into five editions, the last one published posthumously in 2004.

25. Ibid., p. 9.

26. Ibid., p. 10.

27. Quoted from the editors' introduction to the article by G. M. Woodwell, W. M. Malcolm, and R. H. Whittaker, "A-Bombs, Bugbombs, and Us," in *The Subversive Science: Essays toward an Ecology of Man*, ed. Paul Shepard and Daniel McKinley (Boston: Houghton Mifflin, 1969), p. 230.

28. Howard T. Odum, *Environment, Power, and Society* (New York: Wiley, 1971), pp. 9–11.

29. Francis C. Evans, "Ecosystem as the Basic Unit in Ecology," *Science* 123 (1956): 1127–28.

30. Francis C. Evans, "Biology and Urban Areal Research," *Scientific Monthly* 73 (1951): 37–38.

31. Golley, *History of the Ecosystem Concept;* Hagen, *An Entangled Bank;* Stephen Bocking, "Ecosystems, Ecologists, and the Atom: Environmental Research at Oak Ridge National Laboratory," *Journal of the History of Biology* 28 (1995): 1–47; Stephen Bocking, *Ecologists and Environmental Politics: A History of Contemporary Ecology* (New Haven, CT: Yale University Press, 1997), chap. 4.

32. Chunglin Kwa, "Radiation Ecology, Systems Ecology, and the Management of the Environment," in *Science and Nature: Essays in the History of the Environmental Sciences,* ed. Michael Shortland (Stanford-in-the-Vale, Eng.: British Society for the History of Science, 1993), pp. 213–49.

33. Bocking, "Ecosystems, Ecologists, and the Atom"; Bocking, *Ecologists and Environmental Politics*, chaps. 4, 5.

34. Alvin M. Weinberg, "The Promise of Scientific Technology: The New Revolutions," in *Reflections on Big Science,* by Alvin M. Weinberg (Cambridge, MA: MIT Press, 1967), pp. 1–37.

35. Golley, *History of the Ecosystem Concept;* Hagen, *An Entangled Bank;* Kwa, "Radiation Ecology"; Bocking, "Ecosystems, Ecologists, and the Atom."

36. Eugene P. Odum, "Feedback between Radiation Ecology and General Ecology," *Health Physics* 11 (1965): 1257–62.

37. Eugene P. Odum, "Consideration of the Total Environment in Power Reactor Waste Disposal," in *Proceedings of the International Conference on the Peaceful Uses of Atomic Energy*, 13:350–53.

38. Ibid., p. 351.

39. Eugene P. Odum, *Fundamentals of Ecology*, 2nd ed. (Philadelphia: Saunders, 1959), pp. vi, 6–7.

40. Eugene Odum, "The Ecosystem Approach in the Teaching of Ecology Illustrated with Sample Class Data," *Ecology* 38 (1957): 531–35.

41. Odum, *Fundamentals of Ecology*, 2nd ed., p. 467. Radiation ecology is discussed in chap. 14. On Odum's role in the growth of radiation ecology, see Kwa, "Radiation Ecology, Systems Ecology and the Management of the Environment."

42. Howard T. Odum, "Trophic Structure and Productivity of Silver Springs, Florida," *Ecological Monographs* 27 (1955): 55–112; Golley, *History of the Ecosystem Concept*, pp. 69–70.

43. Hagen, *An Entangled Bank*, pp. 122–30; Peter J. Taylor, "Technocratic Optimism, H. T. Odum, and the Partial Transformation of Ecological Metaphor after World War II," *Journal of the History of Biology* 21 (1988): 213–44.

44. Howard T. Odum, "Ecological Potential and the Analogue Circuits for the Ecosystem," *American Scientist* 48 (1960): 1–8.

45. Odum, *Environment, Power, and Society;* Howard T. Odum, "Energy, Value, and Money," in *Ecosystem Modeling in Theory and Practice: An Introduction with Case Histories*, ed. Charles A. S. Hall and John W. Day Jr. (New York: Wiley, 1977), pp. 174–96.

46. Golley, *History of the Ecosystem Concept*, pp. 95–96.

47. Odum, *Fundamentals of Ecology*, 2nd ed., p. 80.

48. Linda Lear, *Rachel Carson, Witness for Nature* (New York: Henry Holt, 1997).

49. Shepard and McKinley, *The Subversive Science*, p. 230, in editors' introduction.

50. Frank E. Egler, "Pesticides—In Our Ecosystem," *American Scientist* 52 (1964): 110–36.

51. Garrett Hardin, "Not Peace, but Ecology," in *Diversity and Stability in Ecological Systems: Report of a Symposium Held May 26–28, 1969* (Upton, NY: Brookhaven National Laboratory, 1969), pp. 151–61, cite p. 151. Hardin's argument for curbing freedoms is in "The Tragedy of the Commons," *Science* 162 (1968): 1243–48, quote p. 1248.

52. Paul B. Sears, "Ecology—A Subversive Subject," *BioScience* 14 (1964): 11–13, quote p. 12.

53. Paul Sears, *Deserts on the March* (Norman: University of Oklahoma Press, 1935; repr., Washington, DC.: Island Press, 1988); Sears, "Ecology," quote p. 11.

54. Lamont C. Cole, "The Impending Emergence of Ecological Thought," *BioScience* 14, no. 7 (1964): 30–32.

55. Sears, "Ecology," p. 13.

56. Ibid., p. 11.

57. Odum, *Environment, Power, and Society.*

58. Odum borrowed this idea from Alfred James Lotka, who considered the world and its inhabitants as though the world were a machine through which energy flowed; he deduced the principle that natural selection should tend to maximize the total energy flux through the system. Alfred J. Lotka, *Elements of Mathematical Biology* (New York: Dover, 1956); Sharon E. Kingsland, *Modeling Nature: Episodes in the History of Population Ecology*, 2nd ed. (Chicago: University of Chicago Press, 1995), chap. 2.

59. W. Michael Kemp, Wade H. B. Smith, Henry N. McKellar, Melvin E. Lehman, Mark Homer, Donald L. Young, and Howard T. Odum, "Energy Cost-Benefit Analysis Applied to Power Plants near Crystal River, Florida," in Hall and Day, *Ecosystem Modeling in Theory and Practice*, pp. 508–43, quote p. 510.

60. Taylor, "Technocratic Optimism," pp. 213–44.

61. Eugene Odum, "The Emergence of Ecology as a New Integrative Discipline," *Science* 195 (1977): 1289–93.

62. Eugene P. Odum, "The Strategy of Ecosystem Development," *Science* 164 (1969): 262–70.

63. Ibid., p. 264.

64. Ibid., p. 268.

65. Alvin M. Weinberg, "Technology and Ecology—Is There a Need for Confronta-

tion?" *BioScience* 23 (1973): 41–45. Weinberg spoke to the American Institute of Biological Sciences and the Ecological Society of America at Minneapolis, August 30, 1972.

66. Ibid., p. 44.

67. Ibid., p. 45.

EIGHT: Defining the Ecosystem

1. Ramón Margalef, *Perspectives in Ecological Theory* (Chicago: University of Chicago Press, 1968), pp. 10–11.

2. John Harte and Robert H. Socolow, eds., *Patient Earth* (New York: Holt, Rinehart, and Winston, 1971), p. 203.

3. Daniel B. Botkin, *Discordant Harmonies: A New Ecology for the Twenty-first Century* (New York: Oxford University Press, 1990).

4. George M. Van Dyne, "Ecosystems, Systems Ecology, and Systems Ecologists," Health Physics Division, Oak Ridge National Laboratory, Oak Ridge, TN, June 1966.

5. Bernard C. Patten, "Systems Ecology: A Course Sequence in Mathematical Ecology," *BioScience* 16 (1966): 593–98.

6. Bernard C. Patten, "An Introduction to the Cybernetics of the Ecosystem: The Trophic-Dynamic Aspect," *Ecology* 40 (1959): 221–31. An interesting feature of this development was the reworking of precybernetic ecological work in order to adapt it to the modern style of mathematical thinking. Patten translated Raymond Lindeman's 1942 article on the ecosystem into the language of cybernetics so that it would fit into the new theoretical context. He thanked Hutchinson for his enthusiastic support of this work.

7. Ramón Margalef, "Information Theory in Ecology," *General Systems* 3 (1958): 36–71.

8. Margalef, *Perspectives*, p. vi.

9. Ibid., p. 17.

10. Ibid., pp. 21, 29.

11. Ibid., p. 30.

12. Ibid., p. 16.

13. Ibid., pp. 26–27; see also Arthur G. Tansley, "The Classification of Vegetation and the Concept of Development," *Journal of Ecology* 8 (1920): 118–44.

14. Margalef, *Perspectives*, pp. 26–30.

15. Philip M. Morse and George E. Kimball, *Methods of Operations Research* (Cambridge, MA: M.I.T. Press, 1963).

16. Bernard C. Patten, ed., *Systems Analysis and Simulation in Ecology*, vol. 1 (New York: Academic Press, 1971), p. 1.

17. George B. Dantzig, *Linear Programming and Extensions* (Princeton, NJ: Princeton University Press, 1963).

18. Gene H. Fisher and Warren E. Walker, "Operations Research and the RAND Corporation," in *Encyclopedia of Operations Research and Management Science*, ed. Saul I. Gass and Carl M. Harris (Boston: Kluwer, 1996), pp. 566–71.

19. Patten, "Systems Ecology," pp. 593–98.

20. Kenneth E. F. Watt, "Use of Mathematics in Population Ecology," *Annual Review of Entomology* 7 (1962): 243–60; Kenneth E. F. Watt, "The Use of Mathematics and Computers to Determine Optimal Strategy and Tactics for a Given Insect Control Problem," *General Systems Yearbook* 9 (1964): 167–80.

21. Kenneth E. F. Watt, ed., *Systems Analysis in Ecology* (New York: Academic Press, 1966), pp. 253–67.

22. Richard Bellman, *Dynamic Programming* (Princeton, NJ: Princeton University Press, 1957).

23. On electronic analog computers, see James E. Stice and Bernet S. Swanson, *Electronic Analog Computer Primer* (New York: Blaisdell, 1965). On analog modeling in ecology, see Howard T. Odum, "Macroscopic Minimodels of Man and Nature," in Bernard C. Patten, *Systems Analysis and Simulation in Ecology*, vol. 4 (New York: Academic Press, 1976), pp. 249–80; J. K. Denmead, "'Accuracy without Precision'—An Introduction to the Analogue Computer," in *Mathematical Models in Ecology*, ed. J. N. R. Jeffries (Oxford: Blackwell Scientific, 1972), pp. 215–35.

24. Frank B. Golley, *A History of the Ecosystem Concept in Ecology: More Than the Sum of the Parts* (New Haven, CT: Yale University Press, 1993), p. 168.

25. Ralph D. Amen, "A Biological Systems Concept," *BioScience* 16 (1966): 396–401, quote p. 401.

26. Charles A. S. Hall and John W. Day Jr., "Systems and Models: Terms and Basic Principles," in *Ecosystem Modeling in Theory and Practice: An Introduction with Case Histories*, ed. Charles A. S. Hall and John W. Day Jr. (New York: Wiley, 1977), pp. 6–36.

27. Bernard C. Patten, preface to Patten, *Systems Analysis*, vol. 1, pp. xiii–xv.

28. Hal Caswell, Herman E. Koenig, James A. Resh, and Quentin E. Ross, "An Introduction to Systems Science for Ecologists," in Bernard C. Patten, ed., *Systems Analysis and Simulation in Ecology*, vol. 2 (New York: Academic Press, 1972), pp. 3–78.

29. Van Dyne, "Ecosystems, Systems Ecology, and Systems Ecologists," p. 7.

30. Sharon E. Kingsland, *Modeling Nature: Episodes in the History of Population Ecology*, 2nd ed. (Chicago: University of Chicago Press, 1995).

31. Margalef, *Perspectives*, p. 49.

32. Garrett Hardin, "The Tragedy of the Commons," *Science* 162 (1968): 1243–48.

33. Paul R. Erlich, *The Population Bomb* (New York: Ballantine Books, 1968).

34. *Diversity and Stability in Ecological Systems: Report of a Symposium Held May 26–28, 1969* (Upton, N.Y.: Brookhaven National Laboratory, 1969).

35. Lynton K. Caldwell, "Health and Homeostasis as Social Concepts: An Exploratory Essay," in *Diversity and Stability*, pp. 206–23, quote p. 206.

36. C. S. Holling, "Stability in Ecological and Social Systems," in *Diversity and Stability*, pp. 128–40; the quote is from the report of discussion, p. 141.

37. Garrett Hardin, "Not Peace, but Ecology," in *Diversity and Stability*, pp. 151–58, with discussion pp. 158–61; quote p. 158.

38. Golley, *History of the Ecosystem Concept in Ecology*, p. 115, quoting from British scientist C. H. Waddington, "The Origin," in E. B. Worthington, *The Evolution of the IBP* (Cambridge: Cambridge University Press, 1975), pp. 4–12.

39. The modeling approaches for each biome study are reviewed in Bernard C. Patten, *Systems Analysis and Simulation in Ecology*, vol. 3 (New York: Academic Press, 1975).

40. Joel B. Hagen, *An Entangled Bank: The Origins of Ecosystem Ecology* (New Brunswick, NJ: Rutgers University Press, 1992), pp. 164–81; Golley, *History of the Ecosystem Concept in Ecology*, pp. 109–40.

41. Eugene P. Odum, "The Emergence of Ecology as a New Integrative Discipline," *Science* 195 (1977): 1289–93.

42. Golley, *History of the Ecosystem Concept in Ecology*, p. 139.

43. Kenneth E. F. Watt, "Critique and Comparison of Biome Ecosystem Modeling," in Patten, *Systems Analysis*, vol. 3, pp. 139–52.

44. Orie Loucks, "Systems Methods in Environmental Court Actions," in Patten, *Systems Analysis*, vol. 2, pp. 419–73.

45. Ibid., p. 441, 421.

46. Orie Loucks, "The Trial of DDT in Wisconsin," in Harte and Socolow, *Patient Earth*, pp. 88–111.

47. Gene E. Likens and F. Herbert Bormann, "Acid Rain: A Serious Regional Environmental Problem," *Science* 184 (1974): 1176–79; F. Herbert Bormann and Gene E. Likens, *Pattern and Process in a Forested Ecosystem: Disturbance, Development, and the Steady State Based on the Hubbard Brook Ecosystem Study* (New York: Springer-Verlag, 1979), pp. 214–15.

48. Ibid., p. 5.

49. Eugene P. Odum, "The Strategy of Ecosystem Development," *Science* 164 (1969): 262–70; Eugene P. Odum, *Fundamentals of Ecology*, 3rd ed. (Philadelphia: Saunders, 1971), pp. 251–57.

50. Bormann and Likens, *Pattern and Process in a Forested Ecosystem*, p. 4.

51. Alexander S. Watt, "Pattern and Process in the Plant Community," *Journal of Ecology* 35 (1947): 1–22.

52. Ibid., p. 2.

53. Bormann and Likens, *Pattern and Process in a Forested Ecosystem*, pp. 174–75.

54. Ibid., p. 178. These questions are complicated by the fact that diversity can be measured in different ways.

55. Ibid., pp. 164–65.

56. Orie L. Loucks, "Evolution of Diversity, Efficiency, and Community Stability," *American Zoologist* 10 (1970): 17–25.

57. Stuart L. Pimm, "The Complexity and Stability of Ecosystems," *Nature* 307 (1984): 321–26.

58. Stuart L. Pimm, *The Balance of Nature? Ecological Issues in the Conservation of Species and Communities* (Chicago: University of Chicago Press, 1991).

59. C. S. Holling, "Resilience of Ecosystems: Local Surprise and Global Change," in *Global Change*, ed. T. F. Malone and J. G. Roederer (Cambridge: Cambridge University Press, 1985), pp. 228–69, cite p. 229. See also C. S. Holling, "Resilience and Stability of Ecological Systems," *Annual Review of Ecology and Systematics* 4 (1973): 1–23.

60. René Dubos, *The Wooing of Earth: New Perspectives on Man's Use of Nature* (New York: Scribner's, 1980), p. 5.

61. Ibid., p. 37.

62. That "nature knows best" had been one of the cardinal rules of Barry Commoner's book *The Closing Circle: Nature, Men, and Technology* (New York: Knopf, 1971), discussed in J. Edward de Steiguer, *The Age of Environmentalism* (New York: McGraw-Hill, 1997), pp. 104–15.

63. Dubos, *Wooing of Earth*, p. 108.

64. Ibid., p. 157.

NINE: New Frontiers

1. Ogden Tanner and Adele Auchincloss, *The New York Botanical Garden: An Illustrated Chronicle of Plants and People* (New York: Walker, 1991), quote p. 156. His talk was published as René Dubos, "To Cultivate Our Garden," in *The World of René Dubos: A Collection from His Writings*, ed. Gerard Piel and Osborn Segerberg Jr. (New York: Henry Holt, 1990), pp. 288–94. Dubos's remark was probably in response to audience reaction, as it is not in the published essay.

2. Information from the Web site of the Henry Morrison Flagler Museum, Palm Beach, Florida, www.flaglermuseum.us/biography.html.

3. Information on the Institute of Ecosystem Studies is taken from its Web site located at www.ecostudies.org.

4. T. F. Malone and J. G. Roederer, *Global Change* (Cambridge: Cambridge University Press, 1985). The volume contains the proceedings of a symposium sponsored by the International Council of Scientific Unions.

5. Stephen Boyden, "The Human Component of Ecosystems," in *Humans as Components of Ecosystems: The Ecology of Subtle Human Effects and Populated Areas*, ed. Mark J. McDonnell and Steward T. A. Pickett (New York: Springer-Verlag, 1993), pp. 72–77, quote p. 73.

6. Billie Lee Turner II, William Clark, Robert Kates, John F. Richards, Jessica T. Mathews, and William B. Meyer, eds., *The Earth as Transformed by Human Action: Global and Regional Changes in the Biosphere over the Past 300 Years* (Cambridge: Cambridge University Press, 1990).

7. Robert M. Adams, "Foreword: The Relativity of Time and Transformation," in ibid., pp. vii–x, cite p. vii.

8. Martin Holdgate, "Postscript," in Turner et al., *Earth as Transformed by Human Action*, pp. 703–6, quotes p. 703.

9. Jane Lubchenco, Annette M. Olson, Linda B. Brubaker, Stephen R. Carpenter, Marjorie M. Holland, Stephen P. Hubbell, Simon A. Levin, James A. MacMahon, Pamela A. Matson, Jerry M. Melillo, Harold A. Mooney, Charles H. Peterson, H. Ronald Pulliam, Leslie A. Real, Philip J. Regal, and Paul G. Risser, "The Sustainable Biosphere Initiative: An Ecological Research Agenda," *Ecology* 72 (1991): 371–412.

10. Peter M. Vitousek, Harold A. Mooney, Jane Lubchenco, and Jerry M. Melillo, "Human Domination of Earth's Ecosystems," *Science* 277 (1997): 494–99.

11. F. Stuart Chapin III, Brian H. Walker, Richard J. Hobbs, David U. Hooper, John H. Lawton, Osvaldo E. Sala, and David Tilman, "Biotic Control over the Functioning of Ecosystems," *Science* 277 (1997): 500–504; P. A. Matson, W. J. Parton, A. G. Power, and M. J. Swift, "Agricultural Intensification and Ecosystem Properties," ibid., pp. 504–9; Louis W. Botsford, Juan Carlos Castilla, and Charles H. Peterson, "The Management of Fisheries and Marine Ecosystems," ibid., pp. 509–15; Andy P. Dobson, A. D. Bradshaw, and A. J. M. Baker, "Hopes for the Future: Restoration Ecology and Conservation Biology," ibid., pp. 515–22; Ian R. Noble and Rodolfo Dirzo, "Forests as Human-Dominated Ecosystems," ibid., pp. 522–25.

12. Gordon G. Whitney, *From Coastal Wilderness to Fruited Plain: A History of Environmental Change in Temperate North America from 1500 to the Present* (Cambridge: Cambridge University Press, 1994).

13. *The International Long Term Ecological Research Network* (Albuquerque, NM: US LTER Network, 1998).

14. Mathis Wackernagel and William Rees, *Our Ecological Footprint: Reducing Human Impact on the Earth* (Gabriola Island, BC: New Society, 1996).

15. Eugene Cittadino, "The Failed Promise of Human Ecology," in *Science and Nature: Essays in the History of the Environmental Sciences*, ed. Michael Shortland (Stanford-in-the-Vale, Eng.: British Society for the History of Science, 1993), pp. 251–83.

16. Francis C. Evans, "Biology and Urban Areal Research," in "Symposium on Viewpoints, Problems, and Methods of Research in Urban Areas," *Scientific Monthly* 73 (1951): 37–50; Evans's contribution pp. 37–38.

17. Lewis Mumford, "The Natural History of Urbanization," in *Man's Role in Changing the Face of the Earth*, ed. William L. Thomas Jr. (Chicago: University of Chicago Press, 1956), pp. 382–98, quote p. 397.

18. Edgar Anderson, "Man as a Maker of New Plants and New Plant Communities," in Thomas, *Man's Role*, pp. 763–77.

19. Edgar Anderson, "The City Is a Garden," *Landscape, Magazine of Human Geography* 7, no. 2 (1957–58): 3–5.

20. Edgar Anderson, "The City Watcher," *Landscape* 8, no. 2 (1958–59): 7–8.

21. Edgar Anderson, "The Country in the City," *Landscape* 5, no. 2 (1956): 32–35, quote p. 35.

22. Edgar Anderson, "The Considered Landscape," *Landscape* 10, no. 1 (1960): 8.

23. Daniel McKinley, "The New Mythology of 'Man in Nature'," *Perspectives in Biology and Medicine* 7 (1964): 93–105; reprinted in *The Subversive Science: Essays toward an Ecology of Man*, ed. Paul Shepard and Daniel McKinley (Boston: Houghton Mifflin, 1969), pp. 351–62, quotes pp. 358, 359.

24. David Ley, *A Social Geography of the City* (New York: HarperCollins, 1983); John R. Borchert, "The Twin Cities Urbanized Area: Past, Present, Future," *Geographical Review* 51 (1961): 47–70; John R. Borchert, *America's Northern Heartland: An Economic and Historical Geography of the Upper Midwest* (Minneapolis: University of Minnesota Press, 1987).

25. Forest Stearns and Tom Montag, eds., *The Urban Ecosystem: A Holistic Approach* (Stroudsburg, PA: Dowden, Hutchinson, and Ross, 1974), quote p. 30.

26. Ibid., pp. 20–21.

27. An example of public communication in keeping with the spirit of this report is a series of informational pamphlets produced in the 1970s by the National Park Service of the Department of the Interior. Written by Theodore W. Sudia, they used concepts from modern ecosystem ecology and evolutionary biology to discuss the features of successful city life. The eight pamphlets, published from 1971 to 1978, are available on the National Park Service Web site at www.cr.nps.gov/history/online_books/urban/.

28. Charles H. Nilon, Alan R. Berkowitz, and Karen S. Hollweg, "Introduction: Ecosystem Understanding Is a Key to Understanding Cities," in *Understanding Urban Ecosystems: A New Frontier for Science and Education*, ed. Alan Berkowitz, Charles Nilon, and Karen S. Hollweg (New York: Springer-Verlag, 2003), pp. 1–13; Frank B. Golley, "Urban Ecosystems and the Twenty-first Century—a Global Imperative," ibid., pp. 401–16; Frank B. Golley, *A History of the Ecosystem Concept: More than the Sum of the Parts* (New Haven, CT: Yale University Press, 1993), pp. 161–64.

29. Kim A. McDonald, "Ecology's Last Frontier: Studying Urban Areas to Monitor

the Impact of Human Activity," *Chronicle of Higher Education*, February 13, 1998, pp. A18–A19; Mary Parlange, "The City as Ecosystem," *BioScience* 48 (1998): 581–85; Charles Nilon, Alan R. Berkowitz, and Karen S. Hollweg, "Understanding Urban Ecosystems: A New Frontier for Science and Education," *Urban Ecosystems* 3 (1999): 3–4.

30. Nancy B. Grimm, J. Morgan Grove, Steward T. A. Pickett, and Charles L. Redman, "Integrated Approaches to Long-Term Studies of Urban Ecological Systems," *BioScience* 50 (2000): 571–84, quote p. 573.

31. Charles L. Redman, "Human Dimensions of Ecosystem Studies," *Ecosystems* 2 (1999): 296–98.

32. S. T. A. Pickett, William R. Burch Jr., and J. Morgan Grove, "Interdisciplinary Research: Maintaining the Constructive Impulse in a Culture of Criticism," *Ecosystems* 2 (1999): 302–7.

33. S .T. A. Pickett and P. S. White, eds., *The Ecology of Natural Disturbance and Patch Dynamics* (Orlando, FL: Academic Press, 1985).

34. Steward T. A. Pickett, William R. Burch Jr., Shawn E. Dalton, Timothy W. Foresman, J. Morgan Grove, and Rowan Rowntree, "A Conceptual Framework for the Study of Human Ecosystems in Urban Areas," *Urban Ecosystems* 1 (1997): 185–99.

35. J. Morgan Grove and William R. Burch Jr., "A Social Ecology Approach and Applications of Urban Ecosystem and Landscape Analyses: A Case Study of Baltimore, Maryland," *Urban Ecosystems* 1 (1997): 259–75; see pp. 264–65.

36. Grimm et al., "Integrated Approaches to Long-Term Studies," p. 579.

37. Charles L. Redman, *Human Impact on Ancient Environments* (Tucson: University of Arizona Press, 1999), esp. chap. 8, quote p. 217.

38. S. T. A. Pickett, "Succession: An Evolutionary Interpretation," *American Naturalist* 110 (1976): 107–19; S. T. A. Pickett, "Non-Equilibrium Coexistence of Plants," *Bulletin of the Torrey Botanical Club* 107 (1980): 238–48; Pickett and White, *Ecology of Natural Disturbance.*

39. M. J. McDonnell and S. T. A. Pickett, "Ecosystem Structure and Function along Urban-Rural Gradients: An Unexploited Opportunity for Ecology," *Ecology* 71 (1990): 1232–37, quote p. 1232.

40. One such experiment began in 1992, when researchers planted fast-growing cottonwood trees in New York City and two rural locations. The urban trees grew faster, suggesting that high ozone levels in the rural areas were responsible for poor growth there. The conclusion was that the rural environment was not necessarily a "safe haven" from urban pollution. See Jillian W. Gregg, Clive G. Jones, and Todd E. Dawson, "Urbanization Effects on Tree Growth in the Vicinity of New York City," *Nature* 424 (2003): 183–87.

41. McDonnell and Pickett, *Humans as Components of Ecosystems.*

42. Peter M. Groffman and Gene Likens, eds., *Integrated Regional Models: Interactions between Humans and Their Environment* (New York: Chapman and Hall, 1994).

43. The project's personnel, collaborations, and research and educational projects are described on the Baltimore Ecosystem Study Web site: http://beslter.org/.

44. Robert Costanza, A. Voinov, R. Boumans, T. Maxwell, F. Villa, L. Wainger, and H. Voinov, "Integrated Ecological Economics of the Patuxent River Watershed, Maryland," *Ecological Monographs* 72 (2002): 203–31. See also Robert Costanza and Alexey Voinov, eds., *Landscape Simulation Modeling: A Spatially Explicit, Dynamic Approach* (New York: Springer, 2004).

45. Grace Brush, "The Chesapeake Bay Estuarine System," in *The Changing Global Environment,* ed. Neil Roberts (Oxford: Blackwell, 1994), pp. 397–416; Grace Brush, "Forests before and after the Colonial Encounter," in *Discovering the Chesapeake: The History of an Ecosystem,* ed. Philip D. Curtin, Grace S. Brush, and George W. Fisher (Baltimore: Johns Hopkins University Press, 2001), pp. 40–59.

46. Grove and Burch, "A Social Ecology Approach," p. 265.

47. Tom Horton, "One Brook's Story of Pollution," *Baltimore Sun,* April 18, 2003.

48. Charles T. Driscoll, David Whitall, John Aber, Elizabeth Boyer, Mark Castro, Christopher Cronen, Christine L. Goodale, Peter Groffman, Charles Hopkinson, Kathleen Lambert, Gregory Lawrence, and Scott Ollinger, "Nitrogen Pollution in the Northeastern United States: Sources, Effects, and Management Options," *BioScience* 53 (2003): 357–74.

49. Mary Gail Hare, "Weeklong Conference Seeks Ways to Protect Reservoir's Watershed," *Baltimore Sun,* April 8, 2003, p. B3.

50. Peter M. Groffman, Danial J. Bain, Lawrence E. Band, Kenneth T. Belt, Grace S. Brush, J. Morgan Grove, Richard V. Pouyat, Ian C. Yesilonis, and Wayne C. Zipperer, "Down by the Riverside: Urban Riparian Ecology," *Frontiers in Ecology and the Environment* 1 (2003): 315–21.

51. Ibid., quote p. 320.

52. Jianguo Wu and Orie L. Loucks, "From Balance of Nature to Hierarchical Patch Dynamics: A Paradigm Shift in Ecology," *Quarterly Review of Biology* 70 (1995): 439–66, quote p. 460.

53. Bryant Smith, "Youths' Perspective on Inner City Environments through the Lens of Neighborhood Science Project," Baltimore Ecosystem Study annual meeting, Baltimore, Maryland, October 29, 2000; "Neighborhood Science Program," Baltimore Ecosystem Study Education meeting, Baltimore, Maryland, July 17, 2001.

54. Michael Anft, "Green in the City," *Baltimore Citypaper,* April 12–18, 2000, online at http://citypaper.com (search at that site for "Green in the City").

55. J. Morgan Grove, Karen E. Hinson, and Robert J. Northrop, "A Social Ecology Approach to Understanding Urban Ecosystems and Landscapes," in Berkowitz, Nilon, and Hollweg, *Understanding Urban Ecosystems,* pp. 167–86.

56. Karen Hinson, presentation at the Baltimore Ecosystem Study Science meeting, focusing on the themes of education, July 17, 2001, Carrie Murray Nature Center, Baltimore, Maryland.

57. Grimm et al., "Integrated Approaches to Long-Term Studies," p. 575.

CONCLUSION: Expanding the Dialogue

1. In 2003–4 the Ecological Society of America established a committee to evaluate and recommend changes that would enhance the effectiveness of ecology in solving environmental problems. The committee's report, "Ecological Science and Sustainability for a Crowded Planet," April 2004, is available online at www.esa.org/ecovisions. See also Margaret Palmer, Emily Bernhardt, Elizabeth Chornesky, Scott Collins, Andrew Dobson, Clifford Duke, Barry Gold, Robert Jacobson, Sharon Kingsland, Rhonda Kranz, Michael Mappin, M. Luisa Martinez, Fiorenza Micheli, Jennifer Morse, Michael Pace, Mercedes Pascual, Stephen Palumbi, O. J. Reichman, Alan Townsend, and Monica Turner, "Ecological Science and Sustainability for the 21st Century," *Frontiers in Ecology*

and the Environment 3 (2005): 4–11. For a parallel discussion within the earth sciences, see the four-part series on science and technology for sustainable development published in the *Proceedings of the National Academy of Sciences (U.S.)* 100 (2003): 8059–91. The introductory essay is William C. Clark and Nancy M. Dickson, "Sustainability Science: The Emerging Research Program," pp. 8059–61.

2. Alan R. Berkowitz, Karen S. Hollweg, and Charles H. Nilon, "Urban Ecosystem Education in the Coming Decade: What Is Possible and How Can We Get There?" in *Understanding Urban Ecosystems: A New Frontier for Science and Education*, ed. Alan Berkowitz, Charles Nilon, and Karen S. Hollweg (New York: Springer-Verlag, 2003), pp. 476–501, quotes pp. 489, 491, 487.

3. Ibid., p. 481.

4. Jane Jacobs, *The Death and Life of Great American Cities* (New York: Random House, 1961), chap. 22, quote p. 433.

5. Ibid., pp. 433–34.

6. M. J. McDonnell and S. T. A. Pickett, "Ecosystem Structure and Function along Urban-Rural Gradients: An Unexploited Opportunity for Ecology," *Ecology* 71 (1990): 1232–37, quote p. 1232.

7. Jane Jacobs, *Death and Life of Great American Cities*, p. 30. See also Jacobs, *The Nature of Economies* (New York: Modern Library, 2000), pp. 127–32, quote p. 30.

8. Steward T. A. Pickett, "Why Is Developing a Broad Understanding of Urban Ecosystems Important to Science and Scientists?" in Berkowitz, Nilon, and Hollweg, *Understanding Urban Ecosystems*, pp. 58–72, quote p. 63.

9. There is a citation in Nancy B. Grimm, J. Morgan Grove, Steward T. A. Pickett, and Charles L. Redman, "Integrated Approaches to Long-Term Studies of Urban Ecological Systems," *BioScience* 50 (2000): 571–84, cite p. 578.

10. Ibid., p. 579.

11. Jane Jacobs, *Dark Age Ahead* (New York: Random House, 2004).

12. Giovanna Di Chiro, "Nature as Community: The Convergence of Environment and Social Justice," in *Uncommon Ground: Toward Reinventing Nature*, ed. William Cronon (New York: Norton, 1995), pp. 298–320.

13. Wangari Maathai, Nobel lecture, delivered in Oslo, Norway, December 10, 2004, available on the Nobel Foundation Web site, http://nobelprize.org/.

Essay on Sources

This essay will guide the reader to the main secondary sources in the history of American ecology, especially those relevant to the themes here explored.

The history of botany in the United States begins with the work of Andrew Denny Rodgers III, author of several books with overlapping themes on nineteenth- and early-twentieth-century botany. The two most relevant here are *John Merle Coulter, Missionary in Science* (Princeton, NJ: Princeton University Press, 1944); and *American Botany, 1873–1892: Decades of Transition* (Princeton, NJ: Princeton University Press, 1944). The lack of full documentation in *American Botany* makes it difficult to identify sources of information and reduces the work's scholarly value. An up-to-date scholarly treatment of the formative years of American botany is Richard A. Overfield, *Science with Practice: Charles E. Bessey and the Maturing of American Botany* (Ames: Iowa State University Press, 1993). For a succinct and well-documented discussion of the rise of the "new botany" and its relation to the history of ecology, see also Eugene Cittadino, "Ecology and the Professionalization of Botany in America, 1890–1905," *Studies in History of Biology* 4 (1980): 171–98. A reissue of A. Hunter Dupree's *Asa Gray: American Scientist, Friend of Darwin* (Baltimore: Johns Hopkins University Press, 1988) makes available the only scholarly biography of the leading American botanist of the nineteenth century. Frank Egerton's edited volumes of the essays of Edward Lee Greene include (in volume 1) an excellent biographical study of Greene by Robert McIntosh: Frank N. Egerton, ed., *Landmarks of Botanical History: Edward Lee Greene*, 2 vols. (Stanford, CA: Stanford University Press, 1983).

For a general history of ecology, see Donald Worster, *Nature's Economy: A History of Ecological Ideas*, 2nd ed. (Cambridge: Cambridge University Press, 1994). Worster provides a wide-ranging and at times provocative survey of ecological ideas in the Anglo-American tradition, with an extended discussion of ecology in the American West. A comprehensive, almost encyclopedic survey of ecological thought may be found in *The Background of Ecology: Concept and Theory* (Cambridge: Cambridge University Press, 1985), by Robert P. McIntosh; it serves as a useful reference work in the history of ecology.

Most histories of American ecology have focused more closely on particular research schools in the western United States. Ronald C. Tobey's *Saving the Prairies: The Life Cycle of the Founding School of American Plant Ecology, 1895–1955* (Berkeley: University of California Press, 1981), offers both a biography of Frederic Clements and a sociological analysis of the Clementsian school. Joel Hagen provides a different perspective on Clements and his school in "Clementsian Ecologists: The Internal Dynamics of a Research School," *Osiris* 8 (1993): 178–95; and in "Organism and Environment: Frederic Clements's Vision of a Unified Physiological Ecology," in *The American Development of Biology*, ed. Ronald Rainger, Keith R. Benson, and Jane Maienschein (Philadelphia: University of Pennsylvania Press, 1988), 257–77. Eugene Cittadino interprets Henry Cowles subtly in "A 'Marvelous Cosmopolitan Preserve': The Dunes, Chicago, and the Dynamic Ecology of Henry Cowles," *Perspectives on Science* 1 (1993): 520–59. J. Ronald Engel devotes a book-length study to Cowles: *Sacred Sands: The Struggle for Community in the Indiana Dunes* (Middletown, CT: Wesleyan University Press, 1983).

Because of his role as a critic of the Clementsian program, Henry A. Gleason has received a degree of attention disproportionate to his ecological productivity. See the detailed studies of his work and its impact by Robert P. McIntosh, "H.A. Gleason, 'Individualistic Ecologist,' 1882–1975: His Contributions to Ecological Theory," *Bulletin of the Torrey Botanical Club* 102 (1975): 253–73; Malcolm Nicolson, "Henry Allan Gleason and the Individualistic Hypothesis," *Botanical Review* 56 (1990): 91–161; and Michael Barbour, "Ecological Fragmentation in the Fifties," in *Uncommon Ground: Toward Reinventing Nature*, ed. William Cronon (New York: Norton, 1995), 233–55.

On ecology in the American Southwest, see Janice E. Bowers's short biography of Forrest Shreve, *A Sense of Place: The Life and Work of Forrest Shreve* (Tucson: University of Arizona Press, 1988). The Desert Botanical Laboratory at Tucson is the subject of several studies describing its range of activities and personnel. Two relatively accessible articles are Janice E. Bowers, "A Debt to the Future: Scientific Achievements of the Desert Laboratory, Tumamoc Hill, Tucson, Arizona," *Desert Plants* 10 (1990): 9–12, 35–47; and Robert P. McIntosh, "Pioneer Support for Ecology," *BioScience* 33 (1983): 107–12.

The zoological side of the history of ecology in the midwestern United States has been investigated by Gregg Mitman, who focused on the Chicago school of ecology in *The State of Nature: Ecology, Community, and American Social Thought, 1900–1950* (Chicago: University of Chicago Press, 1992); and by Robert A. Croker in a biography of Victor Shelford, *Pioneer Ecologist: The Life and Work of Victor Ernest Shelford, 1877–1968* (Washington, DC: Smithsonian Institution Press, 1991). Two well-written and highly recommended studies of American biology, which also have broader geographical and institutional scope, are Mark V. Barrow Jr., *A Passion for Birds: American Ornithology after Audubon* (Princeton, NJ: Princeton University Press, 1998); and Philip J Pauly, *Biologists and the Promise of American Life, from Meriwether Lewis to Alfred Kinsey* (Princeton, NJ: Princeton University Press, 2000). Barrow and Pauly both analyze the professionalization of American biology. Robert Kohler, in his *Landscapes and Labscapes: Exploring the Lab-Field Border in Biology* (Chicago: University of Chicago Press, 2002), samples both botanical and zoological research as he studies the relationship between fieldwork and laboratory science up to the midtwentieth century. He draws on several ecological examples as well as on taxonomic and evolutionary studies.

On postwar ecosystem ecology, the following works are particularly useful sources: Joel B. Hagen, *An Entangled Bank: The Origins of Ecosystem Ecology* (New Brunswick, NJ: Rutgers University Press, 1992); Frank B. Golley, *A History of the Ecosystem Concept: More than the Sum of the Parts* (New Haven, CT: Yale University Press, 1993); Chunglin Kwa, "Radiation Ecology, Systems Ecology and the Management of the Environment," in *Science and Nature: Essays in the History of the Environmental Sciences*, ed. Michael Shortland (Stanford-in-the-Vale, Eng.: British Society for the History of Science, 1993), 213–49; and Stephen Bocking, *Ecologists and Environmental Politics: A History of Contemporary Ecology* (New Haven, CT: Yale University Press, 1997). Golley's analysis is from the standpoint of a scientist who has been a major contributor to ecosystem ecology.

Index